AF572425

Springer Series on Environmental Management

Robert S. DeSanto, Series Editor

Springer Series on Environmental Management
Robert S. DeSanto, Series Editor

Disaster Planning: The Preservation of Life and Property
Harold D. Foster
1980/275 pp./48 illus./cloth
ISBN 0-387-90498-0

Air Pollution and Forests: Interactions between Air Contaminants and Forest Ecosystems
William H. Smith
1981/379 pp./60 illus./cloth
ISBN 0-387-90501-4

Natural Hazard Risk Assessment and Public Policy: Anticipating the Unexpected
William J. Petak
Arthur A. Atkisson
1982/489 pp./89 illus./cloth
ISBN 0-387-90645-2

Environmental Effects of Off-Road Vehicles: Impacts and Management in Arid Regions
R. H. Webb
H. G. Wilshire (Editors)
1983/560 pp./149 illus./cloth
ISBN 0-387-90737-8

Global Fisheries: Perspectives for the '80s
B. J. Rothschild (Editor)
1983/approx. 224 pp./11 illus./cloth
ISBN 0-387-90772-6

Heavy Metals in Natural Waters: Applied Monitoring and Impact Assessment
James W. Moore
S. Ramamoorthy
1984/256 pp./48 illus./cloth
ISBN 0-387-90885-4

Organic Chemicals in Natural Waters: Applied Monitoring and Impact Assessment
James W. Moore
S. Ramamoorthy
1984/282 pp./81 illus./cloth
ISBN 0-387-96034-1

The Hudson River Ecosystem
Karin E. Limburg
Mary Ann Moran
William H. McDowell
1986/344 pp./44 illus./cloth
ISBN 0-387-96220-4

Human System Responses to Disaster: An Inventory of Sociological Findings
Thomas E. Drabek
1986/512 pp./cloth
ISBN 0-387-96323-5

The Changing Environment
James W. Moore
1986/256 pp./40 illus./cloth
ISBN 0-387-96314-6

Balancing the Needs of Water Use
James W. Moore
1988/280 pp./39 illus./cloth
ISBN 0-387-96709-5

The Professional Practice of Environmental Management
Robert S. Dorney
Lindsay Dorney (Editors)
1989/248 pp./23 illus./cloth
ISBN 0-387-96907-1

Landscape Ecology: Theory and Applications
(Student edition)
Zev Naveh
Arthur S. Lieberman
1990/384 pp./78 illus./pbk
ISBN 0-387-97169-6

Long-Term Consequences of Disasters: The Reconstruction of Friuli, Italy, in Its International Context, 1976-1988
Robert Geipel
1991/208 pp./81 illus./cloth
ISBN 0-387-97419-9

Inorganic Contaminants of Surface Water: Research and Monitoring Priorities
James W. Moore
1991/360 pp./13 illus./cloth
ISBN 0-387-97281-1

James W. Moore

Inorganic Contaminants of Surface Water

Research and Monitoring Priorities

With 13 Illustrations

Springer-Verlag
New York Berlin Heidelberg London
Paris Tokyo Hong Kong Barcelona

James W. Moore
Box 42
Vegreville, Alberta
TOB 4LO Canada

Library of Congress Cataloging-in-Publication Data
Moore, James W., 1947–
Inorganic contaminants of surface water : research and monitoring priorities / James W. Moore.
p. cm. – (Springer series on environmental management)
Includes bibliographical references and index.
ISBN 0-387-97281-1 (alk. paper)
1. Inorganic compounds—Environmental aspects. 2. Radioactive pollution of water. I. Title II. II. Series.
TD427.I55M66 1990
628.1'68—dc20 90-9794

Printed on acid-free paper

Typeset by David E. Seham Associates, Metuchen, New Jersey.
Printed and bound by R. R. Donnelley & Sons, Harrisonburg, Virginia.
Printed in the United States of America.

9 8 7 6 5 4 3 2 1

ISBN 0-387-97281-1 Springer-Verlag New York Berlin Heidelberg
ISBN 3-540-97281-1 Springer-Verlag Berlin Heidelberg New York

Series Preface

This series is dedicated to serving the growing community of scholars and practitioners concerned with the principles and applications of environmental management. Each volume is a thorough treatment of a specific topic of importance for proper management practices. A fundamental objective of these books is to help the reader discern and implement man's stewardship of our environment and the world's renewable resources. For we must strive to understand the relationship between man and nature, act to bring harmony to it, and nurture an environment that is both stable and productive.

These objectives have often eluded us because the pursuit of other individual and societal goals has diverted us from a course of living in balance with the environment. At times, therefore, the environmental manager may have to exert restrictive control, which is usually best applied to man, not nature. Attempts to alter or harness nature have often failed or backfired, as exemplified by the results of imprudent use of herbicides, fertilizers, water, and other agents.

Each book in this series will shed light on the fundamental and applied aspects of environmental management. It is hoped that each will help solve a practical and serious environmental problem.

Robert S. DeSanto
East Lyme, Connecticut

Preface

Environmental cycles can be split into two broad categories: those induced by natural processes and those induced by Humankind. Until recently, nature-induced environmental cycles overwhelmingly dominated anything created or caused by humans. However, in a time when the world's population expands at an unparalleled rate, the corresponding burdens placed on the environment and on the process of environmental protection have intensified as well. Today, the impact of Humankind on the environment has overcome certain natural cycles and now demands complex and far-reaching responses.

Over the last twenty years, improvements in environmental protection and modifications in policy have yielded cleaner water, fresher air, an abundance of fish and wildlife, and a higher quality of life in many western nations. Jobs have blossomed as a result of these trends, and entire industries, such as hazardous waste management, have grown dramatically since 1970. Yet for every force in the direction of positive environmental change, a complementary force has labored to oppose it.

As the environment undergoes progressively greater strains, the process of environmental protection must be designed to meet the challenge. Trends in the positive direction should include the diversion of more money to the protection of water and air, the management of hazardous wastes, and the control of nonpoint sources of pollution. But instead we seem to migrate in the opposite direction, toward an era of heightened indifference and lessened control. Conditions in many countries may very well deteriorate in the 1990's as populations grow and money for the environment is compromised.

Only a determined effort by all those concerned about the world they live in, from laymen, educators, and scientists to policy makers and heads of state, will rescue the cycle and reverse the decline in environmental protection. As an important step toward a reversal of these trends, the effective management of inorganic contaminants of surface water has long been the goal of professionals in public health and the environment. Although the greatest progress has been made in controlling some of the sources of these contaminants, the sheer weight of population and economic growth during the 1990's will likely offset many of these gains. The purpose of this book therefore is to provide an information base for the management of inorganic contaminants in an era of reduced environmental commitment. The reader will initially find a prognosis of environmental change during the 1990's, followed by a technical review of current data on inorganic agents, and concluded by recommendations for research and monitoring.

I would like to thank all those who helped with the preparation of this book and, in particular, Greg, Jill, and Inge.

Contents

1
Introduction

The process of environmental protection will come under staggering pressure during the 1990s and for the foreseeable future. It took all of human history to develop an economy of $600 billion by the year 1900. Eighty-five years later, the world economy was expanding by more than that amount every 2 years. In the year 2050, the global economy is expected to reach $13 trillion, more than five times what it is today (Speth, 1988). Resource use will expand enormously, as will the production of waste, the formation of chemical by-products, and the deposition of contaminants into surface waters.

The other staggering inevitability for environmental protection is growth in the world's population: from 3 billion in 1960 to 5.1 billion in 1989 to 10 billion in 2050 (World Resources Institute, 1988; Speth, 1988). All of these people will produce, either directly or indirectly, more waste in an attempt to maintain or increase their standard of living. This will in turn force much of the Western world to use water that does not comply with current quality guidelines, simply because better water will not be available. The same will apply to direct users of surface water, such as fish and aquaculturalists.

Despite the oncoming crush, most Western nations have progressively reduced their expenditures in environmental protection. In the United States, for example, the total amount of money dedicated to research and development by the Environmental Protection Agency has declined from a maximum of $219 million in 1975 to only $90 million in 1985. The most recent budgets have featured a small increase in funding, roughly equivalent to the change in GNP (Livernash, 1988). In Canada, total expenditures by the federal government on the environment have declined by more than 25% since 1980 (Environment Canada, 1980–1988). No country in continental Europe has augmented its environmental budget to keep pace with economic development and population expansion. In the United Kingdom, regulatory agencies are underfunded to the point where their effectiveness is diminished (O'Riordan, 1988).

The relatively low priority placed on environmental protection by many technologically advanced nations is manifest in the adulteration of numerous water supplies as well as drinking water. Some of the best examples come from densely populated nations with an expanding industrial economy. Numerous examples can be cited from Europe, including:

North Sea—titanium, chromium, and PCB (polychlorinated biphenyl) contamination
Mediterranean Sea—mercury, oil, microbial pathogens, eutrophication
Rhine River—copper, zinc, lead, chromium, nickel, organochlorine compounds
Elbe River—heavy metals, low dissolved oxygen levels
Wadden Sea—heavy metals, chlorinated organics (Dethlefsen, 1988; Beukema et al., 1986).

There are numerous other similar examples from other heavily industrialized countries such as Japan (Kimura, 1988) and the Soviet Union (Thompson, 1989), where the multiple use of water has been greatly curtailed by untoward control of anthropogenically derived wastes.

One of the major draws on the water pollution control budget of many nations is the need to replace old sewer systems and other related waterworks. Many billions of dollars need to be spent in Canada, the United States, Europe, and elsewhere on decaying systems that may be 50–80 or more years old (Livernash, 1988). In one sense, then, huge sums of money are being spent by many nations on water pollution control, but this refurbishing activity greatly reduces the budget for research, monitoring, development of pollution control technologies, and other activities.

Objectives

The presence of inorganic contaminants in surface water continues to be one of the most pervasive environmental issues of our time. Although control technologies have been applied to many industrial and municipal sources, the total quantity of these agents released to the environment remains staggering (Table 1.1). Such discharges will, in the 1990s, limit the multiple use of water in many regions of the world, and potentially increase the frequency of chronic disease in the human population (Nriagu, 1988). In fact the annual total toxicity of all the metals mobilized

Table 1.1. Global discharges of trace metals (in 1,000 metric tons/yr).

Metal	Water	Air	Soil
Arsenic	41	19	82
Cadmium	9.4	7.6	22
Chromium	142	30	896
Copper	112	35	954
Lead	138	332	796
Mercury	4.6	3.6	8.3
Nickel	113	56	325
Selenium	41	3.8	41
Tin	ND[a]	6.4	ND
Zinc	226	132	1,372

Sources: Nriagu (1988), Nriagu and Pacyna (1988).
[a]No data.

worldwide exceeds the total activity of all organic wastes generated each year (Nriagu and Pacyna, 1988). What effect this will have on the health of future generations is not known.

The management of inorganic contaminants will become progressively more difficult in future years. Scientists, engineers, technologists, and bureaucrats will need to improve both their effectiveness despite diminishing resources and their ability to identify timely and relevant issues. This then leads to the purpose of this book: to help environmentalists control or otherwise manage inorganic contaminants in surface water during an era of fiscal restraint. The reader will find an initial review of the sources, chemistry, and toxicology of many major pollutants of water, followed by a series of recommendations for research and monitoring. The agents selected for review fall into at least one of the following categories: (1) highly toxic and/or persistent; (2) instrumental in regulating the concentration or hazard posed by more toxic agents, or (3) poorly studied but potentially hazardous. The book concludes with a series of overall recommendations and conclusions.

References

Beukema, A.A., G.P. Hekstra, and C. Venema. 1986. The Netherlands' environmental policy for the North Sea and Wadden Sea. *Environmental Monitoring and Assessment* 7:117–155.

Dethlefsen, V. 1988. Status report on aquatic pollution problems in Europe. *Aquatic Toxicology* 11:259–286.

Dowd, R.M. 1986. Fiscal year 1987. *Environmental Science and Technology* 20:552.

Environment Canada. 1980–1988. *Annual Reports*. Ottawa, Canada.

Kimura, I. 1988. Aquatic pollution problems in Japan. *Aquatic Toxicology* 11:287–301.

Livernash, R. 1988. The shrinking environmental dollar. *Environment* 30(1):7–9.

Nriagu, J.O. 1988. A silent epidemic of environmental metal poisoning? *Environmental Pollution* 50:139–161.

Nriagu, J.O., and J.M. Pacyna. 1988. Quantitative assessment of worldwide contamination of air, water and soils by trace metals. *Nature* 333:134–139.

O'Riordan, T. 1988. The politics of environmental regulation in Great Britain. *Environment* 30(8):4–9.

Speth, J.G. 1988. The greening of technology. *Washington Post* 20 November 1988, p D3.

Thompson, D. 1989. The greening of the U.S.S.R. *Time* 133(1):59–60.

Water Pollution Control Federation. 1986. Is clean water research keeping pace with national pollution control needs? *Journal of the Water Pollution Control Federation* 58:880–885.

World Resources Institute. 1988. *World resources 1988–89*. Basic Books, New York. 372 pp.

2
Aluminum

Aluminum is the third most abundant element in the earth's crust, with an average concentration of approximately 8%. It is chemically reactive, occurring principally as the halide or oxide, usually in complex silicates. Although aluminum is found in essentially all plant and animal species, it is not essential for survival. The environmental significance of aluminum has waxed enormously in recent years for two reasons: (1) increased mobilization due to acidification of surface waters, and (2) potential agent in the genesis of Alzheimer's and related diseases.

Production, Sources, and Residues

Production

Aluminum is produced and consumed in enormous amounts in many nations of the world, including both developed and developing countries. Total world production in recent years has exceeded 15×10^6 metric tons annually (Table 2.1). The United States is the world's largest producer, followed by the USSR, Canada, and Australia. The USA and the USSR also lead the way in consumption, closely followed by Japan and the Federal Republic of Germany (FRG).

Sources and Residues

Water. Because aluminum is abundant in the earth's crust and is produced and consumed in huge amounts, it is inevitable that relatively high

Table 2.1. The world's major producers and consumers of aluminum (1986).

Producing nation	Quantity (1,000 metric tons/yr)	Consuming nation	Quantity (1,000 metric tons/yr)
USA	3,037	USA	4,268
USSR	2,300	USSR	1,885
Canada	1,360	Japan	1,624
Australia	882	China	750
Brazil	762	France	593
Norway	712	Italy	510
Venezuela	424	Brazil	424
China	410	Canada	405
Spain	375	UK	389
All other nations	4,287	All other nations	4,363
World total	15,314	World total	16,396

Sources: World Resources Institute (1988); World Bureau of Metal Statistics (1989).

residues will be found in surface water. Furthermore, the decreasing pH of many surface waters and soils, a result of acid deposition, mobilizes aluminum into dissolved species (see Chemistry section) which are easily transported and are biologically available.

One of the major sources of aluminum in freshwater, surpassing industrial users, is the discharge of alum sludge from municipal water treatment plants. Alum sludges are produced when alum or aluminum sulfate ($Al_2SO_4{\cdot}4H_2O$) is used for coagulation and flocculation of raw water supplies to remove turbidity and/or color. The aluminum content of commercial alum is typically >50,000 mg/L (Cornwell et al., 1987). The floc particles are removed mainly during the sedimentation process in which the flocs slowly settle and form a sludge blanket referred to as alum sludge. This blanket consists mainly of aluminum hydroxide [$Al(OH)_3$], other particles, and flocculated materials. Although the discharge of sludge inevitably increases aluminum residues in surface water, the extent of change does not appear to be great. Cornwell et al. (1987), for example, reported that total Al in the water of the Ohio River rose by a maximum of 0.0008 mg Al/L following the discharge of alum sludge from municipalities.

Aluminum-bearing solid wastes may be dumped in offshore areas, resulting in high but transient aqueous aluminum residues. The North Sea off the coast of Belgium periodically receives 1,000–2,000 metric tons of such waste, causing an initial increase in aluminum to 120 mg/L and pH to 8.5 (Vandelannoote et al., 1987). These can be compared to background levels of 0.007 mg/L and 8.1 pH units, respectively. A hydrotalcite-manasseite-like precipitate ($Mg_6Al_2CO_3(OH)16{\cdot}4H_2O$) is formed on contact with water, which is then highly dispersed.

Evaporation of water is a key factor in controlling aluminum in the water of bogs, ponds, and, to a lesser degree, lakes. As the water evaporates during the summer, aluminum also increases, sometimes to high levels. Urban et al. (1987), working on a series of bogs in eastern North America, found residues of up to 0.5 mg/L in midsummer. The same authors suggested that pH and organic complexation did not control aluminum concentrations in bog water.

Sediment. Relatively high concentrations of Al_2O_3 and related species are found naturally in sediment. Mudroch and Duncan (1986), for example, reported that Al_2O_3 comprised 5.5–16.2% of the dry weight of sediments in the Niagara River (Canada). Similarly elemental Al in the Ganges delta (India) averaged 56,500 mg/kg dry weight of sediment (Subramanian et al., 1988). Such quantities from natural sources are likely to mask most anthropogenic inputs. In fact, Johnson et al. (1986) reported that precipitation loadings to a series of lakes in central Canada accounted for only 2–8% of background loadings. In that study, the concentration of total Al in sediments ranged from 31,000 to 49,500 mg/kg dry weight.

Aluminum from sediments may significantly contaminate surface water if there is turbulence or a decline in pH. Bull and Hall (1986), working on two rivers in the United Kingdom, found that total Al ranged from 0.005 to 0.065 mg/L during periods of low flow but, during moderate flow conditions, the range increased to 0.025 to 0.36 mg/L. The same authors found that, at low pH (<5.5), inorganic Al increased from <0.02 mg/L to a maximum of approximately 0.5 mg/L. Similar findings were reported by Tipping and Hopwood (1988) for monomeric Al and by Sprenger et al. (1987) for total Al in six acidic lakes in New Jersey.

Chemistry

The chemistry of aluminum in water is complex and influenced by pH and the presence of fluoride, sulfate, organic matter, and other ligands. Dissociation of aluminum salts in pure water yields $Al(OH)_6^{3+}$ and other hydrated aluminum ions. Progressive hydrolysis of the aluminum ion produces $Al(OH)_2^+$ and ultimately $Al(OH)_3$, a colloid. If the pH is above neutrality, $Al(OH)_3$ is converted to $Al(OH_4)^-$, the aluminate ion. At extremely low pH (<4), aluminum exists mainly as the trivalent cation; at pH 4.5–6.5, the major species are $Al(OH)^{3+}$, $Al(OH_2)^+$, and $Al(OH_3)$; and at pH >6.5, the main species is $Al(OH_4)^-$. Aluminum solubility is lowest at pH 5.5 to 6.0 (Burrows, 1977).

Aluminum may form relatively stable complexes with fluorides, fulvic acids, and other agents (Plankey and Patterson, 1987, 1988). In addition, monomeric, dimeric, and polymeric Al hydrates may be formed in surface waters, which greatly affect the environmental toxicity of aluminum. The rates of complexation are apparently dependent not only on pH but also on temperature. An example of the mechanism of complexation with fluorine (F) in the presence of fulvic acid (FA) is given below (Plankey and Patterson, 1988):

Al^{3+}	+	F^-	$\rightleftharpoons$	AlF^{2+}			(1)
$AlOH^{2+}$	+	F^-	$\rightleftharpoons$	$AlOHF^+$	+	H_2O	(2)
Al^{3+}	+	HF	$\rightleftharpoons$	AlF^{2+}	+	H^+	(3)
$AlOH^{2+}$	+	HF	$\rightleftharpoons$	AlF^{2+}	+	H_2O	(4)
Al^{3+}	+	FA^-	$\rightleftharpoons$	$Al(FA)^{2+}$			(5)
$AlOH^{2+}$	+	FA^-	$\rightleftharpoons$	$Al(FA)OH^+$			(6)
Al^{3+}	+	HFA	$\rightleftharpoons$	$Al(FA)^{2+}$	+	H^+	(7)
$AlOH^{2+}$	+	HFA	$\rightleftharpoons$	$Al(FA)OH^+$	+	H^+	(8)
$Al(FA)^{2+}$	+	F^-	$\rightleftharpoons$	$Al(FA)F^+$			(9)
$Al(FA)OH^+$	+	F^-	$\rightleftharpoons$	Al(FA)OHF			(10)

where HFA is a fully protonated aluminum binding site and FA^- is a deprotonated site. Reactions 1–3 and 5–8 are important for the complexation of fluoride alone and fulvic acid alone, respectively. The final two reactions (9 and 10) involve fluoride and fulvic acid.

In an experimental watershed in New Hampshire, the speciation of aluminum was determined before and after logging (Lawrence and Driscoll, 1988). The logging operations resulted in an increase in the concentration of nitrates, inorganic Al, and basic cations in the stream, whereas sulfate declined after logging. Precut conditions yielded a predominance of Al–F at the head of the stream followed by $Al(OH)_2^+$ and $AlOH^{2+}$ (Table 2.2). After cutting, the main species was Al^{2+} followed by $AlOH^{2+}$ and Al–F.

Goenaga and Williams (1988) similarly evaluated the impact of changing water chemistry on aluminum species in a Welsh river, particularly in relation to storm episodes. Organic Al complexes increased significantly during high water, reflecting the increase in total organic carbon and ionic Al. Adsorbed Al also increased during floods to the point where it was second in concentration only to Al^{3+}. This means that suspended solids

Table 2.2. Distribution of inorganic Al complexes in a stream[a] in New Hampshire before and after logging.

Precut		Postcut	
Species	Percentage	Species	Percentage
Al-F	46.1	Al^{3+}	39.0
$AlOH^+$	20.1	$AlOH^{2+}$	25.0
$AlOH^{2+}$	19.0	Al–F	22.6
Al^{3+}	14.8	$Al(OH)^+$	13.4

[a]Stream pH approximately 7.
Source: Lawrence and Driscoll (1988).

could act as a sink for inorganic monomeric Al released during storm events.

Another key aspect of aluminum chemistry is the dissolution of aluminum in soil to neutralize acid inputs such as acid rain. Dissolved Al is highly toxic to crop roots and forests, and may wash into surface waters. Mulder et al. (1989) showed that the most soluble fraction is mainly nonsilicate organic Al, formed during the course of soil development. The process of dissolution is relatively rapid and may eventually result in reduced soil acid neutralization.

It is important to note that the toxicity of different aluminum species is poorly understood at the present time. Increased mobilization of aluminum due to acid rain, plus a rise in the frequency of cognitive disorders such as Alzheimer's disease, means that the importance of aluminum species in environmental toxicology and human health will increase. Elucidation of the formation of aluminum species is an important step in managing these conditions.

Bioaccumulation

Plants

Total Al in aquatic and marine plants is often elevated near industrial/municipal waste outfalls. Soderlund et al. (1988) reported residues ranging from 17 to 246 mg/kg dry weight in the growing tips of the macrophyte *Fucus vesiculosus*. These samples were collected in the Baltic Sea near Stockholm. Mason and Macdonald (1988) similarly noted that metal mine drainage resulted in total Al residues of up to 54,000 mg/kg dry weight in aquatic mosses *(Fontinalis squamosa)* in a Welsh river. Less contaminated reaches yielded residues in the 2,000–7,000 mg/kg range. In a study of six acidic lakes (pH 3.6–5.8) in New Jersey, Sprenger and McIntosh (1989) determined total Al in water, sediments, and aquatic plants. The resulting concentration factors (plant concentration/water concentration or sediment concentration) for floating-leafed plants were much lower than for submerged rooted plants, indicating that the primary source of total Al was lake sediment (Table 2.3).

Table 2.3. Summary of concentration factors of total Al in plants from six lakes in New Jersey.

	Concentration factors	
Plant type	Plant/sediment	Plant/water
Submerged-rooted	0.02–0.41	7,100–89,000
Submerged-nonrooted	0.28–0.41	21,600–54,800
Emergent	<0.01–0.17	3,700–11,500
Floating-leafed	<0.01–0.04	180–4,100

Source: Sprenger and McIntosh (1989).

Invertebrates and Fish

Total Al residues are usually lower in invertebrate and fish species, both marine and freshwater. Cleveland et al. (1986) found that concentration factors in juvenile brook trout *Salvelinus fontinalis,* ranging from 50 to 231, were greater after 15 days of exposure than after 30 days. Brumbaugh and Kane (1985), working with smallmouth bass *Micropterus dolomieui* in an acidic (pH 6.3) reservoir in the eastern USA, found average total Al residues of 3.9 mg/kg wet weight in the whole body (less gut contents), compared to 58 mg/kg in the gills. This yielded concentration factors of 4–65 and 60–970, respectively.

Most countries, at present, do not have guidelines for the consumption of fish and related products tainted with aluminum. However, increasing evidence for the role of aluminum in the genesis of cognitive disorders increases the importance of establishing such guidelines.

Toxic Effects to Aquatic Organisms

Plants

Single-celled algae are more sensitive to total Al than macrophytes, reflecting their relative cell volumes:surface areas. The EC (effective concentration)$_{50}$ of the diatom *Cyclotella meneghiniana* was approximately 0.81 mg/L whereas mortality occurred at 6.48 mg/L (reviewed by US Environmental Protection Agency, 1988). Similarly, the EC_{50} for the green algae *Selenastrum capricornutum* ranged from 0.46 to 0.99 mg/L. By contrast the EC_{50} of Eurasian milfoil *Myriophyllum spicatum* and duckweed *Lemna minor* fell within the 2.5 to >45.7 mg/L range (US Environmental Protection Agency, 1988).

Invertebrates and Fish

There are so many Al species, and such little replicated experimental work available, that generalizations about aluminum toxicity must be viewed with caution. Some early studies reported that soluble Al was most toxic to fish at pH 6.8–9.0 (Freeman and Everhart, 1971; Hunter et al., 1980). During the same time period, Driscoll et al. (1980) found that inorganic Al was most toxic to brook trout in slightly acidic water, and a later study concluded that $Al(OH)_2^+$ was the major toxic species (Helliwell et al., 1983). Most recent reports indicate that organic sequestration reduces toxicity to fish (Ramamoorthy, 1988; Seip et al., 1984).

A number of chronic toxic effects have also been observed following exposure to different Al species (Table 2.4). Because most of these studies have been conducted at acidic pH,there is generally an impact on ion flux in test organisms. Although Al^{3+} is often ranked as the most toxic

Table 2.4. Some chronic effects of aluminum to aquatic organisms.

Species	Effect	Conditions
Anodonta grandis[1] (mollusc)	Increase in blood Ca, reduction in blood Na and Cl	pH 4.5, Al (max) 2.2 mg/L
Brown trout[2] *Salmo trutta*	Impaired chloride balance (greatest at pH 5.5)	pH 4–7, Al 0.175 mg/L, Ca 0.35 mEq/L
Brown trout[3]	Impaired development of alevins, decreased Ca uptake and ion deposition in skeleton	pH 4.5–5.4, Al 6–8 μmol/L, Ca 10–50 μmol/L
Brown trout[4]	Retarded growth (greatest at pH <5.5)	pH 4.3–6.5, Al 0–3.7 μmol/L
Rainbow trout[5] *Oncorhynchus mykiss*	Hypoxia, no change in plasma ions	pH 5, Al 2 mg/L, Ca 0.35 mEq/L
Brook trout[5] *Salvelinus fontinalis*	Percent hatch not affected by Al	pH 4.5–7.5, Al 0.3 mg/L, hardness < 9 mg/L

Sources: [1]Malley et al. (1988), [2]Battram (1988), [3]Reader et al. (1988), [4]Sadler and Lynam (1987), [5]Malte (1986), [6]Hunn et al. (1987).

species, Clark and LaZerte (1985) noted that $Al(F)_x$ and $Al(OH)_x$ complexes were as toxic as any other Al species. In that case, toxicity was measured by the success of hatching of amphibian eggs.

To overcome the problem of Al species complex while protecting aquatic organisms, some regulatory agencies have sought to express aluminum only as the acid-soluble form, defined as aluminum that passes through a 0.45-μm membrane after acidification to pH 1.5–2.0 with nitric acid (US Environmental Protection Agency, 1988).

Using acid-soluble data, guidelines for the protection of aquatic species in freshwaters of pH 6.8–9.0 in the USA are (US Environmental Protection Agency, 1988) (1) 0.087 mg/L (4-day average exceeded only once every 3 years), and (2) 0.0750 mg/L (1-hour average exceeded only once every 3 years). The guidelines for the protection of aquatic life in Canada and several European nations are (1) 0.005 mg/L (pH <6.5,Ca <4 mg/L), and (2) 0.1 mg/L (pH >6.5, Ca >4 mg/L).

Health Effects

Intake

The typical Western diet yields a daily aluminum intake of 20–45 mg in a 70-kg reference man. Most of this comes from food and fluids, with only 0.1 mg originating from air. If 2 L of water are consumed daily, and the drinking water guideline of 0.05 mg/L is used, then the maximum daily intake through drinking water is also 0.1 mg/day. Although this represents only a small part of total intake, some water systems carry much more aluminum (>2 mg/L). The total aluminum intake is then augmented by at least 4 mg/day.

Acute Toxicity

Aluminum is relatively nontoxic to humans. In rats, administration of high oral doses leads to lethargy, anorexia, and (eventually) death.

Chronic Toxicity

Interest in the neurotoxic effects of aluminum first appeared during the 1940s, when it was observed that direct application of aluminum paste to the cortex of monkeys produced convulsions with recurrent seizures (Kopeloff et al., 1942, 1947). About 20 years later, it was reported that Holt's adjuvant (which contains aluminum) caused severe convulsions when injected into the cerebrospinal fluid and cerebral hemisphere of rabbits (Klatzo et al., 1965; Wisniewski et al., 1965). Several other studies followed, all noting that the central nervous system of rabbits, cats, dogs, and ferrets was vulnerable to aluminum, reacting with seizures, neurofibrillary changes, and epilepsy. Other species, including mice, rats, guinea pigs, hamsters, and monkeys, showed no such response to aluminum.

As the population of many Western nations ages, the number of cases of senile dementia of the Alzheimer's type increases. Approximately 10% of the population over age 65 develop this condition, which is characterized by intellectual deterioration, memory loss, neurofibrillary changes, and the formation of neuritic (senile) plaques (Sturman and Wisniewski, 1988). It should be noted that the neurofibrillary changes in the spinal cord of humans consist of pairs of filaments helically wound around each other with periodic twists. In rabbits, the changes are composed of disorganized tangles of apparently normal filaments. In both humans and rabbits, the hippocampus region of the cerebrum develops numerous neurofibrillary tangles. Furthermore, neuritic plaques of the Alzheimer's type are not found in aluminum-treated rabbits.

There is an increasing number of studies pointing to a pathogenic role of aluminum in the onset of Alzheimer's disease.

1. Aluminum accumulates in the nucleus of tangle-bearing neurons in patients with Alzheimer's disease (Perl and Good,1988; Krishnan et al., 1988).
2. The performance in behavioral tests of elderly human subjects with no known neurologic disorder is correlated with serum concentrations of aluminum (Bowdler et al., 1979).

3. Visual-motor performance in children is correlated with hair Al/Pb levels (Marlowe et al., 1985).

Furthermore, the relative risk of Alzheimer's disease in England is correlated with aluminum levels in drinking water (Brown, 1989):

Al concentration (mg/L)	<0.01	0.02–0.04	0.05–0.07	0.08–0.11	>0.11
Relative risk	1.0	1.4	1.4	1.6	1.7

Many of the current problems in determining the role of environmental aluminum in the genesis of Alzheimer's disease may reflect differences in the toxic effects of aluminum species. We may in fact be dealing with a number of causative agents and not just the parent element. Careful chemical and basic scientific investigations are required to better define the role of aluminum in the onset of the disease.

Carcinogenicity

There is no replicated evidence available to indicate that aluminum compounds are carcinogenic in humans or experimental animals.

Drinking Water

Residues

Aluminum levels in finished drinking water are highly variable, reflecting differences in raw water total Al concentrations, and the method used to clarify turbid water. Letterman and Driscoll (1988), for example, surveyed 194 water utilities in the United States and found residues of over 0.4 mg/L in raw water and over 2 mg/L in finished water (Figure 2.1). That range in values has been reported in numerous other studies.

Consumption Guidelines

Most regulatory agencies have established guidelines/standards for the control of aluminum in drinking water (Table 2.5). These values are highly variable and reflect differing opinions on the hazard posed by aluminum in drinking water. Utilities have a number of options to reduce aluminum in drinking water to meet the criteria in Table 2.5. The most commonly reported procedures are (1) pH adjustment, (2) reducing the use of alum, (3) replacement of alum with iron salts, (4) replacement of alum with polyelectrolyte flocculents, and (5) optimization of particle removal efficiency. It must be emphasized that all guidelines for the protection of

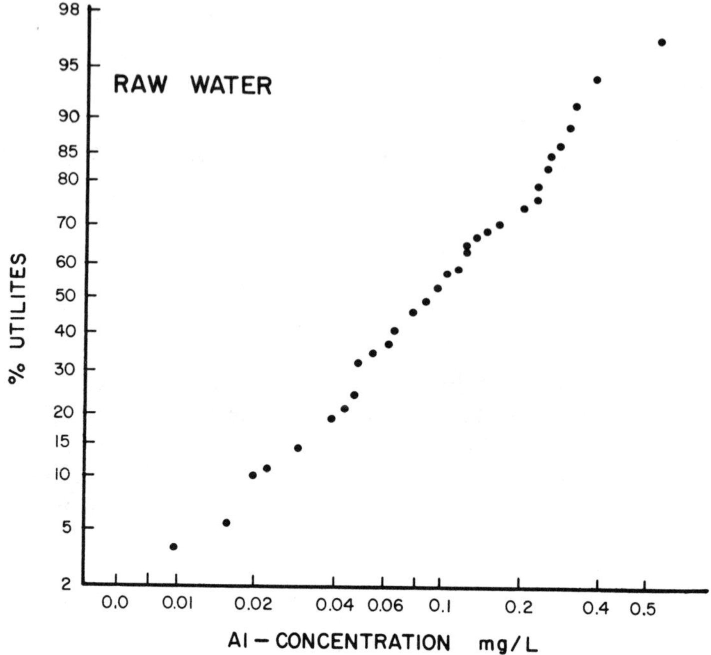

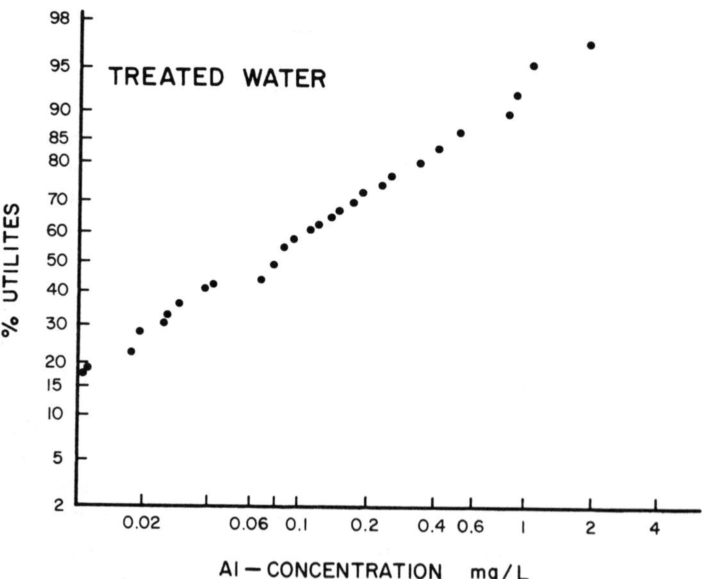

Figure 2.1. Percentage of utilities with aluminum concentration less than or equal to stated concentration. (From Letterman and Driscoll, 1988.) Reprinted from *Journal American Water Works Association*, Vol. 80, No. 4 (April 1988), by permission. Copyright © 1988, American Water Works Association.

Table 2.5. Regulations and guidelines for aluminum in drinking water.

Agency	Concentration (mg/L)
World Health Organization	0.2
US Environmental Protection Agency	0.05
European Economic Community	
Guidance level	0.05
Maximum permissible level	0.2
New York State	
Maximum percent of recorded values	
95	<0.15
75	<0.09
50	<0.05

drinking water refer to total Al. However, considerable differences exist in the toxicity of the different aluminum species, so reports of noncompliance with guidelines may have less toxicologic significance than indicated by total concentration data.

Treatment

The primary source of aluminum in most drinking waters is alum. Hence, excessive residues can be controlled by reducing the amount of alum used in water treatment. Conventional filtration is also effective in removing particulate-bound aluminum from water.

Recommendations

Although aluminum finds so many applications worldwide, and is increasingly mobilized with acid deposition, remarkably little is known about the formation, persistence, and toxicity of Al species to all water users. It has been widely assumed that, for monitoring and regulatory purposes, total Al analysis adequately protects water users, yet there is tremendous variation in the toxic effects of Al species. The following subjects are current research and monitoring priorities.

1. National/international standardization of protocols aimed at studying the Al species complex. This is important because minor differences in experimental conditions can greatly alter the species complex.
2. The formation and/or resolubilization of inorganic and organic complexes under acidic conditions in surface water.

3. The resolubilization of alum sludge and other aluminum-bearing wastes in relation to changes in redox potential and pH of bottom sediments.
4. Routine monitoring of aluminum in drinking water in relation to alum use.
5. The toxicity of Al species to standard test organisms such as rainbow trout and *Daphnia*.
6. Chemical and basic scientific evaluation of the role of Al species in the genesis of Alzheimer's and related diseases.

References

Battram, J.C. 1988. The effects of aluminum and low pH on chloride fluxes in the brown trout, *Salmo trutta* L. *Journal of Fish Biology* 32:937–947.

Bowdler, N.C., D.S. Beasley, E.C. Fritze, A.M. Goulette, J.D. Hatton, J. Hession, D.L. Ostman, D.J. Rugg, and C.J. Schmittdiel. 1979. Behavioral effects of aluminum ingestion on animal and human subjects. *Pharmacology, Biochemistry, and Behavior* 10:505.

Brown, P. 1989. Aluminum in water puts kidney patients at risk. *New Scientist* 121:28.

Brumbaugh, W.G., and D.A. Kane. 1985. Variability of aluminum concentrations in organs and whole bodies of smallmouth bass *(Micropterus dolomieui)*. *Environmental Science and Technology* 19:828–831.

Bull, K.R., and J.R. Hall. 1986. Aluminum in the Rivers Esk and Duddon, Cumbria, and their tributaries. *Environmental Pollution* 12:165–193.

Burrows, W.D. 1977. Aquatic aluminum: chemistry, toxicology and environmental prevalence. *CRC Critical Reviews in Environmental Control* 7:167–216.

Clark, K.L., and B.D. LaZerte. 1985. A laboratory study of the effects of aluminum and pH on amphibian eggs and tadpoles. *Canadian Journal of Fisheries and Aquatic Sciences* 42:1544–1551.

Cleveland, L., E.E. Little, S.J. Hamilton, D.R. Buckler, and J.B. Hunn. 1986. Interactive toxicity of aluminum and acidity to early life stages of brook trout. *Transactions of the American Fisheries Society* 115:610–620.

Cornwell, D.A., J.W. Burmaster, J.L. Francis, J.C. Friedlie, C. Houck, P.H. King, W.R. Knocke, J.T. Novak, A.T. Rolan, and R.S. Giacomo. 1987. Committee report: research needs for alum sludge discharge. *Journal of the American Water Works Association* 79:99–104.

Driscoll, C.T., J.P. Baker, J.J. Bisogni, and C.L. Schofield. 1980. Effect of aluminum speciation on fish in dilute acidified waters. *Nature* 284:161–164.

Freeman, R.A., and W.H. Everhart. 1971. Toxicity of aluminum hydroxide complexes in neutral and basic media to rainbow trout. *Transactions of the American Fisheries Society* 100:644–658.

Goenaga, X., and D.J.A. Williams. 1988. Aluminum speciation in surface waters from a Welsh upland area. *Environmental Pollution* 52:131–149.

Helliwell, S., G.E. Batley, T.M. Florence, and B.G. Lumsden. 1983. Speciation and toxicity of aluminum in a model fresh water. *Environmental Technology Letters* 4:141–144.

Hunn, J.B., L. Cleveland, and E.E. Little. 1987. Influence of pH and aluminum on developing brook trout in a low calcium water. *Environmental Pollution* 43:63–73.

Hunter, J.B., S.L. Ross, and J. Tannahill. 1980. Aluminum pollution and fish toxicity. *Journal of the Water Pollution Control Federation* 79:413–420.

Johnson, M.G., L.R. Culp, and S.E. George. 1986. Temporal and spatial trends in metal loadings to sediments of the Turkey Lakes, Ontario. *Canadian Journal of Fisheries and Aquatic Sciences* 43:754–762.

Klatzo, I., H.M. Wisniewski, and E. Streicher. 1965. Experimental production of neurofibrillary degeneration. I. Light microscope observations. *Journal of Neuropathology and Experimental Neurology* 24:187.

Kopeloff, L.M., S.E. Barrera, and N. Kopeloff. 1942. Recurrent convulsive seizures in animals produced by immunologic and chemical means. *American Journal of Psychiatry* 98:881.

Kopeloff, N., L.M. Kopeloff, and B.L. Pacella. 1947. The experimental production of epilepsy in animals. *In: Epilepsy,* eds. P.H. Hoch and R.P. Knight, 163. Grune and Stratton, Orlando, FL.

Krishnan, S., D.R. McLachlan, B. Krishnan, S.S.A. Fenton, and J.E. Harrison. 1988. Aluminum toxicity to the brain. *Science of the Total Environment* 71:59–64.

Lawrence, G.B., and C.T. Driscoll. 1988. Aluminum chemistry downstream of a whole-tree-harvested watershed. *Environmental Science and Technology* 22:1293–1299.

Letterman, R.D., and C.T. Driscoll. 1988. Survey of residual aluminum in filtered water. *Journal of the American Water Works Association* 80:154–158.

Malley, D.F., J.D. Huebner, and K. Donkersloot. 1988. Effects on ionic composition of blood and tissues of *Anodonta grandis grandis* (Bivalvia) of an addition of aluminum and acid to a lake. *Archives of Environmental Contamination and Toxicology* 17:479–491.

Malte, H. 1986. Effects of aluminum in hard, acid water on metabolic rate, blood gas tensions and ionic status in the rainbow trout. *Journal of Fish Biology* 29:187–198.

Marlowe, M., J. Stellern, J. Errera, and C. Moon. 1985. Main and interaction effects of metal pollutants on visual-motor performance. *Archives of Environmental Health* 40:221–225.

Mason, C.F., and S.M. Macdonald. 1988. Metal contamination in mosses and otter distribution in a rural Welsh river receiving mine drainage. *Chemosphere* 17:1159–1166.

Mudroch, A., and G.A. Duncan. 1986. Distribution of metals in different size fractions of sediment from the Niagara River. *Journal of Great Lakes Research* 12:117–126.

Mulder, J., N. van Breeman, and H.C. Eijck. 1989. Depletion of soil aluminum by acid deposition and implications for acid neutralization. *Nature* 337:247–249.

Perl, D.P., and P.F. Good. 1988. Aluminum, environment and central nervous system disease. *Environmental Technology Letters* 9:901–906.

Plankey, B.J., and H.H. Patterson. 1987. Kinetics of aluminum–fulvic acid complexation in acidic waters. *Environmental Science and Technology* 21:595–601.

Plankey, B.J., and H.H. Patterson. 1988. Effect of fulvic acid on the kinetics of aluminum fluoride complexation in acidic waters. *Environmental Science and Technology* 22:1454–1459.

Ramamoorthy, S. 1988. Effect of pH on speciation and toxicity of aluminum to rainbow trout *(Salmo gairdneri). Canadian Journal of Fisheries and Aquatic Science* 45:634–642.

Reader, J.P., T.R.K. Dalziel, and R. Morris. 1988. Growth, mineral uptake and skeletal calcium deposition in brown trout, *Salmo trutta* L., yolk-sac fry exposed to aluminum and manganese in soft acid water. *Journal of Fish Biology* 32:607–624.

Sadler, K., and S. Lynam. 1987. Some effects on the growth of brown trout from exposure to aluminium at different pH levels. *Journal of Fish Biology* 31:209–219.

Seip, H.M., L. Muller, and A. Naas. 1984. Aluminum speciation: comparison of two spectrophotometric analytical methods and observed concentrations in some acidic aquatic systems in southern Norway. *Water, Air and Soil Pollution* 23:81–95.

Simpson, A.M., W. Hatton, and M. Brockbank. 1988. Aluminum, its use and control, in potable water. *Environmental Technology Letters* 9:907–916.

Soderlund, S., A. Forsberg, and M. Pedersen. 1988. Concentrations of cadmium and other metals in *Fucus vesiculosus* L. and *Fontinalis dalecarlica* Br. Eur. from the northern Baltic Sea and the Southern Bothnian Sea. *Environmental Pollution* 51:197–212.

Sprenger, M., and A. McIntosh. 1989. Relationship between concentrations of aluminum, cadmium, lead, and zinc in water, sediments, and aquatic macrophytes in six acidic lakes. *Archives of Environmental Contamination and Toxicology* 18:225–231.

Sprenger, M., A. McIntosh, and T. Lewis. 1987. Variability in concentrations of selected trace elements in water and sediment in six acidic lakes. *Archives of Environmental Contamination and Toxicology* 16:383–390.

Sturman, J.A., and H.M. Wisniewski. 1988. Aluminum. *In: Metal neurotoxicity,* eds. S.C. Bondy and K.N. Prasad, 61–85. CRC Press, Boca Raton, FL.

Subramanian, V., P.K. Jha, and R. van Grieken. 1988. Heavy metals in the Ganges Estuary. *Marine Pollution Bulletin* 19:290–293.

Tipping, E., and J. Hopwood. 1988. Estimating streamwater concentrations of aluminum released from streambeds during acid episodes. *Environmental Technology Letters* 9:703–712.

Urban, N.R., S.J. Eisenreich, and E. Gorham. 1987. Aluminum, iron, zinc, and lead in bog waters of northeastern North America. *Canadian Journal of Fisheries and Aquatic Sciences* 44:1165–1172.

US Environmental Protection Agency. 1988. Ambient water quality criteria for aluminum. 1988. US Environmental Protection Agency, EPA 440/5-86-008, Washington, DC. 47 pp.

Vandelannoote, R., L. van't Dack, and R. van Grieken. 1987. Effects of alkaline aluminate waste dumping on seawater chemistry. *Marine Environmental Research* 21:275–288.

Wisniewski, H.M., R.D. Terry, C. Pena, E. Streicher, and I. Klatzo. 1965. Experimental production of neurofibrillary degeneration. *Journal of Neuropathology and Experimental Neurology* 24:139.

World Bureau of Metal Statistics. 1989. *World metals statistics yearbook.* WBMS, London.

World Resources Institute. 1988. *World resources 1988–1989*. Basic Books, New York. 372 pp.

3
Arsenic

Arsenic, a rare but ubiquitous element, cycles rapidly through water, land, air, and living systems. It occurs in the Earth's crust at an average concentration of 2–5 mg/kg, principally as sulfide and oxide complexes. Arsenic forms a variety of inorganic and organic compounds of different toxicity, reflecting the physicochemical properties of arsenicals of different valency.

Production, Sources, and Residues

Production

Worldwide production of arsenic was approximately 40×10^3 metric tons in 1930, increasing to $49–50 \times 10^3$ metric tons in 1970, but then falling to approximately 45×10^3 metric tons/yr in more recent years (US Minerals Yearbooks, 1930–1988). The major uses of arsenic are insecticides, rodenticides, plant desiccants, pharmaceuticals, textile dying, and detergents.

Sources and Residues

Worldwide anthropogenic input of arsenic to freshwaters is approximately 41×10^3 metric tons/yr (Table 3.1). The most important source is domestic waste water, reflecting the use of arsenic in household preparations and in small industries that discharge effluents to municipal waste systems, particularly in developing nations. Other major sources include sewage sludge, manufacturing processes, and smelting and refining.

Table 3.1. Worldwide anthropogenic input of arsenic to freshwater.

Source	Input (thousand metric tons/yr)
Domestic wastewater	3.0–15.3
Sewage sludge	0.4–6.7
Manufacturing processes	
metals	0.3–1.5
chemicals	0.6–7.0
pulp and paper	0.4–4.2
petroleum products	0–0.1
Smelting and refining	1.0–13
Base metal mining and dressing	0–0.75
Steam electrical production	2.4–14
Atmospheric production	3.6–7.7
Total input	12–70
Median value	41

Source: Nriagu and Pacyna (1988).

In a study of municipal waste treatment plants in the state of Washington, Nevissi et al. (1988) found that primary sludge contained total As residues of 0.06–0.34 mg/L. This resulted in a daily discharge of 0.09 kg total As for primary treated sludge and 0.32 kg total As for digested sludge. The same study showed that total As in the interstitial water of digested sludge ranged from <0.04 to 1.48 mg/L, depending on pH. In another investigation, Sung et al. (1986) determined total As in the effluent of 25 sewage treatment systems in the USA (Table 3.2). Remarkably high residues, up to 3.2 mg/L, were observed in some waste products. In the UK, approximately 15 metric tons of arsenic are deposited each year in marine waters, a result of sewage disposal (Hutton and Symon, 1986).

Manufacturing and nonferrous metal mining and smelting are primary agents in the release of arsenic into surface waters. In the UK, 650 metric tons of total As are deposited in coastal and marine waters each year from the nonferrous metal industry (Hutton and Symon, 1986). This rate is typical of many nations, and results in significant contamination of sediments, periodically in excess of 1,000 mg/kg. Manufacturing discharges are typically less than those of mining and smelting. For example, the UK rate of deposition from the manufacturing sector is 250 metric tons per year (Hutton and Symon, 1986). It should be pointed out that arsenic can be easily removed from the liquid waste of the manufacturing and mining sectors. High deposition rates can be greatly reduced to meet rigid environmental standards using relatively simple control technologies (Environment Canada, 1988).

Atmospheric deposition of arsenic comes mainly from the burning of coal and other fossil fuels for power production, the smelting of minerals

Table 3.2. Concentration of total As in effluents from sewage treatment systems in the USA.

Sample	Number of plants	Concentration (mg/L) Mean	Range
Raw sewage	24	0.006	ND[a]–0.017
Primary effluent	22	0.004	ND–0.021
Secondary effluent	22	0.003	ND–0.019
Discharge	23	0.003	ND–0.020
Primary sludge	22	0.438	0.004–3.20
Secondary sludge	8	0.070	0.004–0.20

[a]Not detected.
Source: Sung et al. (1986).

containing arsenic, and the manufacturing sector where control technologies are not vigorously implemented. Worldwide anthropogenic sources produce 12–26 $\times$ 10^3 metric tons of total As annually, compared to 0.9–23 $\times$ 10^3 metric tons per year for natural sources (Nriagu, 1989). Although the anthropogenic sources of arsenic continue to grow in many developing nations, there has been some decline in emission patterns in developed nations. The following data refer to anthropogenic emissions (metric tons per year) in the Hudson-Raritan Basin, USA (Ayres and Rod, 1986):

Year	1880	1900	1920	1940	1960	1980
Emission	60	250	830	1700	960	680

Arsenic emissions inevitably yield relatively high residues in rainfall, which then contaminates surface water. Scudlark and Church (1988), in their review of a number of studies, reported that total As residues in rainfall near a copper smelter ranged from 0.98 to 17.0 mg/L. For comparison, total As residues in rainfall in Hawaii and other noncontaminated areas were as low as 0.005 mg/L. Scudlark and Church (1988) also noted that total As in rainfall was positively correlated with acidity of the same rainfall, suggesting common atmospheric sources and fates from fossil fuel combustion.

Chemistry

Arsenic potentially exists in four states in surface water (+5, +3, 0, −3) for a total of eight electron reductions. As^{3-} is found only at low Eh, and the metal is extremely rare (Figure 3.1). Under mildly reducing conditions, arsenites predominate, whereas under well-oxygenated conditions

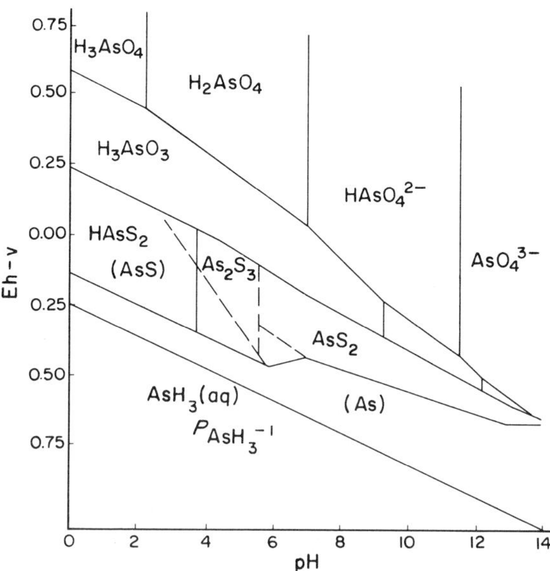

Figure 3.1. Speciation of arsenic as a function of pH and Eh. (From Ferguson and Gavis, 1972.)

and high Eh, arsenates are most common. Oxidation of arsenite to arsenate is slow in most surface waters, increasing at pH extremes. As^{3+} is considered a hard acid and preferentially complexes with oxides and nitrogen, As^{5+}, on the other hand, behaves like a soft acid and preferentially complexes with sulfides. As^{3+} and As^{5+} also undergo a series of biological transformations in aquatic systems, yielding a large number of compounds (Figure 3.2).

Arsenic is predominantly bound to sediments in most water courses. Chunguo and Zihui (1988) showed that the main species in the sediments of the Xiangjiang River (China) were aluminum arsenate, iron arsenate, calcium arsenate, and iron occluded arsenic. Lesser species were soluble As^{3+}, soluble organic As, and organic As. Similarly, Brannon and Patrick (1987) reported that As^{5+} was relatively immobile when combined with iron and aluminum compounds, as also noted for As^{3+} (Sakata, 1987). Sorption of As^{3+} and As^{5+} by humic acids and other organic matter is also possible, a process mediated by pH, absorbate concentration, and ash content of the substrate (Thanabalasingam and Pickering, 1986; Faust et al., 1987).

Desorption of arsenic from sediments is controlled by pH, arsenic concentration in the interstitial water of sediments, and changes in the total iron, extractable iron, and calcium carbonate equivalent in sediments (Xu

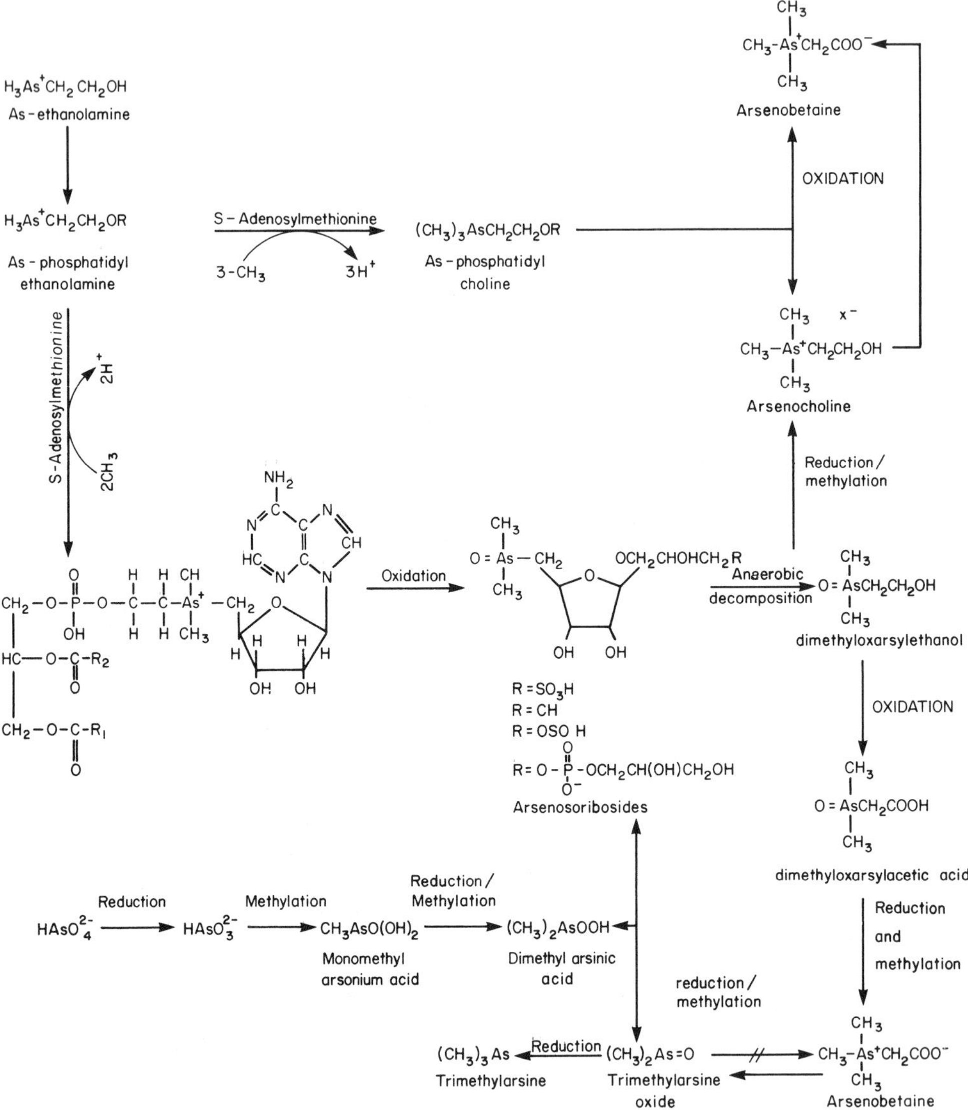

Figure 3.2. Potential biological transformations of arsenic. (From Maher, 1988.)

et al., 1988; Brannon and Patrick, 1987). Studies by Moore et al. (1988) on contaminated sediments in a Montana reservoir indicated that diagenetic sulfides were important sinks for arsenic in reduced, sulfidic sediments, and that diagenetic sulfides controlled the distribution of arsenic. During reduction, oxyhydroxides of iron and manganese dissolved, releasing As^{3+} to the groundwater.

Bioaccumulation

Plants

Arsenic is not a major contaminant of aquatic plants, except in cases of extreme ambient pollution. Some of the highest reported residues (up to 3,700 mg/kg dry weight) for total As came from a series of lakes in northern Canada contaminated by mine and smelter wastes (Wagemann et al., 1978). The aquatic weeds *(Myriophyllum exalbescens)* from another lake in northern Canada contaminated by emissions from a smelter had average total As residues of 263 mg/kg dry weight (Franzin and McFarlane, 1980). Studies in the Calasieu Estuary (USA) indicated that benthic diatoms contained total As at a concentration of 3.8–12.7 mg/kg dry weight, giving a concentration factor of 300 (Ramelow et al., 1987).

It is widely assumed that many algal species can partition inorganic As into sugar moieties, producing arsenobetaine and other organic compounds, which are then available for passage through the food chain. In some cases, the extent of transformation from inorganic to organic compounds is enormous. For example, Klumpp and Peterson (1979) reported that dimethyl arsenate accounted for 86–97% of total As in five species of marine algae whereas the major metabolites in the marine phytoplankter *Dunaliella tertiolecta* were monomethlyarsonic acid, dimethylarsinic acid, arsenite, and arsenate. It should be emphasized that most of the organic As in algae is the result of metabolic transformations and not uptake from the environment.

Invertebrates

Although soft-tissue residues in invertebrates from unpolluted freshwaters generally range from 0.5 to 20 mg/kg, >1,000 mg As/kg dry weight was reported by Wagemann et al. (1978) for several arthropod species from contaminated lakes in northern Canada. Total As in estuarine invertebrates is also often high, exceeding 30 mg/kg in several reports. This is of considerable significance because (1) some invertebrate species are consumed by humans, and (2) relatively little is known about the species of arsenic within invertebrate tissues, particularly those consumed by humans.

Fish

Total As does not normally concentrate in either freshwater or marine fish, so there are only a few reports of noncompliance of tissue residues with health consumption guidelines. Although these guidelines assume that all of the arsenic in fish is inorganic, it has been repeatedly noted that inorganic arsenicals are rapidly transformed to organic compounds in fish. For example, Edmonds and Francesconi (1987) found that oral administration of sodium arsenate to the estuary catfish *Cnidoglanis macrocephalus* and school whiting *Sillago bassensis* resulted in the accumulation of trimethylarsine oxide in their tissues. Since most organic arsenicals exhibit relatively low mammalian toxicity and are rapidly excreted, the consumption guidelines (often given as 0.5 mg/kg) may in fact be too strict.

It has also been noted that the primary arsenical contributing to arsenic burdens in marine fish is arsenobetaine. This compound apparently originates from a food chain source based on algal-derived arsenosugars. Hence, arsenobetaine is not formed in fish tissues from ingested organic As, by bacterial action in the gut, or by other means, and is not related to the formation of trimethylarsine oxide.

It is apparent that, in cases of extreme ambient pollution or other cases where fisheries utilization is limited, analysis of major arsenic species would be useful in defining the threat to consumers.

Toxic Effects to Aquatic Organisms

Arsenate is generally excreted at a faster rate than arsenite, reflecting its poor affinity with thiol groups, and is therefore less toxic than arsenite. The toxic mechanism of arsenite includes inhibition of thiol-dependent enzymes retained in proteins. Arsenate, on the other hand, inhibits ATP synthesis by uncoupling oxidative phosphorylation and replacing the stable phosphoryl group.

Plants, Invertebrates, and Fish

No definitive description is available on the acute toxicity of arsenicals to a cross section of aquatic plants and animals, both marine and freshwater. This reflects, to a large degree, the difficulty in working with the large number of arsenic species present in the environment and the lack of standard test protocols. The work conducted to date indicates that large doses of arsenic, typically greater than 1 mg As/L, are required to induce acute toxic effects in most plants and invertebrates, regardless of As species. Almost no information is available on the effects of temperature, salinity, and water hardness on acute toxicity.

Although more data are available on the toxicity of arsenicals to fish, the lack of standard protocols once again precludes a definitive comparison among the different arsenic species. It appears, however, that relatively large doses, >1 mg As/L, are once again required to elicit acute toxic effects in most fish (Mance, 1987). Water hardness does not appear to ameliorate arsenic-induced toxicosis, and little is known about the effects of temperature and salinity on toxicity.

A wide range of chronic effects has been reported for both invertebrates and fish exposed to inorganic As in water (Table 3.3). In all cases, the dose required to elicit a toxic response is relatively large—typically >5 mg As/L, and occasionally much more. Oral administration of arsenic also leads to relatively mild symptoms (Table 3.3).

Table 3.3. Examples of some long-term effects of arsenicals to aquatic invertebrates and fish.

Species	Effect	Minimum effect level
Mud snail[1] *Nassarius obsoletus*	Reduced oxygen metabolism	Aqueous sodium arsenite, 2 mg As/L
Crab[2] *Scylla serrata*	Histopathologic changes to gills and hepatopancreas	Aqueous arsenic trioxide, 32 mg As/L
Cladoceran[3] *Daphnia magna*	Young per adult decreased, body size decreased	Aqueous sodium arsenite, 1.3 mg As/L
Fathead minnow[3] *Pimephales promelas*	Egg hatch not affected, fry development not affected, fry survival decreased	Aqueous sodium arsenite, 16.5 mg As/L
Flagfish[3] *Jordanella floridae*	Egg hatch not affected, fry development not affected, fry survival not affected	Aqueous sodium arsenite, 16.3 mg As/L
Murrel[4] *Channa punctatus*	Growth decreased, muscle RNA and protein decreased	Aqueous sodium arsenite, 7 mg As/L
Rainbow trout[5] *Oncorhynchus mykiss*	No observed effect on growth	Dietary[a]
Rainbow trout[6]	Hemoglobin decreased, growth decreased, mean corpuscular hemoglobin decreased	Dietary sodium arsenite, 30 mg As/L

[a]Arsenic trioxide: 1–180 mg As/kg; arsanilic acid: >1,497 mg As/kg; dimethylarsenic acid: >1,497 mg As/kg; disodium arsenate: 1–137 mg As/kg.

Sources: [1]MacInnes and Thurberg (1973), [2]Krishnaja et al. (1987), [3]Lima et al. (1984), [4]Shukla et al. (1987), [5]Cockwell and Hilton (1988), [6]Oladimeji et al. (1984).

Several nations have promulgated guidelines for the protection of marine and freshwater organisms:

Canada (total As)
  0.05 mg/L
USA (acid-extractable As)
  0.19 mg/L (4-day average exceeded only once every 3 years)
  0.36 mg/L (1-hour average exceeded only once every 3 years)
European Community (dissolved As)
  0.05 mg/L (freshwater fish, annual average)
  0.15 mg/L (other freshwater life, annual average)
  0.025 mg/L (marine fish and shellfish, annual average)

Health Effects

Intake

Respiratory intake of arsenic by nonoccupationally exposed populations amounts to only 0.0001 mg/day in a 70-kg reference male (Federal Register, 1985). Of this, 0.00003 mg/day is absorbed, assuming 30% absorption and a daily ventilation rate of 20 m^3. Food is a much more significant source: The Food and Drug Administration estimates that the per-capita daily intake in the United States is 0.062 mg, again for the nonoccupationally exposed population. This is equivalent to the daily intake through drinking water.

Acute Toxicity

Acute poisoning in humans is characterized by central nervous system effects, leading to coma and eventual death. The respiratory tract, gastrointestinal tract, and skin are also affected during bouts of severe poisoning. Chronic intoxication results in neurological disorders, muscular weakness, loss of appetite, nausea, and skin disorders such as hyperpigmentation and keratosis. Low-level ingestion of arsenic (1–10 mg As/L drinking water) over extended periods may lead to the onset of acute conditions (World Health Organization, 1984).

Chronic Toxicity

People living near arsenic sources generally carry a much larger body burden of arsenic than those in remote locations. Binder et al. (1987) estimated that children playing on dirt near an abandoned copper smelter ingested up to 1.3 mg As/day. This produced urine As residues of 0.066 mg/L, compared to 0.011 mg/L for control subjects. Arsenic was also found, albeit in relatively low concentrations, in the dust in houses near the smelter.

Inorganic As^{3+} is methylated in the liver of most mammals, and As^{5+} is reduced in the blood prior to methylation in the liver. Dimethylarsinic acid is a major metabolite in mammals, followed by methylarsonic acid and lesser amounts of trimethylarsenic compounds (Yamato, 1988). Approximately 70% of the daily dose of arsenic, which has a half-life of 10–30 h, is excreted in the urine. It appears that the arsenic species present in human urine reflect in vivo metabolism to reduce toxicity, and the nature of the contaminating source (Yamato, 1988).

Arsenic is readily transported to the fetus in pregnant mammals. Both arsenate and arsenite have potential embryolethal effects and teratogenic potential in several mammalian species. Administration of inorganic As to the mother leads to rapid methylation of arsenic in the fetus (Hood et al., 1988), thereby reducing toxicity. The primary metabolites in the fetus of several rodent species are dimethylarsinic acid and methylarsonic acid.

Carcinogenicity

Inorganic As was associated with the genesis of skin cancer as early as 1888 (Pershagen, 1981). The International Agency for Research on Cancer currently classifies inorganic As in Group 1: inadequate evidence for carcinogenicity in animals and sufficient evidence for carcinogenicity (skin and lung) in humans. Although organic As sources such as carbarsone and arsanilic acid have not been tested, it is widely assumed that the carcinogenic potency of these compounds is less than that of inorganic As.

Drinking Water

Residues

Although arsenic is routinely detected in drinking water, residues are generally low and of little toxicologic significance. In a survey by the American Water Works Association (1985) of municipal water systems in 39 states and 3 territories, there were only 46 cases of noncompliance with the Maximum Contaminant Limit of 0.05 mg/L. For comparison, fluoride and nitrates were in noncompliance with guidelines in 907 and 369 cases, respectively. The AWWA survey found that the average concentration of arsenic was 0.08 mg/L (range 0.05–0.19 mg/L) in cases of noncompliance. Most of these values were reported for groundwater samples, reflecting the mobilization of arsenic from geologic formations.

Consumption Guidelines

Many nations, plus the World Health Organization, have adopted a recommended guideline/standard of 0.05 mg/L for arsenic in drinking water. This value was calculated based on an animal study in which adolescent and infant rhesus monkeys were exposed to arsenic for 1 year (Heywood and Sortwell, 1979). A No-Observed-Effect Level of 3.74 mg arsenate/kg/day (2.8 mg As/kg/day) was found. Using an uncertainty factor of 1,000 and a daily water consumption of 2 L per adult, the acceptable daily intake amounted to 0.10 mg As/L. However, the study by Heywood and Sortwell (1979) was based on a relatively small number of animals (four per treatment group), so the recommended level was adjusted downward to 0.05 mg/L (Federal Register, 1985). This lower value gives a daily arsenic intake of 0.10 mg, similar to that from diet and other sources (World Health Organization, 1984).

Although the current drinking water guideline does not consider potential carcinogenic effects, consumption of water at 0.05 mg As/L is projected to increase the number of skin cancers by 2.5 per million. Since the extrapolation model used for the calculation assumes a linear fit, cutting the exposure limit by half (as has been proposed in several nations) will reduce the number of excess skin cancers by 50%. If such changes to the guideline are implemented, it will be useful to conduct epidemiological studies to determine the actual reduction in the number of skin cancers.

Treatment

Arsenic can be effectively removed from drinking water using activated alumina with pH adjustment to approximately 5.5 (Hathaway and Rubel, 1987). Activated alumina without pH adjustment, and strong-base anion exchange resins without pH adjustment are less effective in removal. Another method, end-of- point reverse osmosis, resulted in a 73% drop in arsenic in one study, so that procedure can be used in households and small-scale municipal works (Fox and Sorg, 1987).

The major operational factors affecting arsenic removal are flow rate, media clogging, and downtime, whereas the primary water quality considerations are pH of the feed water, arsenic concentration, sulfate concentration, and alkalinity. The use of activated alumina produces a waste product high in dissolved solids, aluminum, and soluble arsenic. If the waste is acidified, aluminum and arsenic form precipitates that can be separated from the remaining fraction by mechanical dewatering. The solid cake can then be deposited in a landfill.

Recommendations

Arsenic does not constitute a threat to most users of water, so the number of priority areas for research and monitoring is relatively small.

1. Physicochemical studies on the mobilization of arsenic into surface and groundwater.
2. Speciation of arsenic in algae, invertebrates, and fish, particularly those used for human consumption.
3. Epidemiological studies on the occurrence of skin cancer in populations exposed to arsenic in water.
4. The role of arsenic's carcinogenic potential in setting drinking water guidelines.

Studies on the toxicity of arsenicals to marine and aquatic organisms do not appear to be a priority, given the low toxicity of all agents studied to date. In addition, the routine determination of total As residues in algae, invertebrates, and fish should be discouraged in favor of speciation studies.

References

American Water Works Association. 1985. An AWWA survey of inorganic contaminants in water supplies. *Journal of the American Water Works Association* 77:67–72.

Ayres, R.U., and S.R. Rod. 1986. Patterns of pollution in the Hudson-Raritan Basin. *Environment* 28:14–20.

Binder, S., D. Forney, W. Kaye, and D. Paschal. 1987. Arsenic exposure in children living near a former copper smelter. *Bulletin of Environmental Contamination and Toxicology* 39:114–121.

Brannon, J.M., and W.H. Patrick. 1987. Fixation, transformation, and mobilization of arsenic in sediments. *Environmental Science and Technology* 21:450–459.

Chunguo, C., and L. Zihui. 1988. Chemical speciation and distribution of arsenic in water, suspended solids and sediment of Xiangjiang River, China. *Science of the Total Environment* 77:69–82.

Cockwell, K.A., and J.W. Hilton. 1988. Preliminary investigations on the comparative chronic toxicity of four dietary arsenicals to juvenile rainbow trout (*Salmo gairdneri* R.). *Aquatic Toxicology* 12:73–82.

Edmonds, J.S., and K.A. Franesconi. 1987. Trimethylarsine oxide in estuary catfish *(Cnidoglanis macrocephalus)* and school whiting *(Sillago bassensis)* after oral administration of sodium arsenate; and as a natural component of estuary catfish. *Science of the Total Environment* 64:317–323.

Environment Canada. 1988. *Status report on water pollution control in the Canadian metal mining industry (1986)*. Report EPS 1/MM/3, Environment Canada, Ottawa. 29 pp.

Faust, S.D., A.J. Winka, and T. Belton. 1987. An assessment of chemical and biological significance of arsenical species in the Maurice River drainage basin (N.J.). II. Partitioning of arsenic into bottom sediments. *Journal of Environmental Science and Health* A22:239–262.

Federal Register. 1985. National primary drinking water regulations. *Federal Register* 50:46936–47022.

Ferguson, J.F., and J. Gavis. 1972. A review of the arsenic cycle in natural waters. *Water Research* 6:1259–1274.

Fox, K.R., and T.J. Sorg. 1987. Controlling arsenic, fluoride, and uranium by point-of-use treatment. *Journal of the American Water Works Association* 79:81–84.

Franzin, W.G., and G.A. McFarlane. 1980. An analysis of the aquatic macrophyte, *Myriophyllum exalbescens,* as an indicator of metal contamination of aquatic ecosystems near a base metal smelter. *Bulletin of Environmental Contamination and Toxicology* 24:597–605.

Hathaway, S.W., and F. Rubel. 1987. Removing arsenic from drinking water. *Journal of the American Water Works Asssociation* 79:61–65.

Heywood, R., and R.J. Sortwell. 1979. Arsenic intoxication in the rhesus monkey. *Toxicology Letters* 3:137–144.

Hood, R.D., G.C. Vedel, M.J. Zaworotko, F.M. Tatum, and R.G. Meeks. 1988. Uptake, distribution, and metabolism of trivalent arsenic in the pregnant mouse. *Journal of Toxicology and Environmental Health* 25:423–434.

Hutton, M., and C. Symon. 1986. The quantities of cadmium, lead, mercury and arsenic entering the U.K. environment from human activities. *Science of the Total Environment* 57:129–150.

Klumpp, D.W., and P.J. Peterson. 1979. Arsenic and other trace elements in the waters and organisms of an estuary in SW England. *Environmental Pollution* 19:11–20.

Krishnaja, A.P., M.S. Rege, and A.G. Joshi. 1987. Toxic effects of certain heavy metals (Hg, Cd, Pb, As and Se) on the intertidal crab *Scylla serrata. Marine Environmental Research* 21:109–119.

Lima, A.R., C. Curtis, D.E. Hammermeister, T.P. Markee, C.E. Northcott, and L.T. Brooke. 1984. Acute and chronic toxicities of arsenic (III) to fathead minnows, flagfish, daphnids, and an amphipod. *Archives of Environmental Contamination and Toxicology* 13:595–601.

MacInness, J.R., and F.P. Thurberg. 1973. Effects of metals on the behavior and oxygen consumption of the mud snail. *Marine Pollution Bulletin* 4:185–186.

Maher, W.A. 1988. Arsenic in the marine environment of south Australia. *In: The biological alkylation of heavy metals,* eds. P.J. Craig and F. Glockling, 120–126. Royal Society of Chemistry, London.

Mance, G. 1987. *Pollution threat of heavy metals in aquatic environments.* Elsevier Applied Science, New York. 373 pp.

Moore, J.N., W.H. Ficklin, and C. Johns. 1988. Partitioning of arsenic and metals in reducing sulfidic sediments. *Environmental Science and Technology* 22:432–437.

Mudroch, A., L. Sarazin, and T. Lomas. 1988. Summary of surface and background concentrations of selected elements in the Great Lakes sediments. *Journal of Great Lakes Research* 14:241–251.

Nevissi, A.E., F.B. DeWalle, J.F.C. Sung, K. Mayer, and R. Dalsey. 1988. Heavy metal variability of different municipal sludges as measured by atomic absorption and inductively coupled plasma emission spectroscopy. *Journal of Environmental Science and Health* A23:823–841.

Nriagu, J.O. 1988. A silent epidemic of environmental metal poisoning. *Environmental Pollution* 50:139–161.

Nriagu, J.O. 1989. A global assessment of natural sources of atmospheric trace metals. *Nature* 338:47–49.

Nriagu, J.O., and J.M. Pacyna. 1988. Quantitative assessment of worldwide contamination of air, water and soils by trace metals. *Nature* 333:134–139.

Oladimeji, A.A., S.U. Qadri, and A.S.W. deFreitas. 1984. Long-term effects of arsenic accumulation in rainbow trout, *Salmo gairdneri. Bulletin of Environmental Contamination and Toxicology* 32:732–741.

Pershagen, G. 1981. The carcinogenicity of arsenic. *Environmental Health Perspectives* 40:93–100.

Ramelow, G.J., R.S. Maples, R.L. Thompsom, C.S. Mueller, C. Webre, and J.N. Beck. 1987. Periphyton as monitors of heavy metal pollution in the Calcasieu River estuary. *Environmental Pollution* 43:247–261.

Sakata, M. 1987. Relationship between adsorption of arsenic (III) and boron by soil and soil properties. *Environmental Science and Technology* 21:1126–1130.

Scudlark, J.R., and T.M. Church. 1988. The atmospheric deposition of arsenic and association with acid precipitation. *Atmospheric Environment* 22:937–943.

Shukla, J.P., K.N. Shukla, and U.N. Dwivedi. 1987. Survivality and impaired growth in arsenic treated fingerlings of *Channa punctatus*, a fresh water murrel. *Archives of Hydrochemistry and Hydrobiology* 15:307–311.

Sung, J.F.C., A.E. Nevissi, and F.B. DeWalle. 1986. Concentration and removal of major trace elements in municipal wastewater. *Journal of Environmental Science and Health* A21:435–448.

Thanabalasingam, P., and W.F. Pickering. 1986. Arsenic sorption by humic acids. *Environmental Pollution* 12:233–246.

US Environmental Protection Agency. 1985. *Ambient water quality criteria for arsenic—1984*. US Environmental Protection Agency, EPA 440/5-84-033, Washington, DC. 66 pp.

US Minerals Yearbooks. 1930–1988. Bureau of mines, US Department of the Interior, Washington, DC.

Wagemann, R., N.B. Snow, D.M. Rosenberg, and A. Lutz. 1978. Arsenic in sediments, water and aquatic biota from lakes in the vicinity of Yellowknife, Northwest Territories, Canada. *Archives of Environmental Contamination and Toxicology* 7:169–191.

World Health Organization. 1984. *Guidelines for drinking-water quality*. WHO, Geneva.

Xu, H., B. Allard, and A. Grimwall. 1988. Influence of pH and organic substance on the adsorption of As (V) on geologic materials. *Water, Air, and Soil Pollution* 40:293–305.

Yamato, N. 1988. Concentrations and chemical species of arsenic in human urine and hair. *Bulletin of Environmental Contamination and Toxicology* 40:633–640.

4
Asbestos

Asbestos is a generic term for a variety of hydrated silicate minerals that have been crystallized to form long, flexible fibers and can in turn be separated into bundles of fibrils. Asbestos is distinguished from non-asbestiform analogs by the presence of easily separated fibers, typically measuring >5 μm in length. There are six common types of asbestiform minerals and an equal number of asbestiform analogs (Table 4.1). Chrysotile is generally the most common form found in water.

Production, Sources, and Residues

Production

Annual world production of asbestos has amounted to approximately 4.2×10^6 metric tons in recent years, making it the 28th most important mineral in terms of production weight (Noetstaller, 1988). Canada is the world's largest exporter of asbestos, and Japan is the world's largest importer (Table 4.2). Continuing concern over the potential carcinogenicity of asbestos will likely erode demand for asbestos, particularly in developed nations. The main uses of asbestos are friction products, floor products, coatings and compounds, asbestos cement, packing and gaskets, asbestos cement sheet, and roofing products.

Table 4.1. Asbestos minerals and nonasbestiform analogs.

Asbestiform	Nonasbestiform
Serpentine	
Chrysotile	Antigorite
Amphibole	
Anthophyllite asbestos	Anthophyllite
Amosite	Cummingtonite–grunerite
Actinolite asbestos	Actinolite
Tremolite asbestos	Tremolite
Crocidolite	Riebeckite

Sources and Residues

Natural and, to a lesser degree, anthropogenic discharges contribute to asbestos in surface water. Concentrations of more than 100 million fibers/L have been observed in surface water where bedrock formations contribute to asbestos loadings (Table 4.3). Some of the aqueducts in California, built on serpentine deposits, contain sediments with a chrysotile asbestos load of up to 2.6% by weight (Jones and McGuire, 1987). In another case, a major landslide in the Sumas River, Washington, exposed serpentinitic rocks, resulting in the mobilization of large amounts of asbestos (Schreier, 1987). Below the slide, asbestos fiber concentrations ranged from 380 million/L to 340,000 million/L. Fiber concentration in that study was related to the seasonal pattern of rainfall and river discharge. Extremely high fiber residues also resulted in increased concentrations of nickel, chromium, cobalt, and manganese in river water.

Rainfall is a potentially significant source of asbestos in surface water, particularly in areas of asbestos-bearing geologic formations. Bacon et al. (1986) found that rainfall in southeastern Quebec carried asbestos residues of 14.6 million fibers/L (range 1.9–23.7 million fibers/L). That area

Table 4.2. World's largest exporters and importers of asbestos.

Exporter	Quantity (1,000 metric tons/yr)	Importer	Quantity (1,000 metric tons/yr)
Canada	753	Japan	237
USSR	276	FRG	155
South Africa	171	USA	141
Zimbabwe	150	Republic of Korea	113
Italy	20	France	73

Source: Noetstaller (1988).

Table 4.3. Examples of asbestos concentrations (million fibers/L) in surface waters.

Location	Concentration	Source
Lakes, southeast Quebec (Canada)[1]	1.7–147.8	Mainly geologic plus rail ballast and rainfall
Lakes, Staten Island, New York[2]	15–86	Geologic serpentine formation
Reservoirs, southern California[3]	100–200,000	Geologic serpentine formation
Reservoirs, New Jersey[4]	0–4.7	Crocidolite fault zone
Stream, New Jersey[4]	0.3–3.5	Tremolite-actinolite quarry
River, northern British Columbia[5]	max 10,000	Metamorphic rocks

Sources: [1]Bacon et al. (1986), [2]Maresca et al. (1984), [3]Hayward (1984), [4]Puffer et al. (1983), [5]Schreier and Taylor (1980).

of Quebec contains some of the world's largest deposits of asbestos. Hallenbeck (1977) reported much lower levels in rainfall (0.1–1 million fibers/L) over Chicago, where the only sources were motor vehicles and construction sites.

The other major source for surface waters is industrial/ municipal effluents. In one study, Stewart et al. (1976) sampled effluent from 22 industrial plants which used about 45% of all asbestos in the USA. Asbestos residues in the effluents of the plants ranged from nondetectable to over 100,000 million fibers/L, much more than that associated with natural sources. Asbestos paper plants were the greatest source of fibers. Similarly, Patel-Mandlik et al. (1988) reported the presence of asbestos in 9 of 15 municipal sludges from communities in the USA. The major forms were chrysotile and amosite, accounting for 1–5% by volume of the sludge ash. In another study, Monaro et al. (1983) determined asbestos residues in the Becancour River (Quebec), which receives waste waters from asbestos mines. Maximum concentrations in the river exceeded 5,000 million fibers/L during periods of high discharge, but decreased rapidly in lakes downstream of the mines.

Chemistry

Asbestos, as a mineral group, is resistant to many of the processes of chemical speciation such as photolysis, hydrolysis, volatilization, oxidation, and reduction. Although this effectively eliminates the existence of species based on such processes, surface charge and sorption can be used to delineate some asbestos forms. For example, Bales and Morgan (1985) showed that chrysotile freshly suspended in an inorganic electrolyte had a positive surface charge below pH 8.9. However, as the fibers aged over 2 weeks, the outer magnesium hydroxide surface broke down, exposing silica and reversing the charge. There was no absorption of inorganic anions such as nitrate, chloride, and bicarbonate, but natural organic material was absorbed.

Asbestos can also be speciated based on its heavy metal content. This is of considerable toxicologic significance because the deposit of large amounts of asbestos into surface waters also introduces potentially toxic metals to the same system. Schreier (1987) found that a slide of asbestos-bearing rock into the Sumas River (Washington) resulted in chromium and nickel residues of up to 0.034 and 0.525 mg/L in the water. For comparison, maximum residues at the control site were 0.006 and 0.040 mg/L, respectively. Other studies have shown that, in addition to chromium and nickel, metals such as copper and cobalt may be present in relatively high levels. These are important observations that must be considered during routine monitoring in areas where asbestos residues are thought to be high.

Bioaccumulation

Plants, Invertebrates, and Fish

Algae, macrophytes, microcrustaceans, molluscs, and fish can accumulate chrysotile. Uptake is particularly pronounced in filter feeders, and concentration factors (fibers in tissues/fibers in water) of up to 100 have been found in the viscera of the clam *Corbicula*, and up to 500 in whole clam homogenates (Belanger et al., 1987). Such high concentration factors are due to the filtering of the large volumes of water used in respiration. This is of considerable significance to human health if the clams are ultimately used for food.

Table 4.4. Concentration (fibers/mg wet weight) of asbestos in lake trout caught from a contaminated site (Split Rock) and a control site (Huron Bay) in Lake Superior.

	Contaminated site		Control site	
Fish tissue	Amphibole	Chrysotile	Amphibole	Chrysotile
Muscle	1.0	27.0	1.4	0.7
Kidney	1,247	<6.2	35	5.4
Liver	20.5	20.5	<6.5	6.5

Source: Batterman and Cook (1981).

Since fish sorb asbestos primarily through food and, to a lesser degree, the gills, residues in tissues are generally low. For example, Batterman and Cook (1981) examined the tissues of lake trout *Salvelinus namaycush* collected from a highly contaminated site in western Lake Superior and from a control site in the central portion of the lake (Table 4.4). Both amphibole and chrysotile were elevated in fish from the contaminated site, but the concentration factors were <0.01.

Overall, asbestos does not appear to pose a health risk to consumers of fish, including those caught in contaminated waters.

Toxic Effects to Aquatic Organisms

Although acute toxic effects of asbestos are rarely reported, even in highly contaminated waters, chronic intoxication occurs in a number of species when residues fall in the 1–100 million fibers/L range. For example, Belanger et al. (1986a,b) found that exposure of the clam *Corbicula* to those levels caused damage to gill tissues and that such changes (plus shell and tissue growth) were most pronounced in the winter, when food supply was limited.

In another study, Belanger et al. (1986c) reported that behavioral stress effects such as rheotaxis were observed in coho salmon *Oncorhynchus kisutch* at 3 million fibers/L and green sunfish *Lepomis cyanellus* at 1.5–3 million fibers/L. The ventral epidermis of both species contained necrotic cells, as did the lateral line of coho salmon.

Health Effects

Intake

Assuming a daily water consumption of 2 L, the number of fibers ingested annually per person is 0.9 million to well over 10,000 million (Rowe, 1983). Daily exposure through inhalation of ambient air is approximately 3,000 fibers at a ventilation rate of 20 m^3 for the adult males, giving a yearly total of 1.1 million (Federal Register, 1985). Although data are highly variable, diet contributes at least as much asbestos as water, and periodically much more.

Acute Toxicity

Asbestos is generally not acutely toxic to humans or experimental animals (except at very high doses).

Chronic Toxicity and Carcinogenicity

Numerous effects have been observed following inhalation of asbestos fibers, including the development of lung tumors and mesothelioma in laboratory animals and occupationally exposed workers. As a result, the International Agency for Research on Cancer classifies asbestos as Group 1: sufficient evidence for carcinogenicity in animals and humans. This classification is based on inhalation of asbestos and is independent of fiber length. It has been estimated that, by the year 2000, cancers attributable to inhalation of asbestos will number from 550 to 233,000 in occupationally exposed workers in the USA (Mauskopf, 1987). The large discrepancy between these estimates reflects differences in projected exposure and the predictive model used to generate the data.

Several epidemiological studies have been conducted on potential associations between ingestion of asbestos fibers and gastrointestinal cancer. Marsh (1983), having reviewed 13 such studies, concluded that, although one or more studies found associations between exposure and cancer mortality/incidence, no study exists that would establish risk levels for ingested asbestos. Similarly, the majority of ingestion studies have failed to produce a carcinogenic effect in laboratory animals.

Drinking Water

Residues

Asbestos occurs in drinking water as a result of contaminated raw water supplies and as a result of corrosion of asbestos cement pipes in the distribution system. In a survey of drinking water in 406 cities in the USA, asbestos was found in 289 systems, as outlined below (Federal Register, 1985):

Asbestos (million fibers/L)	<1	1–10	>10	not detected
Number of cities	216	33	40	117

Another report listed asbestos in 12 of 100 systems, with residues of 0.4 million to 1,071 million fibers/L (Federal Register, 1985).

One of the most dramatic episodes of contamination of drinking water occurred in Woodstock, New York, a result of deteriorating asbestos cement (Webber et al., 1988). The pipe had eroded over a number of years and caused the clogging of faucet strainers. Contamination in excess of 10,000 million fibers/L was detected in some samples. The residues were so high that household air also became contaminated compared to controls (0.12 vs. 0.04 fibers/cm^3). Potential health impacts of the episode have yet to be determined.

It should be pointed out that most developed nations prohibit the installation of new asbestos cement pipe in delivery systems, and that current problems solely reflect the deterioration of old pipes.

Consumption Guidelines

Although there is no firm evidence that asbestos in drinking water is a threat to consumers, the association between occupational fiber inhalation and the genesis of lung cancer and mesothelioma suggests that drinking water guidelines should be developed. To date, however, many nations plus the World Health Organization have not yet finalized any guideline. The US Environmental Protection Agency proposed the use of a one hit model, predicting that 7.1×10^7, 7.1×10^6, and 7.1×10^5 fibers/L would result in a lifetime cancer risk of 10^{-5}, 10^{-6}, and 10^{-7}, respectively. These levels were calculated for the intermediate length of chrysotile fibers only. If a risk of 10^{-6} is accepted, the recommended guideline is 7.1×10^6 fibers/L.

Treatment

Asbestos in drinking water can be readily controlled using both conventional and special coagulation-filtration treatment process trains. Existing facilities have achieved reductions of 18- to 300-fold, whereas pilot-scale processes have achieved removals of more than 10,000-fold. Good process control in many plants can yield residues in finished water of 0.1–1.1 million fibers/L.

Recommendations

Because asbestos has achieved such notoriety as an airborne carcinogen, it is likely that interest in waterborne asbestos will also remain high for some time. Certainly, the presence of enormous amounts of asbestos cement in water delivery systems means that relatively high residues will be found in drinking water for many years to come. However, no health problem has been clearly related to the ingestion of asbestos in water or food products such as fish. The following priorities reflect these considerations.

1. Mobilization of heavy metals in surface waters when ambient asbestos residues are high.
2. Asbestos residues in clams and other filter feeders that may be used for human food.
3. Chronic toxicity of asbestos at environmentally relevant concentrations to freshwater and marine organisms.

4. Sensitive epidemiological surveys relating exposure to neoplasia in the stomach, pancreas, esophagus, and small intestine.
5. Establishment of a widely accepted guideline or maximum recommended contaminant level for asbestos in drinking water.

References

Bacon, D.W., O.T. Coomes, A.A. Marsan, and N. Rowlands. 1986. Assessing potential sources of asbestos fibers in water supplies of S.E. Quebec. *Water Resources Bulletin* 22:29–38.

Bales, R.C., and J.J. Morgan. 1985. Surface charge and adsorption properties of chrysotile asbestos in natural waters. *Environmental Science and Technology* 19:1213–1219.

Bales, R.C., D.D. Newkirk, and S.B. Hayward. 1984. Chrysotile asbestos in California surface waters: from upstream rivers through water treatment. *Journal of the American Water Works Association* 76:66–74.

Batterman, A.R., and P.M. Cook. 1981. Determination of mineral fiber concentrations in fish tissues. *Canadian Journal of Fisheries and Aquatic Sciences* 38:952–959.

Belanger, S.E., D.S. Cherry, and J. Cairns. 1986a. Uptake of chrysotile asbestos fibers alters growth and reproduction of Asiatic clams. *Canadian Journal of Fisheries and Aquatic Sciences* 43:43–52.

Belanger, S.E., D.S. Cherry, and J. Cairns. 1986b. Seasonal, behavioral and growth changes of juvenile *Corbicula fluminea* exposed to chrysotile asbestos. *Water Research* 20:1243–1250.

Belanger, S.E., K. Schurr, D.J. Allen, and A.F. Gohara 1986c. Effects of chrysotile asbestos on coho salmon and green sunfish: evidence of behavioral and pathological stress. *Environmental Research* 39:74–85.

Belanger, S.E., D.S. Cherry, J. Cairns, and M.J. McGuire. 1987. Using Asiatic clams as a biomonitor for chrysotile asbestos in public water supplies. *Journal of the American Water Works Association* 79:69–74.

Federal Register. 1985. National primary drinking water regulations. *Federal Register* 50:46937–47022.

Hallenbeck, W.H. 1977. *Asbestos in potable water*. National Technical Information Service, PB-265-389, Springfield, VA. 78 pp.

Hayward, S.B. 1984. Field monitoring of chrysotile asbestos in California waters. *Journal of the American Water Works Association* 76:66–73.

Jones, J., and M.J. McGuire. 1987. Dredging to reduce asbestos concentrations in the California aqueduct. *Journal of the American Water Works Association* 79:30–37.

Maresca, G.P., J.H. Puffer, and M. Germine. 1984. Asbestos in lake and reservoir waters of Staten Island, New York: source, concentration, mineralogy and size distribution. *Environmental Geology and Water Science* 6:201–210.

Marsh, G.M. 1983. Critical review of epidemiological studies related to ingested asbestos. *Environmental Health Perspectives* 53:49–56.

Mauskopf, J.A. 1987. Projections of cancer risks attributable to future exposure to asbestos. *Risk Analysis* 7:477–486.

Monaro, S., S. Landsberger, R. Lecomte, and P. Paradis. 1983. Asbestos pollution levels in river water measured by proton-induced X-ray emission (Pixie) techniques. *Environmental Pollution* 5:83–90.

Noetstaller, R. 1988. *Industrial minerals*. World Bank Technical Paper Number 76, Washington, DC 117 pp.

Patel-Mandlik, K.J., C.G. Manos, and D.J. Lisk. 1988. Identification of asbestos and glass fibers in sewage sludges of small New York State cities. *Chemosphere* 17:1025–1032.

Puffer, J.H., G.P. Maresca, and M. Germine. 1983. *Asbestos in water supplies of the northern New Jersey area: source, concentration, mineralogy, and size distribution*. National Technical Information Service, PB84-136811, Springfield, VA. 22 pp.

Rowe, J.N. 1983. Relative source contributions of diet and air to ingested asbestos exposure. *Environmental Health Perspectives* 53:115–120.

Schreier, H. 1987. Asbestos fibers introduce trace metals into streamwater and sediments. *Environmental Pollution* 43:229–242.

Schreier, H., and J. Taylor. 1980. *Asbestos fibers in receiving waters*. Department of Environment, Technical Bulletin 117, Vancouver, BC, Canada. 18 pp.

Stewart, I.M., R.E. Putscher, H.J. Humecki, and R.J. Shimps. 1976. *Asbestos fibers in natural runoff and discharges from sources manufacturing paper products. II. Non-point sources and point sources manufacturing asbestos products*. National Technical Information Service, PB-263-746, Springfield, VA. 166 pp.

Webber, J.S., S. Syrotynski, and M.V. King. 1988. Asbestos-contaminated drinking water: its impact on household air. *Environmental Research* 46:153–167.

5
Barium

Barium is a reactive metal, existing in nature mainly as the mineral barite (barium sulfate), with lesser amounts of witherite (barium carbonate). Although the concentration of barium in the Earth's crust generally ranges from 300 to 500 mg/kg, certain geologic formations, such as fossil fuel deposits, yield concentrations in excess of 100,000 mg/kg. Barium and calcium are metabolized together in most animal species, and are predominantly deposited in the skeleton.

Production, Sources, and Residues

Production

World production of barite in recent years has amounted to 5.5×10^6 metric tons, making it the 25th most important mineral in terms of production weight (Noetstaller, 1988). China leads the way in world exports, followed by Morocco and Ireland, whereas the main importers are the USA, the UK, and the GDR (Table 5.1).

The major application of barite is as a drilling fluid in oil and gas wells. Barite is circulated down the well to cool and lubricate the bit and string, removing cuttings from the hole and controlling formation pressures. This accounts for over 90% of barium use, the remainder going to the production of lithopone (zinc sulfide + barite) and barium chemicals.

Sources and Residues

Most episodes of barium contamination of water come from natural geologic deposits and untoward spills of drilling fluids, rather than from in-

Table 5.1. World's largest exporters and importers of barite.

Exporter	Quantity (1,000 metric tons/yr)	Importer	Quantity (1,000 metric tons/yr)
China	753	USA	1,246
Morocco	276	UK	128
Ireland	171	GDR	103
Thailand	150	Norway	90
Turkey	20	The Netherlands	81

Source: Noetstaller (1988).

dustrial and municipal discharges. Although residues in surface water are generally <0.1 mg/L, some underground sources contain levels of up to 10 mg/L in geothermal brines (World Health Organization, 1984; Brenniman et al., 1979). Vastly higher residues can be detected around offshore drilling platforms when drilling fluids are discharged to the surrounding water. Barium in sediments is typically <100 mg/kg dry weight, except in cases of geologic deposits, where much higher levels can once again be observed (Byrne and DeLeon, 1987; Forstner and Wittmann, 1979).

Barium in the atmosphere comes mainly from wind erosion of soil particles. In their review of seven studies, Wiersma and Davidson (1986) noted that barium over Antarctica amounted to only 0.02 ng/m^3, but in Montana concentrations were 8–10 ng/m^3. Another review (Cass and McRae, 1986) noted that emissions of particulate barium in the Los Angeles area came to 193 kg/day, compared to 35,000 kg/day for aluminum and 8,800 kg/day for lead. Overall, therefore, atmospheric deposition does not constitute a significant source of barium in surface waters, except in areas where geologic formations contribute substantially to the flux of barium.

Chemistry

Although barium is often detected in water, relatively little is known about its environmental chemistry and biological fate. The principal form of barium in natural freshwaters is Ba^{2+} (American Water Works Association, 1988). Although the halide, nitrate, and acetate salts of barium are soluble in water, the carbonate, sulfate, chromate, fluoride, oxalate, and phosphate salts are relatively insoluble. This means that excess carbonate and sulfate in freshwaters limits the concentration of barium through precipitation, whereas in estuaries and coastal waters, excess chloride enhances solubility.

A major fate process at circumneutral pH is sorption, but the rate of desorption increases greatly with declining pH. For example, the concentration of barium in leachate from fly ash was approximately 2.8 mg/L at pH 7, increasing to 5.4 mg/L at pH 5 (Robertson et al., 1986). This finding is of considerable environmental relevance because declining pH in many surface waters (a result of acid deposition) is likely to mobilize increasing amounts of barium. Other environmental fate processes, including photolysis, hydrolysis, and biological transformation, do not appear to significantly affect barium in either freshwater or saltwater.

Bioaccumulation

Barium is accumulated by freshwater and marine organisms. Residues in the brown algae *Laminaria saccharina* along the coast of New England averaged 11.1 mg/kg wet weight, compared to 0.8–46.4 mg/kg for the red algae *Ptilota serrata* from the same location (Sears et al., 1985). Submerged plants along the shore of Lake Huron contained residues ranging from nondetectable to 160 mg/kg whereas plants from small Michigan lakes contained barium in the range nondetectable–256 mg/kg (Estabrook et al., 1985).

Barium is passed through the food chain but is not concentrated; hence, residues in invertebrates and higher species seldom exceed 10 mg/kg and are generally much lower (Howard and Brown, 1986, 1987; Phillips et al., 1987). There appears to be no instance in which high barium levels have caused a limitation on human consumption of fish and other food products.

Toxic Effects to Aquatic Organisms

Relatively little is known about the toxicity of barium in marine and freshwater species. It is generally believed that barium is, at most, moderately acutely toxic, elicting responses comparable to those of chromium and manganese (Wang, 1986a). Furthermore, the toxicity of drilling fluids, consisting mainly of barite, is also low to a wide range of species. For example, Petrazzuolo (1981) calculated the 96-h LC_{50} values for 45 marine animal species using a total of 116 different fluids. He found that the lethal concentration was generally more than 1,000 mg fluid/L, as listed below:

Number of species	Number of fluids	Number of 96-h LC_{50} values (mg/L)			
		<100	100–999	1,000–99,999	>100,000
45	116	1	8	129	125

Little information is available on the factors affecting the toxicity of barium, apart from the fact the ligands significantly reduce toxic effects to aquatic plants and presumably animal species (Wang, 1986b). Some jurisdictions use a guideline of 5 mg Ba/L for the protection of aquatic life (Wang, 1988), but this is preliminary and based on limited information.

Health Effects

Intake

Adult intake of barium from dietary sources ranges from 0.4 to 1.2 mg/day in Western nations (World Health Organization, 1984; Federal Register, 1985). Respiratory intake is much lower in the nonoccupationally exposed population, amounting to only 0.0004–0.026 mg/day. It is estimated that a 70-kg male contains a total barium burden of 22 mg, of which 20 mg is stored in the skeleton. Excretion of barium occurs primarily (>75%) through the feces, rather than in the urine or sweat.

Acute and Chronic Toxicity

Acute exposure in humans and experimental animals results in gastrointestinal effects (nausea, vomiting, diarrhea), followed by myocardial and general muscular stimulation and, eventually, respiratory arrest. Chronic exposure via the respiratory route may result in a benign pneumoconiosis, known as baritosis. Subcutaneous exposure of experimental animals has caused spleen, liver, and bone marrow lesions and subsequent effects on blood.

Carcinogenicity

The International Agency for Research on Cancer has not evaluated barium for possible carcinogenic effects because of inadequate data from animal and human studies.

Drinking Water

Residues

Barium is regularly found in drinking water, often in relatively high concentrations. In a major survey of water systems in the United States, barium residues were in noncompliance with the recommended contaminant limit (1.0 mg/L) in 39 cases out of a total of 1,583 episodes of noncompliance for all other contaminants (American Water Works Association, 1985). The average concentration of barium in these cases was 3.0 mg/L, with a range of 1.1–9.6 mg/L. Another survey of 132 groundwater systems showed that residues in 18 systems exceeded 0.25 mg/L and that two systems exceeded 1.0 mg/L (American Water Works Association, 1985).

Consumption Guidelines

The current guideline/standard for the protection of drinking water in several nations is 1.0 mg barium/L. This value was derived by using the adjusted acceptable daily intake (AADI) of 1.8 mg/L derived from animal studies, and a survey in Illinois in which adults consumed water with a barium content of up to 10 mg/L for 10 years (Brenniman et al., 1981). Since no hypertensive effects were noted in the latter study, it is assumed that the use of a relatively low safety factor (10) protects consumers over a 70-year life span. The US Environmental Protection Agency (1989) recently reduced the safety factor to 2, thereby giving a lifetime health advisory of 5 mg/L. It should be noted, however, that no epidemiological data are available to evaluate this concentration.

Treatment

Barium is easily removed from drinking water by lime-softening to pH 10–11, followed by ion exchange (American Water Works Association, 1988). Since the recommended contaminant limit is relatively high, removal targets can be effectively achieved using this procedure. Some resins have a greater affinity for barium than either calcium or magnesium (Snoeyink et al., 1987). The ion exchange resin can be regenerated with reclaimed brine, a procedure that reduces disposal problems (Myers et al., 1985).

Recommendations

It is widely assumed that barium poses a relatively low environmental and health risk, and certainly the small number of studies conducted to date reflect that point of view. However, increased mobilization of barium, as acidification of surface waters increases in many nations, may also increase residues, potentially to the point where chronic toxic effects will be observed in some water users. Detailed studies on the environmental chemistry of barium in coastal and freshwaters is a good first step in resolving these issues. If barium mobilization is significant at environmentally relevant pH, follow-up studies on chronic effects in freshwater and coastal species, both plant and animal, should be initiated.

The revision of the drinking water guideline to 5.0 mg/L needs to be carefully evaluated, considering the unique reduction in the safety factor and the insufficient data on long-term (>10 years) effects. The following recommendations reflect these considerations:

1. Environmental fate of barium in coastal and freshwaters.
2. Chronic effects of barium exposure to marine and freshwater species, both plant and animal.
3. Revision of drinking water guidelines.

References

American Water Works Association. 1985. An AWWA survey of inorganic contaminants in water supplies. *Journal of the American Water Works Association* 77:67–72.

American Water Works Association. 1988. A review of solid-solution interactions and implications for the control of trace inorganic materials in water treatment. *Journal of the American Water Works Association* 80:56–64.

Brenniman, G.R., T. Namekata, W.H. Kojola, B.W. Carnow, and P.S. Levy. 1979. Cardiovascular disease death rates in communities with elevated levels of barium in drinking water. *Environmental Research* 20:318–324.

Brenniman, G.R., W.H. Kojola, P.S. Levy, B.W. Carnow, and T. Namekata. 1981. High barium levels in public drinking water and its association with elevated blood pressure. *Archives of Environmental Health* 36:28–32.

Byrne, C.J., and I.R. DeLeon. 1987. Contributions of heavy metals from municipal runoff to the sediments of Lake Pontchartrain, Louisiana. *Chemosphere* 16:2579–2583.

Cass, G.R., and G.J. McRae. 1986. Emissions and air quality relationships for atmospheric trace metals. *In: Toxic metals in the atmosphere,* eds. J.O. Nriagu and C.I. Davidson, 145–171. Wiley, New York.

Estabrook, G.F., D.W. Burk, D.R. Inman, P.B. Kaufman, J.R. Wells, J.D. Jones, and N. Ghosheh. 1985. Comparison of heavy metals in aquatic plants on Charity Island, Saginaw Bay, Lake Huron, U.S.A., with plants along the shoreline of Saginaw Bay. *American Journal of Botany* 72:209–216.

Federal Register. 1985. National primary drinking water regulations. *Federal Register* 50:46935–47022.

Forstner, U., and G.T.W. Wittmann. 1979. *Metal pollution in the aquatic environment.* Springer-Verlag, Berlin. 486 pp.

Howard, L.S., and B.E. Brown. 1986. Metals in tissues and skeleton of *Fungia fungites* from Phuket, Thailand. *Marine Pollution Bulletin* 17:569–570.

Howard, L.S., and B.E. Brown. 1987. Metals in *Pocillopora damicornis* exposed to tin smelter effluent. *Marine Pollution Bulletin* 18:451–454.

Myers, A.G., V.L. Snoeyink, and D.W. Snyder. 1985. Removing barium and radium through calcium cation exchange. *Journal of the American Water Works Association* 77:60–66.

Noetstaller, R. 1988. *Industrial minerals.* World Bank Technical Paper 76, World Bank, Washington, D.C. 117 pp.

Petrazzuolo, G. 1981. Preliminary report on an environmental assessment of drilling fluids and cuttings released onto the outer continental shelf of the Gulf of Mexico. Office of Water Enforcement and Waste Management, US Environmental Protection Agency, Washington, DC.

Phillips, C.R., J.R. Payne, J.L. Lambach, G.H. Farmer, and R.R. Sims. 1987. Georges Bank monitoring program: hydrocarbons in bottom sediments and hydrocarbons and trace metals in tissues. *Marine Environmental Research* 22:33–74.

Robertson, J.B., J.T. Braya, and C. Schoonmaker. 1986. The effect of pH on the metal concentrations found in the leachates of wood fly ash. *Hazardous Waste and Hazardous Materials* 3:161–166.

Sears, J.R., K.J. Pecci, and R.A. Cooper. 1985. Trace metal concentrations in offshore, deep-water seaweeds in the western North Atlantic Ocean. *Marine Pollution Bulletin* 16:325–328.

Snoeyink, V.L., C. Cairns-Chambers, and J.L. Pfeffer. 1987. Strong-acid ion exchange for removing barium, radium, and hardness. *Journal of the American Water Works Association* 79:66–72.

US Environmental Protection Agency. 1989. Office of drinking water health advisories. *Reviews of Environmental Contamination and Toxicology* 107:1–184.

Wang, W. 1986a. Toxicity tests of aquatic pollutants by using common duckweed. *Environmental Pollution* (Series B) 11:1–14.

Wang, W. 1986b. The effect of river water on phytotoxicity of Ba, Cd and Cr. *Environmental Pollution* (Series B) 11:193–204.

Wang, W. 1988. Site-specific barium toxicity to common duckweed, *Lemna minor*. *Aquatic Toxicology* 12:203–212.

Wiersma, G.B., and C.I. Davidson. 1986. Trace metals in the atmosphere of remote areas. *In: Toxic metals in the atmosphere,* eds. J.O. Nriagu and C.I. Davidson, 201–266. Wiley, New York.

World Health Organization. 1984. *Guidelines for drinking-water quality*. World Health Organization, Geneva.

6
Beryllium

Beryllium is found in the Earth's crust at an average concentration of 2.5 mg/kg, principally in association with silicate minerals rather than sulfides. Approximately 90% of the total crustal beryllium is bound to feldspar and related structures in the form of beryl, bromellite, chrysoberyl, and beryllonite. Although beryllium is absorbed by most plant and animal species, it is not normally metabolized.

Production, Sources, and Residues

Production

Beryllium finds its main application in fatigue-resistant alloys, nuclear reactors and weapons, aerospace and missile components, and specialized electronics. Because these applications are relatively limited, world production of the refined metal is also small, <300 metric tons per year. The Soviet Union and the United States account for most of that total, largely reflecting demand from the aerospace and weapons industries.

Sources

The overwhelming source of beryllium in the air is coal combustion. It has been estimated that total emissions in the USA are 193 metric tons Be per year, of which 180 metric tons come from coal (US Environmental Protection Agency, 1984). The other sources, in order of declining importance, are combustion of fuel oil, windblown dust, beryllium ore processing, and volcanic emissions.

Residues

Beryllium residues in freshwaters are generally <0.001 mg/L, whereas marine waters contain much lower levels, <1 ng/L (reviewed by International Agency for Research on Cancer, 1980). No detailed inventories are available on the sources of beryllium in freshwater, but it is widely assumed that desorption and runoff from coal slag and ashes are important in local areas. For example, Kubiznakova (1987) found that coal burned at some power plants in Czechoslovakia contained residues of 1.5 to 30 mg/kg. This produced waste water with a beryllium content of 0.0032 mg/L, compared to 0.0005 mg/L for the receiving river.

Any other coal-using industry is a potential source of beryllium in surface waters. Itoh (1986) found that residues in the sediments of Nagoya Harbor (Japan) carried beryllium levels as high as 34 mg/kg dry weight near an abandoned smelter, falling to 2–3 mg/kg in a control area. In contrast, municipal effluents are not normally a significant source of contamination in most waters, reflecting the limited use of beryllium in most households and industries. Byrne and DeLeon (1986) reported that residues in the sediments of a Louisiana estuary (serving 1.5 million people) ranged from 0.07 to 0.48 mg/kg.

It is widely assumed that atmospheric emissions from coal-burning plants, and subsequent deposition, are major sources of contamination in surface waters. There are, however, remarkably few studies supporting this idea and, in some cases, data are available to the contrary. As an example, the total annual emission of beryllium in Europe from all sources is approximately 50 metric tons, much lower than any other trace element, as listed below (Pacyna, 1986):

Element	As	Cd	Cr	Cu	Ni	Pb	Se	Zn
Emission (thousand metric tons/yr)	6.5	2.7	18.9	15.5	16.0	123.0	0.4	80.0

This limited rate of deposition is generally found in industrialized areas, so the extent of contamination introduced to surface waters is also probably small.

Chemistry

Beryllium exhibits a complicated coordination chemistry, forming complexes, chelates, and oxycarboxylates with a variety of materials. The major compound in surface waters appears to be beryllium hydroxide, which is practically insoluble. Hence, most reports of beryllium in water are of particulate rather than dissolved residues. This is of considerable toxicologic significance because insoluble particulate compounds are not easily transported across the gill membrane in fish and other aquatic species, thereby reducing accumulation in edible tissues.

Volatilization, photolysis, and biotransformations are not significant fate processes. Although sorption/desorption probably play a key role in determining the fate and partitioning of beryllium, there are insufficient data available to definitively describe such processes.

Bioaccumulation

Bioaccumulation does occur, but concentration factors are always low, often <1. For example, Byrne and DeLeon (1986), working on oysters *Crassostrea virginica* from an estuary in Louisiana, found whole-body residues averaging 0.17 mg/kg dry weight (range 0.05–0.38 mg/kg), compared to sediment residues of 0.29 (range 0.07–0.48) mg/kg. This gives a concentration factor of only 0.59. Byrne et al. (1985), studying bowhead whale *Balaena mysticetus* from the Bering, Chukchi, and Beaufort Seas, found that residues in blubber, kidney, and muscle were <0.01 mg/kg wet weight but that the liver contained 0.24 mg/kg.

Overall, beryllium does not appear to threaten the consumption of fish and other food products in any location.

Toxic Effects to Aquatic Organisms

The acute and chronic toxicity of beryllium to aquatic organisms has not been adequately documented. The limited information available to date indicates that acute toxicity to freshwater organisms is strongly affected

Table 6.1. Acute toxicity of beryllium and other metals to three species of aquatic organisms.

	LC_{50} (mg/L)		
Element	Fathead minnow	Bluegill	*Daphnia*
Beryllium (soft water)	0.2	1.3	ND[a]
Beryllium (hard water)	15.0	12.0	1.0
Arsenic	15.7	16.2	5.3
Cadmium	0.4	1.2	0.07
Copper	0.4	1.2	0.07
Lead	5.7	23.8	0.5
Nickel	17.0	40.0	2.3
Selenium	5.2	28.5	0.4
Zinc	4.7	4.7	0.1

[a]No data.
Source: LeBlanc (1984).

by water hardness. The LC_{50} (lethal concentration) for fathead minnow *Pimephales promelas* in soft water is 0.2 mg/L compared to 15 mg/L in hard water (Table 6.1). The corresponding values for bluegill *Lepomis machrochirus* are 1.3 and 12 mg/L, respectively.

The toxicity of beryllium in soft water is comparable to that of cadmium and copper, using both fathead minnow and bluegill as test species (Table 6.1). The cladoceran *Daphnia magna* appears to be much more sensitive to beryllium than fish; for example, in soft water, the LC_{50} for *Daphnia magna* is only 1.0 mg/L, more than an order of magnitude lower than that recorded for fish.

Because relatively little is known about potential acute and chronic effects to freshwater and marine organisms, most jurisdictions have not finalized corresponding water quality guidelines for beryllium. Some agencies, such as the US Environmental Protection Agency and the Ontario Department of Environment, have recommended concentrations of 0.11 and 1.10 mg/L for soft water and hard water, respectively. In addition, the US National Academy of Sciences recommends a maximum level of 0.0015 mg/L for the protection of marine life, and a minimal risk value of 0.0001 mg/L. However, these guidelines are based on insufficient data and are probably not reliable. A clear environmental priority is to establish guidelines for the protection of freshwater and marine life.

Health Effects

Intake

Intake of beryllium for a nonoccupationally exposed 70-kg reference man is approximately 0.012 mg/day through food and water and <0.001 mg/day through air (Carson et al., 1987; US Environmental Protection Agency, 1984). No significant uptake occurs through the skin, and, in fact, ingested beryllium is also inefficiently absorbed in the gut. The total estimated body burden of beryllium is 0.027 mg, of which 0.009 mg is in the skeleton. Elimination of beryllium from the body is extremely slow and occurs mainly through the feces.

Acute Toxicity

Occupational exposure provides the only examples of acute exposure to beryllium. Under extreme cases, a fatal pulmonary edema may develop.

Chronic Toxicity

Only soluble beryllium salts induce chronic toxic effects in humans. These effects include contact dermatitis (edematous papulovesicular lesions) and conjunctivitis. Inhalation of beryllium produces irritation of the respiratory tract including pharyngitis, tracheobronchitis, and chemical pneumonitis potentially leading to pulmonary edema.

A pulmonary granulomatosis known as berylliosis may develop following chronic exposure to beryllium. The major symptoms include pneumonitis, chest pain, and pulmonary dysfunction. Other chronic effects include alveolar metaplasia (noted in experimental animals but not in humans), macrocytic anemia, osteosarcoma, and pulmonary cancer. Beryllium is also readily transported from pregnant females to the developing fetus in laboratory rats (Mathur et al., 1987).

Carcinogenicity

The International Agency for Research on Cancer (1980) has evaluated beryllium as follows:

> There is sufficient evidence that beryllium metal and several beryllium compounds are carcinogenic to three experimental animal species. The epidemiological evidence that occupational exposure to beryllium may lead to an increase in lung cancer risk is limited. Taken together, the experimental and human data indicate that beryllium should be considered suspect of being carcinogenic to humans.

It should be noted that this evaluation is based mainly on inhalation exposure, rather than ingestion as either food or water (Kuschner, 1981).

Drinking Water

Residues

Only a few studies have determined the concentration of beryllium in drinking water. Residues are generally <0.001 mg/L, and are occasionally more than an order of magnitude lower than that value (World Health Organization, 1984; International Agency for Research on Cancer, 1980). Assuming an average daily water consumption of 2 L, the corresponding intake of beryllium is <0.002 mg/L, and generally much lower (US Environmental Protection Agency, 1984).

Consumption Guidelines

Most nations, plus the World Health Organization, have yet to establish a recommended contaminant guideline or limit for drinking water. It is assumed that absorption is inefficient and that most beryllium comes from food rather than water. The US Environmental Protection Agency has established criteria of 37, 3.7, and 0.37 ng/L, which may result in incremental increases of cancer risk over a 70-year lifetime of 10^{-5}, 10^{-6}, and 10^{-7}, respectively.

Treatment

Relatively little is known about the removal of beryllium from drinking water. It is likely that flocculation with alum or related agent, plus conventional filtration, will cause a significant reduction in residues.

Recommendations

Relatively few environmental studies have been conducted on beryllium, reflecting its scarcity in water and relatively low toxicity. Beryllium has not been implicated in the adulteration of fish tissues and other food products, and has not been responsible for any known fish kills or decline in productivity in surface waters. The human population at risk appears to be limited to occupational environments and possibly people living near beryllium sources. Contamination of drinking water is essentially unknown, except where spills occur.

Overall, therefore, beryllium does not appear to constitute a significant environmental or health risk, so the number of recommendations for further work is relatively small:

1. Atmospheric deposition of beryllium near coal-fired power plants and concomitant effects on surface water loading.
2. Sorption/desorption from sediments and suspended solids under different environmental conditions.
3. Chronic toxic effects to several aquatic species.
4. Guidelines for the protection of freshwater and marine life.

References

Byrne, C., R. Balasubramanian, E. Overton, and T.F. Albert. 1985. Concentrations of trace metals in the bowhead whale. *Marine Pollution Bulletin* 16:497–498.

Byrne, C.J., and I.R. DeLeon. 1986. Trace metal residues in biota and sediments from Lake Pontchartrain, Louisiana. *Bulletin of Environmental Contamination and Toxicology* 37:151–158.

Carson, B.L., H.V. Ellis, and J.L. McCann. 1987. *Toxicology and biological monitoring of metals in humans*. Lewis Publishing Inc., Chelsea, MI. 328 pp.

DeLeon, I.R., C.J. Byrne, E.A. Peuler, S.R. Antoine, J. Schaeffer, and R.C. Murphy. 1986. Trace organic and heavy metal pollutants in the Mississippi River. *Chemosphere* 15:795–805.

International Agency for Research on Cancer. 1980. *Evaluation of the carcinogenic risk of chemicals to humans: some metals and metallic compounds*. 23:1–438. Lyon, France.

Itoh, K. 1986. Beryllium in sediments of Nagoya Harbor estuaries. *Bulletin of Environmental Contamination and Toxicology* 36:935–941.

Kubiznakova, J. 1987. Beryllium pollution from slag and ashes from thermal power stations. *Water, Air, and Soil Pollution* 34:363–367.

Kuschner, M. 1981. The carcinogenicity of beryllium. *Environmental Health Perspectives* 40:101–105.

LeBlanc, G.A. 1984. Interspecies relationships in acute toxicity of chemicals to aquatic organisms. *Environmental Toxicology and Chemistry* 3:47–60.

Mathur, S., S. Sharma, and A.O. Prakash. 1987. Effect of beryllium nitrate on early and late pregnancy in rats. *Bulletin of Environmental Contamination and Toxicology* 38:73–77.

Pacyna, J.M. 1986. Atmospheric trace elements from natural and anthropogenic sources. *In: Toxic metals in the atmosphere,* eds. J.O. Nriagu and C.I. Davidson, 33–52. Wiley, New York.

US Environmental Protection Agency. 1984. *Health assessment document for beryllium.* US Environmental Protection Agency, EPA 600/8-84-026A. Washington, DC.

World Health Organization. 1984. *Guidelines for drinking-water quality.* World Health Organization, Geneva.

7
Boron

Boron is a metalloid that occurs in more than 80 minerals, the most common being tourmaline, a complex silicate mineral present in igneous rocks and some sedimentary rocks. Boron in this form is largely inert and is released into the environment at an extremely slow rate through natural weathering processes. However, once released from silicate structures, boron forms highly soluble species that are readily absorbed by most plants and animals.

Production, Sources, and Residues

Production

World production of boron was $2{,}200 \times 10^3$ metric tons in 1983, and is anticipated to grow to $2{,}570 \times 10^3$ metric tons/yr in 1990 and $3{,}100 \times 10^3$ metric tons/yr by the year 2000 (Noetstaller, 1988). Almost all of the world's production comes from bedded deposits and lake brines in Turkey and California, whereas the major importers of boron are Italy, France, and the FRG (Table 7.1). It is anticipated that current world capacity will meet demand until at least the next century.

The No. 1 use of boron, often accounting for more than 50% of total production in many nations, is in the manufacture of ceramics and glass, including thermal shock-resistant products and fiber glass. Soaps and detergents occupy the second most important category, accounting for 8–

Table 7.1. World's leading exporters and importers of boron.

Exporting nation	Quantity (10^3 metric tons/yr)	Importing nation	Quantity (10^3 metric tons/yr)
Turkey	630	Italy	152
USA	607	France	112
USSR	12	FRG	96
Peru	5	Belgium	76
		Japan	49

Source: Noetstaller (1988).

40% of production (depending on nation), followed by use in agricultural chemicals, and coating and plating products.

Sources and Residues

Although the amount of boron mobilized annually from anthropogenic sources has not been accurately determined, it is widely assumed that the most important sources to surface waters are sewage, sewage sludge, and user industries such as detergent manufacturers. Natural weathering releases approximately 360×10^3 metric tons of elemental boron worldwide each year. That value is substantially lower than those reported for most other elements including aluminum, iron, manganese, and lead, and is within an order of magnitude of those reported for silver and molybdenum (Westall and Stumm, 1980).

Table 7.2. Concentration (mg/L) of total boron in freshwaters.

Location	Range	Number of samples
Pacific coast rivers, Canada[1]	0.01–2.00	418
Prairie rivers, Canada[1]	0.01–0.59	166
Central lakes and rivers, Canada[1]	<0.01–3.69	324
Atlantic coast rivers, Canada[1]	<0.01–2.30	417
Rivers, USSR[2]	0.01–0.02	NR[a]
Rivers, UK[3]	0.7–0.9	NR
Rivers, Sweden[4]	0.06–0.65	20
Alaskan rivers[5]	0.01–0.12	28

[a]NR not reported.
Sources: [1]Naquadat (1985), [2]Schoeller (1981), [3]Matthews (1974), [4]HMSO (1980), [5]Kopp and Kroner (1970).

Boron concentrations in freshwaters are typically <0.5 mg/L and may be as low as 0.005 mg/L or less (Table 7.2). In a nationwide Canadian study, for example, over 90% of the 1,325 samples that were collected had median levels of <0.05 mg/L (Naquadat, 1985). Although several studies in Europe and the United States have also shown that boron is generally <0.1 mg/L in freshwater, residues in volcanic areas and in the vicinity of borax lakes may range up to 9 mg/L (Canadian Water Quality Guidelines, 1987; Forstner and Wittmann, 1979). Seawater contains relatively large amounts of boron, averaging 4.5 mg/L, a reflection of the solubility of the dominant species—boric acid (H_3BO_3).

Chemistry

Boron has two oxidation states, B^o and B^{3+}, that form various boranes (hydrides) and organoboron complexes. The elemental form is not normally found in surface waters, while sodium and calcium borate are relatively rare. Boric acid, H_3BO_3, is the most important species in freshwaters, is moderately soluble, and does not readily dissociate. Boric acid is also the predominant form in saltwater, and is followed in importance by the borate anion, $B(OH)_4^-$. Although relatively little is known about the environmental fate of boron complexes, it is widely assumed that sorption plays a key role in controlling concentrations in freshwaters.

Bioaccumulation

Boron is accumulated by most species of aquatic plants and animals, both marine and freshwater. Residues do not increase through the food chain, so the corresponding concentration factors are relatively low, <50. Saiki and May (1988), working in the heavily irrigated San Joaquin River (California), noted that residues in bluegill *Lepomis machrochirus* ranged from nondetectable to 7.8 mg/kg dry weight, whereas carp *Cyprinus carpio* carried burdens from nondetectable to 18.3 mg/kg. These relatively high levels are due to the elevation of boron in irrigation water, a result of tile drainage.

It should be noted that most studies of inorganic residues in fish and other aquatic organisms do not consider boron as part of the analysis suite. At present, however, the data base is inadequate to determine regional or seasonal trends in residues, so any potential threat posed to fisheries cannot be determined at this time. It is therefore recommended that routine monitoring of inorganic residues in tissues be expanded to include boron.

Toxic Effects to Aquatic Organisms

Plants

Boron, particularly undissociated boric acid, is only mildly toxic to most species of aquatic plants. Antia and Cheng (1975), for example, showed that marine phytoplankton could readily tolerate concentrations of up to 10 mg B/L, and that growth of 14 of the 19 species studied was not affected by levels of 50 mg/L. In a similar study on nanoplankton, Subba Rao (1981) demonstrated that 50% of the species studied could tolerate 30 mg/L with no reduction in photosynthesis. Similarly, the EC_{50} (effective concentration) for duckweed *Lemna minor* is high, >60 mg/L (Wang, 1986).

Invertebrates

Although only a few studies are available, it appears that boron is not acutely toxic to most invertebrate species. The LC_{50} for the cladoceran *Daphnia magna* is approximately 13 mg/L, whereas the No Observed Effect Level is 6 mg/L (reviewed by Butterwick et al., 1989). Using the mean brood size and the mean total young as end points, Gersich et al. (1985) showed that the Maximum Acceptable Toxicant Concentration was 9.3 mg/L. No definitive data are available on the toxicity of boron to marine invertebrates, but it is widely assumed that most species are relatively insensitive to undissociated boric acid, the predominant species in marine waters.

Fish

Of all the species studied to date, the early life stages of rainbow trout *Oncorhynchus mykiss* appear to be most sensitive, with a Lowest Observed Effect Concentration of approximately 0.1 mg/L (Butterwick *et al.*, 1989). For comparison, other species such as the dab *Limanda limanda* are relatively insensitive (LC_{50} 74 mg/L) (Taylor et al., 1985). Water hardness, although affecting the toxicity of many heavy metals, does not consistently show any affect on the acute or chronic toxicity of boron to fish.

There do not appear to be any guidelines or standards aimed at protecting marine or freshwater organisms from elevated boron levels.

Health Effects

Intake

Total intake of boron in a 70-kg reference man ranges from 1.3 to 4.5 mg/day (reviewed by Carson et al., 1987; Canadian Water Quality Guidelines, 1987). Assuming that the daily consumption of water is 2 L, and the average boron concentration in drinking water is 0.1 mg/L, intake from water comes to only 0.2 mg/day, leaving food as the primary source. Approximately 75% of ingested boron is eliminated in the urine, the rest leaving in the feces. The total body burden of boron in a 70-kg reference man is 200 mg, of which 140 mg is in the soft tissues. Boron is not an essential part of the human diet (but is required for most plant species).

Acute Toxicity

Acute exposure to boric acid, whether through ingestion or skin absorption, causes nausea, vomiting, other gastrointestinal ailments, and erythematous lesions. In addition, circulatory collapse, tachycardia, cyanosis, and central nervous system disorders may occur in extreme cases.

Chronic Toxicity

Chronic effects in occupationally exposed workers have included dry skin and gastric disorders.

Carcinogenicity

No information is available on the carcinogenicity of boron and related compounds, but they are widely assumed to have little if any carcinogenic activity.

Drinking Water

Residues and Consumption Guidelines

Boron does not constitute a threat to most drinking water supplies. The maximum acceptable concentration for the protection of drinking water in many countries is 5 mg/L. This value is based on limiting the daily intake from water to 10 mg. Such high levels can only be found in areas of extreme ambient contamination and in the absence of water treatment facilities.

Treatment

No special treatment is usually used to remove boron, simply because boron is not considered a threat to most water systems. Granular activated carbon is capable of 90% removal from raw waters, provided boron concentrations do not exceed 5 mg/L (Choi and Chen, 1979). Demineralization using cation exchange resin or by reverse osmosis has also been used for removal in special cases.

Recommendations

Although the world's use of boron will continue to grow, the concomitant production and discharge of waste material to the environment will not keep pace, at least for the foreseeable future. Most of the growth will, in fact, be in the production of glass products, a relatively waste-free industry. Irrigation, on the other hand, inevitably leads to increased boron residues in water, particularly in arid areas of the world. Although the release of these enriched irrigation waters produces some elevation in residues in lakes and rivers, no toxic effects to aquatic organisms have ever been attributed to boron.

Because boron is not regarded as a significant contaminant of surface water at this time, relatively little research or monitoring has been directed to the area of residue dynamics in water, sediments, and biological tissues. Such baseline work is, however, important, particularly in heavily irrigated areas. The only recommendation at this time centers around that area:

1. Concentration of boron in a range of aquatic species, both plant and animal.

References

Antia, N.J., and J.Y. Cheng. 1975. Culture studies on the effects from borate pollution on the growth of marine phytoplankton. *Journal of the Fisheries Research Board of Canada* 32:2487–2492.

Butterwick, L., N. de Oude, and K. Raymond. 1989. Safety assessment of boron in aquatic and terrestrial environments. *Ecotoxicology and Environmental Safety* 17:339–371.

Canadian Water Quality Guidelines. 1987. Canadian Council of Resource and Environment Ministers. Environment Canada, Ottawa.

Carson, B.L., H.V. Ellis, and J.L. McCann. 1987. *Toxicology and biological monitoring of metals in humans*. Lewis Publishers, Chelsea, MI. 328 pp.

Choi, W.W., and K.Y. Chen. 1979. Evaluation of boron removal by adsorption on solids. *Environmental Science and Technology* 13:189–196.

Forstner, U., and G.T.W. Wittmann. 1979. *Metal pollution in the aquatic environment*. Springer-Verlag, Berlin.

Gersich, F.M., D.L. Hopkins, S.L. Applegath, C.G. Mendoza, and D.P. Milazzo. 1985. The sensitivity of chronic endpoints used in *Daphnia magna* Straus life-cycle tests. *In: Aquatic toxicology and hazard assessment*. Eighth Symposium, American Society for Testing and Materials, Philadelphia, pp. 245–252.

HMSO. 1980. *Twentieth and final report on the standing technical committee on synthetic detergents*. Her Majesty's Stationery Office, London.

Kopp, J.F., and R.C. Kroner. 1970. *Trace metals in waters of the United States*. US Department of the Interior, Cincinnati, OH. 123 pp.

Matthews, P.J. 1974. Survey of boron content in certain waters of the Greater London area using a novel analytical method. *Water Research* 8:1021–1028.

Naquadat. 1985. *National water quality data bank*. Environment Canada, Ottawa.

Noetstaller, R. 1988. *Industrial minerals*. World Bank Technical Paper 76. World Bank, Washington, DC. 117 pp.

Saiki, M.K., and T.W. May. 1988. Trace element residues in bluegills and common carp from the lower San Joaquin River, California, and its tributaries. *Science of the Total Environment* 74:199–217.

Schoeller, F. 1981. Boron content of lower Austria's waters. *Oesterreich Wasserwirtsch*. 33:9–15.

Subba Rao, D.V. 1981. Effect of boron on primary production of nanoplankton. *Canadian Journal of Fisheries and Aquatic Sciences* 38:52–58.

Taylor, D., B.G. Maddock, and G. Mance. 1985. The acute toxicity of nine "grey list" metals (arsenic, boron, chromium, copper, lead, nickel, tin, vanadium and zinc) to two marine species: dab (*Limanda limanda*) and grey mullet (*Chelon labrosus*). *Aquatic Toxicology* 7:135–144.

Wang, W. 1986. Toxicity tests of aquatic pollutants by using common duckweed. *Environmental Pollution* (Series B) 11:1–14.

Westall, J., and W. Stumm. 1980. The hydrosphere. *In: The handbook of environmental chemistry,* ed. O. Hutzinger, 17–49. Springer-Verlag, New York.

8
Cadmium

Cadmium is the 64th most abundant element, occurring in the earth's crust at an average concentration of 0.2 mg/kg. Cadmium is classified as a soft acid, preferentially complexing with sulfides, often as greenockite (a hexagonal crystalline form of cadmium sulfide), hawleyite (a cubic crystalline form of cadmium sulfide), sphalerite (zinc sulfide), and otavite (a mineralized form of calcium carbonate). Cadmium is of considerable environmental and health significance because of its increasing mobilization and toxicity to many life forms.

Production, Sources, and Residues

Production

World production of refined cadmium increased from 6,000 metric tons per year in 1950 to 15,000 metric tons per year in 1980, and now amounts to approximately 19,000 metric tons per year (Table 8.1). The leading producers are currently the USSR, Japan, Canada, and the USA, and Belgium; the leading consumers are the USA, the USSR, Japan, Belgium, and the UK. The primary uses of cadmium are in electroplating other metals or alloys for protection against corrosion, and in the manufacture of storage batteries, glass ceramics, and some biocides. At one time, cadmium was used in large quantities as a pigment in paint, other coatings, and plastics. However, occupational health concerns have forced a significant cutback in usage, and also forced increased research on alternative pigment formulations (Chemical Marketing Reporter, 1989).

Table 8.1. World production and consumption of refined cadmium (1986).

Country	Production (thousand metric tons)	Country	Consumption (thousand metric tons)
USSR	>2,700	USA	3,673
Japan	2,581	USSR	2,200
Canada	1,712	Japan	2,089
USA	1,678	Belgium	1,925
Belgium	1,293	UK	1,391
FRG	1,095	FRG	1,372
Australia	879	France	1,010
Mexico	851	Mexico	573
Netherlands	598	Sweden	268
Finland	564	Brazil	224
World total	19,000	World total	18,000

Source: World Bureau of Metal Statistics (1989).

Sources

Estimates of the worldwide anthropogenic input of cadmium to freshwater range from 2.1 to 17 $\times$ 10^3 metric tons per year (Table 8.2). The major specific sources on a worldwide basis are atmospheric deposition, smelting and refining of nonferrous metals, manufacturing processes related to chemicals and metals, and domestic waste water. Only about 15% of the atmospheric deposition comes from natural sources, such as volcanoes, windborne soil particles, and biogenic particles (Nriagu, 1989).

Table 8.2. Worldwide anthropogenic input of cadmium to freshwaters.

Source	Input (thousand metric tons/yr)
Atmospheric deposition	0.9–3.6
Smelting and refining nonferrous metals	0.01– 3.6
Manufacturing processes	
chemicals	0.1–2.5
metals	0.5–1.8
Domestic wastewater	
central	0.2–1.8
noncentral	0.3–1.2
Discharge of sewage sludge	0.1–1.3
Steam electricity production	0.01–0.24
Base metal mining and dressing	0–0.3
Total input	2.1–17

Source: Nriagu and Pacyna (1988).

The dominant worldwide source in oceans is atmospheric precipitation, giving 4.0×10^{14} metric tons/yr, followed by river water discharge (3.2×10^{13} metric tons/yr), river-suspended particulate discharge (1.8×10^{10} metric tons/yr), and atmospheric particulate deposition (5×10^{8} metric tons/yr) (reviewed by Yeats and Bewers, 1987). Hutton and Symon (1986) noted that approximately 43 metric tons of cadmium per year entered UK coastal waters. Of this, 33.3 metric tons came from sewage and sewage sludge, 5.7 metric tons from the production and use of phosphate fertilizers, 2.6 metric tons from nonferrous metal production, and 1.7 metric tons from iron and steel production.

Residues

Cadmium is routinely detected in most surface waters, both as dissolved Cd and particulate Cd. Dissolved Cd residues can be extremely low (2 ng/L) or range up to 400 ng/L in heavily industrialized areas (reviewed by Yeats and Bewers, 1987). Particulate Cd is often highly variable, largely reflecting the mobilization of contaminated bottom sediments under high flow conditions. Examples of the cadmium content of suspended matter in rivers are as follows: 120–840 mg/kg in the Van River, Europe (Whitehead et al., 1988); 0.02–11.3 mg/kg in the Roya River, Europe (Whitehead et al., 1988); and <0.1–1.7 mg/kg in the Rhone River, Monaco (Huynh-Ngoc et al., 1988). On a worldwide basis, the particulate Cd/dissolved Cd distribution coefficients range from 1×10^{4} to 1×10^{5} (Yeats and Bewers, 1987).

Cadmium in uncontaminated freshwater sediments is generally detected at 0.1–1.0 mg/kg dry weight, increasing to 4–10 mg/kg in contaminated areas (e.g., Mohapatra, 1988; Mudroch et al., 1988). In the Great Lakes of North America, residues in >90% of Lake Michigan are <2 mg/kg, and there are only a few places where cadmium exceeds 6 mg/kg (Figure 8.1). For comparison, more than half of Lake Erie (which has a heavily industrialized basin) contains residues in the range 4–>6 mg/kg (Figure 8.1). Cadmium in the sediments of the Rhine River (The Netherlands) averaged 20 mg/kg in 1980, up from 1–2 mg/kg in 1900 (Beukema et al., 1986).

Cadmium is appreciably enriched in coastal marine waters and sediments. There is in fact a strong negative correlation between pH and dissolved Cd for the North Atlantic, expressed by the relationship (Yeats and Bewers, 1987):

$$\text{Cd (ng/L)} = 137 - 3.76 \times \text{salinity (‰)}$$

where $r = 0.99$.

Hence, in inshore areas where salinity is low (28 ‰), cadmium residues come to 32 ng/L, whereas in the open ocean (salinity 36 ‰), cadmium comes to 2 ng/L. Similarly, cadmium in inshore areas of the North Sea

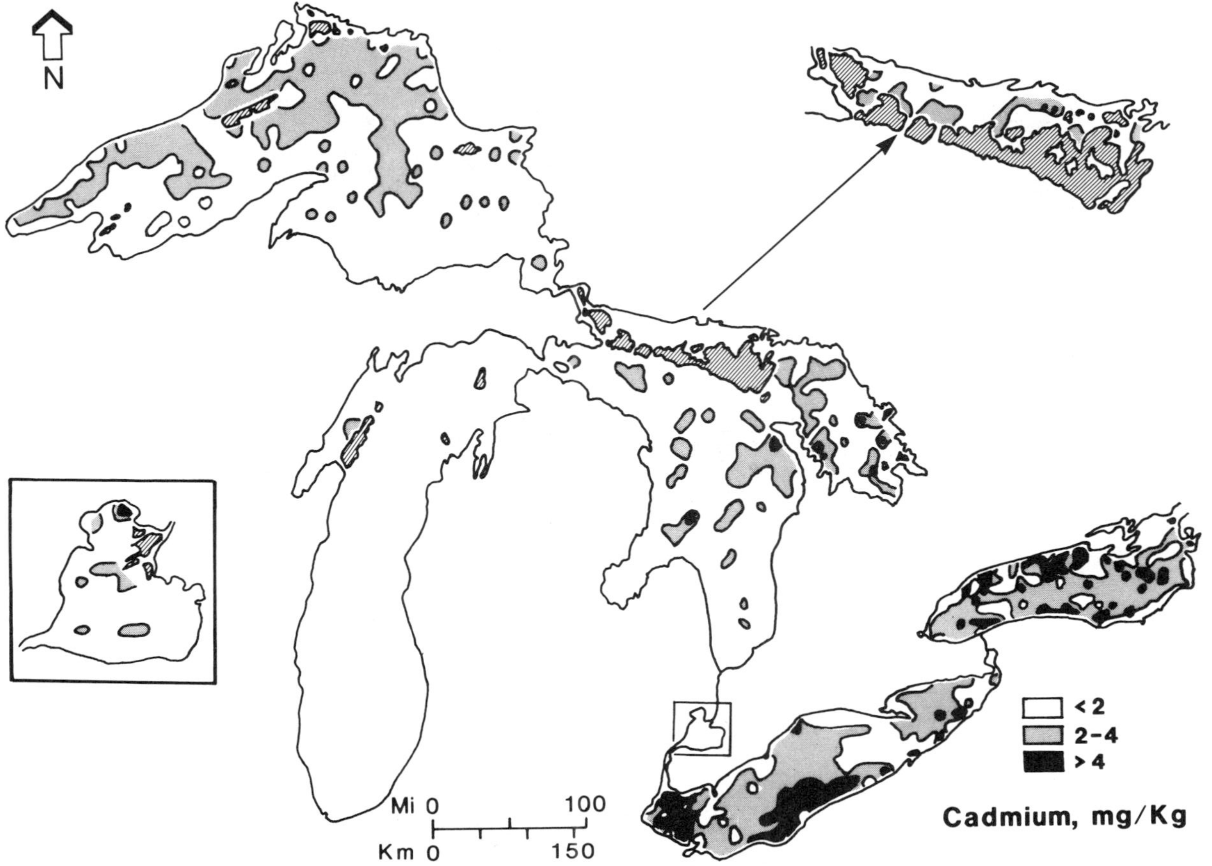

Figure 8.1. Distribution of cadmium in the sediments of the Great Lakes. (From Thomas and Mudroch, 1979.)

near The Netherlands averages 600–700 ng/L, decreasing linearly to 20 ng/L in the main body of the North Sea (Beukema et al., 1986).

It is widely assumed that the influx of anthropogenically derived cadmium will have no significant impact on open-ocean residues for the foreseeable future. For example, total influx from the continents is only 1×10^4 metric tons per year, about 50 times less than the amount derived from internal transport of sediments. However, as the rate of input to the oceans continues to increase, the cadmium/phosphate ratio will change, leading to a buildup of cadmium in the area of phosphate regeneration. Some increase in residues is likely to occur in pelagic sediments over the long term—15,000 years (the residence time of cadmium in the ocean).

Chemistry

Cadmium is relatively mobile in aquatic systems, existing as Cd^{2+}, $Cd(OH)_2$ (aq.), $Cd(OH)_3^-$, $Cd(OH)_4^{2-}$, and $CdCO_3$, and in various other organic and inorganic complexes. In many freshwaters, the affinity of ligands to complex with cadmium follows the order: humic acids, CO_3^{2-}, OH^-, Cl^-, SO_4^{2-}. The solubility of calcium hydroxide complexes decreases as pH increases, owing to the formation of solid $Cd(OH)_2$ according to the reaction:

$$Cd^{2+} + 2\ OH^- \rightleftharpoons Cd(OH)_2$$

Since the 2+ valency state predominates in freshwater, redox potential has little effect on speciation.

As salinity increases moving into coastal and marine waters, cadmium complexation with the chloride ion also increases until the dominant complexes all contain chloride. Nuernberg and Valenta (1983) listed the dominant species in seawater with low biological productivity as follows:

Species	Distribution (%)
Cd^{2+}	1.9
$CdCl^+$	29.1
$CdCl_2$	37.2
$CdCl_3$	31.0
Other	0.8

In marine waters where dissolved organic matter is high, this speciation scheme would shift to some degree, favoring the formation of strong organic complexes.

Sorption to suspended solids such as clay is an important, often dominant fate process in freshwaters. Coprecipitation with hydrous iron, aluminum–manganese oxides, and carbonate materials also occurs, and periodically dominates fate processes. For example, Samanidou and Fytianos (1987) found that the main species in the sediments of the Axios River (Greece) were in association with Fe–Mn hydrous oxides, accounting for 22–43% of total Cd, followed by associations with carbonate at 14–35%. Several studies have reported large-scale desorption from particulates as rivers mix with seawater (Comans and Van Dijk, 1988). Apparently, dissolved chloride competes effectively for Cd^{2+}, then causing Cd^{2+} to complex with marine particles.

Other processes, such as photolysis and volatilization, are not widely regarded as important in the environmental fate of cadmium.

Bioaccumulation

Plants

Total residues in aquatic plants are generally low, <5 mg/kg dry weight, except near polluting sources, where concentrations of up to 342 mg/kg have been observed (Table 8.3). The corresponding concentration factors (tissue residue/water residue) generally fall in the 500–5,000 range (e.g., Melhuus et al., 1978). Adsorption of cadmium (in those species studied) is rapid, with a steady state in residues occurring in <30 min (Skowronski, 1986; Skowronski and Przytocka-Jusiak, 1986). Uptake of cadmium by the alga *Scenedesmus pannonicus*, and other species, is generally greatest at neutral or basic pH (Demon et al., 1988). The presence of manganese and iron in water significantly inhibits uptake, a reflection of competition for uptake sites.

Table 8.3. Concentration (mg/kg dry weight) of cadmium in marine and freshwater plants.

Species	Average (range)	Location
Tropical seagrass[1] (9 species)	0.42 (0.05–1.20)	Flores Sea, Indonesia
Deep-water seaweeds[2] (4 species)	0.22 (0.04–2.35)	North Atlantic
Submerged macrophytes[3] (8 species)	—[a] (ND[b]–5.7)	Saginaw Bay, Lake Michigan
Submerged macrophytes[3] (8 species)	1.3 (0.29–2.01)	Unnamed lakes, Michigan
Lemanea fluviatilis[4]	22.7 (2–342)	35 rivers, Europe

[a]Not determined.
[b]Not detected.
Sources: [1]Nienhuis (1986), [2]Sears et al. (1985), [3]Estabrook et al. (1985), [4]Harding and Whitton (1981).

Invertebrates

Invertebrates, particularly marine mussels, have been widely used to monitor cadmium residues on a worldwide basis. *Mytilus* spp. are sessile and ubiquitous, and have the ability to concentrate many elements, making them particularly well suited for programs in coastal waters. Cossa (1988) reviewed data for *Mytilus* spp. from 591 stations from 13 regions, and found maximum residues along the Pacific coast of the USA, the Baltic Sea, and the Irish Sea (Table 8.4). Concentration factors (tissue concentration/water concentration) ranged from 88,000 to 96,000 (Table 8.4). Although Lima et al. (1986) similarly showed that the oyster *Crassostrea brasiliana* from Sepetiba Bay (Brazil) was an effective monitor of environmental cadmium levels, Uthe and Chou (1987) reported that over 90% of total Cd in the mollusc *Placopecten magellanicus* was in the digestive gland. This indicated that the observed contamination was a reflection of feeding and nutritional inadequacies, rather than ambient residues.

It must be emphasized that the debate concerning the source of cadmium is relatively esoteric when people continue to consume contaminated molluscs. McKenzie-Parnell et al. (1988), for example, found that the concentration of cadmium in the New Zealand dredge oyster *Tiostrea lutaria* was sufficiently high that the ingestion of just one oyster more than doubled the normal dietary intake of adults. In fact, the highest daily individual fecal excretion of cadmium observed in a population of fishermen was 0.58 mg, equivalent to more than 10 times the weekly intake guideline. It is apparent that edible mussels must be routinely monitored for elevated cadmium levels. The contamination of freshwater invertebrates is not nearly so controversial, largely because the consumption of mollusks and other species is relatively small. Furthermore, residues are also often low, <20 mg/kg in most cases. Pugsley et al. (1988) reported an average concentration of 6.4 mg/kg (range 1.9–19.0 mg/kg) in the clam *Lampsilis radiata* from Lake St. Clair and the Detroit River (a highly industrialized and urbanized part of North America). Similarly, the gas-

Table 8.4. Concentration (mg/kg dry weight) of cadmium in the soft tissues of mussel from different regions, cadmium concentration (ng/L) in the overlying water, and the corresponding concentration factors.

Region	Mussel average (range)	Water average (range)	Concentration factor
Pacific coast, USA	3.3 (1.4–7.7)	no data	no data
Baltic Sea	2.7 (0.7–12.9)	28 (16–58)	96,400
Irish Sea	2.2 (0.4–15.0)	25 (16–33)	88,000
St. Lawrence Gulf	1.9 (1.0–4.2)	20 (<10–31)	95,000
North Sea	1.5 (0.3–11.0)	16 (8–25)	93,800

Source: Cossa (1988).

tropod *Lymnaea stagnalis* from Lake Balaton (Hungary) contained residues of 2–3 mg/kg (V.-Balogh et al., 1988). Abaychi and Mustafa (1988), working in the Shatt al-Arab River, Iraq, reported a relatively high concentration of 35 mg/kg (range 2.2–70.0 mg/kg) in soft tissues of the Asiatic clam *Corbicula fluminea*. Maximum residues were recorded in winter, corresponding to the period of maximum discharge of domestic and industrial wastes.

Uptake of cadmium by invertebrates occurs through both water and food, and is temperature dependent (Baumer and Zauke, 1987). The presence of cations such as lead, zinc, and calcium may reduce uptake in some species. Metal binding proteins, of both high and low molecular weight and containing metallothionein, have been identified in many species (Harrison et al., 1988). Cadmium depuration is remarkably slow, often requiring more than 50 days for elimination of 50% of the total Cd burden (reviewed by Ray and McLeese, 1987). This is one of the reasons why concentration factors are so high in some species.

Fish

Although cadmium in fish tissues is generally <0.1 mg/kg wet weight, there are occasional reports of levels 1 to 2 orders of magnitude higher in fish from contaminated areas. A survey of 67 sites in Great Britain gave the following results on a wet-weight basis (Mason, 1987):

Species	Average (mg/kg)	Range (mg/kg)
Brown trout *(Salmo trutta)*	0.06	<0.001–0.27
Gudgeon *(Gobio gobio)*	0.08	<0.001–0.20
Perch *(Perca fluviatilis)*	0.04	<0.001–0.35
Pike *(Esox lucius)*	<0.001	<0.001

Similarly, residues in 50 species of Australian fish from the Great Barrier Reef were almost always below the detection limit of 0.1 mg/kg dry weight (Denton and Burdon-Jones, 1986), as also noted for 12 species from the Mediterranean (Hornung and Ramelow, 1987) and for sharks from the Atlantic near the British Isles (Vas and Gordon, 1988). Such low residues are significant because limits do not usually need to be placed on the consumption of fish inhabiting cadmium-polluted waters.

In most cases, cadmium is concentrated in the kidney, liver, and gill of fish. Concentration factors in the 6,000–19,000 range have been reported for kidney, whereas the corresponding values for liver and gills are 2,600–6,500 and 1,100–2,400, respectively (reviewed by McCracken, 1987). In a study of fish from the Australian Great Barrier Reef, cadmium in liver often exceeded 20 mg/kg and periodically topped 100 mg/kg (Denton and Burdon-Jones, 1986). Similarly, liver residues in the eel *Anguilla anguilla* from the Urola River (Spain) averaged 2.3 mg/kg dry weight, compared to 0.09 mg/kg for gill and 0.03 mg/kg for muscle (Legorburu et al., 1988).

The routes of uptake by adult fish are through both food and water; fish eggs apparently sorb cadmium faster through water than alevins (Beattie and Pascoe, 1978). Uptake increases with low dissolved oxygen levels, owing to increased gill ventilation, and availability of the free Cd^{2+} ion. Joensen and Korsgaard (1986) noted that cadmium was rapidly passed to the embryos of the viviparous fish *Zoarces viviparus,* following intraovarian loading in the mother. Although residues in the embryo declined rapidly near the end of the experiment, cadmium continuously concentrated in the liver and kidney of the mother fish. Many species of fish contain metal-binding proteins including metallothionein.

Toxic Effects to Aquatic Organisms

Plants

Cadmium and, in particular, the free cadmium ion are highly toxic to most plant and animal species. The EC (effective concentration)$_{50}$ for duckweed *Lemna minor* is only 0.2 mg/L, compared to 0.45 mg/L for nickel and 1.1 mg/L for copper (Wang, 1986a,b). Similarly, the EC_{50} for the green alga *Selenastrum capricornutum* was only 0.006 mg/L (Sedlacek et al., 1983). Although higher tolerances are periodically recorded, it is known that some species can adapt to relatively high cadmium levels (Klerks and Weis, 1987). Acute and chronic toxicity to aquatic plants is ameliorated by changing pH and the concomitant effect on the availability of the free cadmium ion (Campbell and Stokes, 1985; Peterson et al., 1984). Similarly, marine algae are generally much more tolerant of cadmium than freshwater species, owing to the binding of the free cadmium ion with chloride.

Invertebrates

The acute toxicity of cadmium to freshwater invertebrates varies enormously among the different taxonomic groups. The LC_{50} for one of the most sensitive species, the cladoceran *Daphnia magna*, is approximately 0.005 mg/L, whereas the LC_{50} for the amphipod *Gammarus pulex* is 0.7 mg/L (reviewed by Martin and Holdich, 1986; Mance, 1987). Comparably high values (0.1–0.9 mg/L) have been reported for a number of oligochaete and dipertan species. Much higher LC_{50}s have been reported for marine invertebrate species (Mance, 1987). For example, the LC_{50} for the crustacean *Penaeus indicus* was 2.0 mg/L and the corresponding concentration for the amphipod *Marinogammarus obstusatus* ranged from 3.5 to 13.3 mg/L.

Numerous subacute effects have been reported following exposure of both freshwater and marine invertebrates to cadmium. These include (1) reduction in emergence of dipteran larvae (*Polypedilum nubifer*) following oral administration at 220 mg/kg in food (Hatakeyama, 1987); (2) increase in oxygen affinity of blood in the polychaete *Arenicola marina* (Everaarts and Reichert, 1988); (3) extensive hyperplasia of secondary lamellae and necrosis in the gills of the crab *Scylla serrata* following exposure at 2.4 mg/L (Krishnaja et al., 1987); and (4) reduction in follicle development in both male and female gonads in the mussel *Mytilus edulis* exposed to 0.1 mg Cd/L (Kluytmans et al., 1988).

In freshwaters, cadmium toxicity is usually greatest in soft water, and is also pH-dependent. Although the presence of chelating agents in the water, both organic and inorganic, may reduce toxicity, low-molecular-weight humic acids often have little or no impact on toxicity. Elevated temperatures, and any other factor that may increase the metabolic rate of invertebrates, usually increases both the uptake and toxicity of cadmium. High salinity greatly reduces potential toxic effects to most species. This is of considerable importance to estuarine species, which are subjected to rapidly changing water quality conditions on a twice-daily basis.

Fish

The acute toxicity of cadmium to freshwater fish is highly variable. Salmonids seem to be particularly sensitive, with LC_{50}s of 0.001 to 0.03 mg/L being reported. Hardness plays a key role in determining toxicity. Sprague (1987) described such affects in salmonids by the equation:

$$\log_{10}(LC_{50}) = 0.8334\ [\log_{10}(\text{hardness})]$$

where hardness is expressed in mg/L. This equation indicates that the LC_{50} for salmonids is 2.5 μg/L in water of hardness 20 mg/L, and 27 μg/L at a hardness of 350 mg/L. LC_{50}s for nonsalmonid fishes range from 0.1 to 23 mg/L.

Marine species similarly show considerable variability in their response to cadmium. For example, the LC_{50} for the mummichog *Fundulus heteroclitus* was 60 mg/L, while for the Atlantic silverside *Menidia menidia* and the flounder *Pseudopleuronectes americanus*, it was 6.4 and 0.6 mg/L, respectively (reviewed by Mance, 1987). Much of this variability is due to differences in salinity favored by the various species. Other factors that affect toxicity include the life stage of the fish (eggs and fry are generally more sensitive than juveniles or adults), presence of competing ions including other metals, and any factor (such as temperature) that increases the metabolic rate of the fish.

Numerous subacute effects have been reported for both freshwater and marine fish following exposure to cadmium. These include (1) decrease in plasma calcium levels in the freshwater teleost *Oreochromis mossam-*

bicus following exposure at 0.01 mg/L (Fu et al., 1989); (2) decrease in plasma sodium, potassium, calcium, and chloride and elevation in magnesium in rainbow trout *Oncorhynchus mykiss* following exposure to 0.006 mg Cd/L (Giles, 1984); (3) inhibition of calcium influx in rainbow trout following acclimation to high and low levels of calcium in the water (Reid and McDonald, 1988); (4) morphologic change in the gut of rainbow trout following oral administration at 5 mg Cd/kg fish (Crespo et al., 1986); and (5) a 16% decrease in growth of the freshwater murrel *Ophiocephalus punctatus* following 13 days of exposure at 3.5 mg Cd/L (Shukla and Pandey, 1988).

Many nations have promulgated guidelines or standards for the protection of freshwater and marine life. Examples of some of these are listed:

Canada
- 0.0002 mg/L (water hardness 0 to 60 mg/L)
- 0.0008 mg/L (water hardness 61 to 120 mg/L)
- 0.0013 mg/L (water hardness 121 to 180 mg/L)
- 0.0018 mg/L (water hardness >180 mg/L)

United States
- 0.00066 mg/L (water hardness 50 mg/L; 4-day average)
- 0.0011 mg/L (water hardness 100 mg/L; 4-day average)
- 0.002 mg/L (water hardness 200 mg/L; 4-day average)

United Kingdom
- 0.002 mg/L (dissolved Cd)
- 0.003 mg/L (saltwater; total Cd).

Health Effects

Intake

Intake of cadmium for a nonoccupationally exposed 70-kg reference man amounts to approximately 0.15 mg/day from food and fluids and <0.001 mg/day through inhalation in nonsmokers (reviewed by Carson et al., 1987). About 0.15 mg are lost per day, mainly though the urine (0.10 mg) and, to a lesser degree, the feces (0.05 mg). The estimated body burden of cadmium is 50 mg, of which 38 mg is located in the soft tissues. These values are doubled for tobacco smokers. Pocock et al. (1988) reported that heavy smoking caused a four fold increase in blood Cd levels of British middle-aged men.

Acute Toxicity

In humans, acute exposure to cadmium leads to nausea, vomiting, salivation, diarrhea, and muscular cramps (US Environmental Protection

Agency, 1989). Severe to fatal cases may show the following symptoms: liver injury, convulsions, shock, renal failure, and cardiopulmonary depression.

Chronic Toxicity

Renal toxicity, such as proteinuria, is the most common symptom of chronic exposure to cadmium. Renal dysfunction is likely to be displayed in 10% of the population at concentrations of 0.18–0.22 mg Cd/g renal cortex. Any individual with a tissue residue in excess of 0.285 mg/g usually suffers from renal dysfunction. Scott et al. (1987), working on human kidneys from the United Kingdom, found that maximum cortex residues, 0.02 mg/g, occurred in 50- to 59-year-olds, followed by those in the 60- to 69- and 40- to 49-year-old categories. In addition, heavy smokers had about 33% more cadmium in their kidneys than nonsmokers. Nogawa et al. (1986) reported that Cd residues in 173 autopsied Japanese reached 0.20 mg/g in the kidney cortex. These individuals came from a highly polluted area and showed kidney damage on autopsy.

Carcinogenicity

The carcinogenicity of cadmium following oral exposure has not been definitively described because there is no or little evidence that cadmium is carcinogenic via the oral route (US Environmental Protection Agency, 1989). The International Agency for Research on Cancer has classified cadmium and some cadmium compounds in Group 2B: limited evidence for carcinogenicity in humans and sufficient evidence of carcinogenicity in animals. This classification is based on exposure by inhalation.

Drinking Water

Residues

Cadmium is routinely detected in finished drinking water, albeit at relatively low concentrations. In a survey by the American Water Works Association (1985) of drinking water in 39 states and three territories, there were 14 episodes of noncompliance with guidelines and/or standards. For comparison, fluoride and nitrates were in noncompliance in 907 and 369 episodes, respectively. Cadmium concentration in the episodes of noncompliance averaged 0.017 mg/L, with a range of 0.011–0.03 mg/L. Similarly, Azcue et al. (1988) reported that cadmium concentrations in drinking water from Rio de Janeiro (Brazil) averaged only 0.0002 mg/L (range 0.0001–0.0004 mg/L), while Ajmal and Uddin (1986a,b) found that drinking water for the City of Aligarh (India) always contained cadmium below 0.005 mg/L.

Consumption Guidelines

Many nations, plus the World Health Organization, use a maximum guideline/standard of 0.005 mg Cd/L for the protection of drinking water. This was derived by initially assuming that the critical concentration of cadmium in the renal cortex of humans associated with renal dysfunction is 0.20 mg/g (US Environmental Protection Agency, 1989). The daily intake of cadmium needed to obtain 0.20 mg/g in the renal cortex of a 70-kg reference man is approximately 0.352 mg for 50 years, assuming 4.5% absorption of the daily oral dose and 0.01% excretion per day of the total body burden. This gives a daily cadmium intake of 0.005 mg/kg associated with renal dysfunction, a value considered to be the Lowest-Observed-Adverse-Effect Level (LOAEL).

Using 0.005 mg/kg/day as the LOAEL, the drinking water guideline was established as follows:

1. Determination of the Reference Dose (RD)

   $$\text{RD} = (0.005 \text{ mg/kg/day})/10 = 0.0005 \text{ mg/kg/day}$$

   The number 10 refers to the uncertainty factor, which in this case is relatively low because of the good data base on the toxic effects of cadmium.
2. Determination of the Drinking Water Equivalent (DWEL)

   $$\text{DWEL} = (0.0005 \text{ mg/kg/day})(70 \text{ kg})/(2 \text{ L/day}) = 0.018 \text{ mg/L}$$
3. Determination of the Guideline

   $$\text{Guideline} = (0.018 \text{ mg/L})(25\%) = 0.005 \text{ mg/L}$$

   The value of 25% comes from the theoretical allocation to drinking water of the Acceptable Daily Intake of cadmium.

Treatment

Cadmium can be effectively removed from raw water by destablization and aggregation of preexisting solids that contain cadmium, or by adsorption onto amorphous iron or aluminum hydroxides (American Water Works Association, 1988). In order to improve removal efficiency it is necessary to increase pH above neutrality. In one study (American Water Works Association, 1988), the predicted removal of cadmium by adsorption onto iron hydroxide increased from 27.4% at pH 7 to 88.3% at pH 8. Reverse osmosis is an effective, albeit small-scale, treatment process. Removal efficiencies of 90–98% have been reported using that technique (reviewed by US Environmental Protection Agency, 1989). The extent and type of treatment technology selected depend on the characteristics of the raw water, and compatibility with existing systems.

Recommendations

An enormous body of literature is available on the chemistry and the environmental and health effects of cadmium. Many areas (such as bioaccumulation and toxic effects to freshwater organisms) have been adequately investigated, and further research cannot be considered high priority at this time.

Because most regulatory agencies recognize the environmental significance of cadmium, discharge from point sources will probably come under progressively tighter control for the foreseeable future. However, nonpoint source discharges (particularly atmospheric deposition) will gain in importance, and total discharges to the environment will either increase or remain at current levels.

Another area of concern is the flux in cadmium in estuarine waters as salinity shifts with the tides and freshwater inputs. There have been only a few studies conducted on the toxic effects to estuarine/coastal organisms of cycling levels of cadmium.

The health effects and drinking water guidelines of cadmium have been firmly established, so further work in these areas cannot be considered a priority at this time.

The following recommendations for research and monitoring reflect these considerations.

1. Nonpoint sources of contamination, particularly atmospheric deposition.
2. Toxic effects to estuarine/coastal species under variable cadmium regimens.
3. Routine monitoring of surface and drinking water, and fish and other food products.

References

Abaychi, J.K., and Y.Z. Mustafa. 1988. The Asiatic clam, *Corbicula fluminea:* an indicator of trace metal pollution in the Shatt al-Arab River, Iraq. *Environmental Pollution* 54:109–122.

Ajmal, M., and R. Uddin. 1986a. Studies on heavy metals in the ground waters of the City of Aligarh U.P. (India). *Environmental Monitoring and Assessment* 6:181–194.

Ajmal, M., and R. Uddin. 1986b. Quality of drinking water in the Aligarh Muslim University campus, Aligarh, U.P. (India) with respect to heavy metals. *Environmental Monitoring and Assessment* 6:195–205.

American Water Works Association. 1985. An AWWA survey of inorganic contaminants in water supplies. *Journal of the American Water Works Association* 77:67–72.

American Water Works Association. 1988. A review of solid-solution interactions and implications for the control of trace inorganic materials in water treatment. *Journal of the American Water Works Association* 80:56–64.

Azcue, J.M.P., W.C. Pfeiffer, M. Fiszman, and O. Malm. 1988. Heavy metal removal by different water treatment plants, in Rio de Janeiro State, Brazil. *Environmental Technology Letters* 9:429–436.

Baumer, H.P., and G.P. Zauke. 1987. Modelling the dynamic relationship between Cd-concentrations in *Gammarus tigrinus* and water temperature. *Environmental Technology Letters* 8:529–544.

Beattie, J.H., and D. Pascoe. 1978. Cadmium uptake by rainbow trout, *Salmo gairdneri* eggs and alevins. *Journal of Fish Biology* 13:631–637.

Beukema, A.A., G.P. Hekstra, and C. Venema. 1986. The Netherlands' environmental policy for the North Sea and Wadden Sea. *Environmental Monitoring and Assessment* 7:117–155.

Campbell, P.G.C., and P.M. Stokes. 1985. Acidification and toxicity of metals to aquatic biota. *Canadian Journal of Fisheries and Aquatic Sciences* 42:2034–2049.

Carson, B.L., H.V. Ellis, and J.L. McCann. 1987. *Toxicology and biological monitoring of metals in humans*. Lewis Publishing, Chelsea, MI. 328 pp.

Chemical Marketing Reporter. 1989. Cadmium pigment future dim; search for alternatives is on. *Chemical Marketing Reporter* 15 May 1989, p. 26.

Comans, R.N.J., and C.P.J. van Dijk. 1988. Role of complexation processes in cadmium mobilization during estuarine mixing. *Nature* 336:151–154.

Cossa, D. 1988. Cadmium in *Mytilus* spp.: worldwide survey and relationship between seawater and mussel content. *Marine Environmental Research* 26:265–284.

Crespo, S., G. Nonnotte, D.A. Colin, C. Leray, L. Nonnette, and A. Aubree. 1986. Morphological and functional alterations induced in trout intestine by dietary cadmium and lead. *Journal of Fish Biology* 28:69–80.

Demon, A., M. de Bruin, and H.T. Wolterbeek. 1988. The influence of pH on trace element uptake by an alga (*Scenedesmus pannonicus* subsp. *berlin*) and fungus (*Aureobasidium pullulans*). *Environmental Monitoring and Assessment* 10:165–173.

Denton, G.R.W., and C. Burdon-Jones. 1986. Trace metals in fish from the Great Barrier Reef. *Marine Pollution Bulletin* 17:201–209.

Estabrook, G.F., D.W. Burk, D.R. Inman, P.B. Kaufman, J.R. Wells, J.D. Jones, and N. Ghosheh. 1985. Comparison of heavy metals in aquatic plants on Charity Island, Saginaw Bay, Lake Huron, U.S.A., with plants along the shoreline of Saginaw Bay. *American Journal of Botany* 72:209–216.

Everaarts, J.M., and M.J.M. Reichert. 1988. The effect of cadmium on some oxygen-binding properties of the blood pigment of the lugworm *Arenicola marina* (Annelida, Polychaeta). *Marine Environmental Research* 25:275–289.

Fu, H., R.A.C. Lock, and S.E.W. Bonga. 1989. Effect of cadmium on prolactin cell activity and plasma electrolytes in the freshwater teleost *Oreochromis mossambicus*. *Aquatic Toxicology* 14:295–306.

Giles, M.A. 1984. Electrolyte and water balance in plasma and urine of rainbow trout (*Salmo gairdneri*) during chronic exposure to cadmium. *Canadian Journal of Fisheries and Aquatic Sciences* 41:1678–1685.

Harding, J.P.C., and B.A. Whitton. 1981. Accumulation of zinc, cadmium, and lead by field populations of *Lemanea*. *Water Research* 15:301–319.

Harrison, F.L., J.R. Lam, and J. Novacek. 1988. Partitioning of metals among metal-binding proteins in the bay mussel, *Mytilus edulis*. *Environmental Research* 24:167–170.

Hatakeyama, S. 1987. Chronic effects of Cd on reproduction of *Polypedilum nubifer* (Chironomidae) through water and food. *Environmental Pollution* 48:249–261.

Hornung, H., and G.J. Ramelow. 1987. Distribution of Cd, Cr, Cu, and Zn in eastern Mediterranean fishes. *Marine Pollution Bulletin* 18:45–49.

Hutton, M., and C. Symon. 1986. The quantities of cadmium, lead, mercury and arsenic entering the U.K. environment from human activities. *Science of the Total Environment* 57:129–150.

Huynh-Ngoc, L., N.E. Whitehead, and B. Oregioni. 1988. Cadmium in the Rhone River. *Water Research* 22:571–576.

Joensen, J.E., and B. Korsgaard. 1986. Uptake, time-course distribution and elimination of cadmium in embryos and tissues of the pregnant *Zoarces viviparus* (L.) after intraovarian loading. *Journal of Fish Biology* 28:61–68.

Klerks, P.L., and J.S. Weis. 1987. Genetic adaptation to heavy metals in aquatic organisms: a review. *Environmental Pollution* 45:173–205.

Kluytmans, F., F. Brands, and D.I. Zandee. 1988. Interactions of cadmium with the reproductive cycle of *Mytilus edulis* L. *Marine Environmental Research* 24:189–192.

Krishnaja, A.P., M.S. Rege, and A.G. Joshi. 1987. Toxic effects of certain heavy metals (Hg, Cd, Pb, As and Se) on the intertidal crab *Scylla serrata*. *Marine Environmental Research* 21:109–119.

Legorburu, I., L. Canton, E. Millan, and A. Casado. 1988. Trace metal levels in fish from Urola River (Spain): Anguillidae, Mugillidae and Salmonidae. *Environmental Technology Letters* 9:1373–1378.

Lima, N.R.W., L.D. de Lacerda, W.C. Pfeiffer, and M. Fiszman. 1986. Temporal and spatial variability in Zn, Cr, Cd and Fe concentrations in oyster tissues (*Crassostrea brasiliana* Lamarck, 1819) from Sepetiba Bay, Brazil. *Environmental Technology Letters* 7:453–460.

Mance, G. 1987. *Pollution threat of heavy metals in aquatic environments*. Elsevier, New York. 372 pp.

Martin, T.R., and D.M. Holdich. 1986. The acute lethal concentration of heavy metals to peracarid crustaceans (with particular reference to fresh-water asellids and gammarids). *Water Research* 20:1137–1147.

Mason, C.F. 1987. A survey of mercury, lead and cadmium in muscle of British freshwater fish. *Chemosphere* 16:901–906.

McCracken, I.R. 1987. Biological cycling of cadmium in fresh water. *In: Cadmium in the aquatic environment,* eds. J.O. Nriagu and J.B. Sprague, 89–116. Wiley, New York.

McKenzie-Parnell, J.M., T.E. Kjellstrom, R.P. Sharma, and M.F. Robinson. 1988. Unusually high intake and fecal output of cadmium, and fecal output of other trace elements in New Zealand adults consuming dredge oysters. *Environmental Research* 46:1–14.

Melhuus, A., K.L. Seip, H.P. Seip, and S. Myklestad. 1978. A preliminary study of the use of benthic algae as biological indicators of heavy metal pollution in Sørfjorden, Norway. *Environmental Pollution* 15:101–107.

Mohapatra, S.P. 1988. Distribution of heavy metals in polluted creek sediment. *Environmental Monitoring and Assessment* 10:157–163.

Mudroch, A., L. Arazin, and T. Lomas. 1988. Summary of surface and background concentrations of selected elements in the Great Lakes sediments. *Journal of Great Lakes Research* 14:241–251.

Nienhuis, P.H. 1986. Background levels of heavy metals in nine tropical seagrass species in Indonesia. *Marine Pollution Bulletin* 17:508–511.

Nogawa, K., R. Honda, Y. Yamada, T. Kido, I. Tsuritani, M. Ishizaki, and H. Yamaya. 1986. Critical concentration of cadmium in kidney cortex of humans exposed to environmental cadmium. *Environmental Research* 40:251–260.

Nriagu, J.O. 1989. A global assessment of natural sources of atmospheric trace metals. *Nature* 338:47–49.

Nriagu, J.O., and J.M. Pacyna. 1988. Quantitative assessment of worldwide contamination of air, water and soils by trace metals. *Nature* 333:134–139.

Nuernberg, H.W., and P. Valenta. 1983. Potentialities and applications of voltammetry in chemical speciation of trace metals in the sea. *In: Trace metals in seawater,* eds. C.S. Wong, E. Boyle, K.N. Bruland, J.D. Burton, and E.D. Goldberg, 671–697. Plenum, New York.

Peterson, H.G., F.P. Healey, and R. Wagemann. 1984. Metal toxicity to algae: a highly pH dependent phenomenon. *Canadian Journal of Fisheries and Aquatic Sciences* 41:974–979.

Pocock, S.J., H.T. Delves, D. Ashby, A.G. Shaper, and B.E. Clayton. 1988. Blood cadmium concentrations in the general population of British middle-aged men. *Human Toxicology* 7:95–103.

Pugsley, C.W., P.D.N. Hebert, and P.M. McQuarrie. 1988. Distribution of contaminants in clams and sediments from the Huron-Erie corridor. II. Lead and cadmium. *Journal of Great Lakes Research* 14:356–368.

Ray, S., and D.W. McLeese. 1987. Biological cycling of cadmium in marine environment. *In: Cadmium in the aquatic environment,* eds. J.O. Nriagu and J.B. Sprague, 199–229. Wiley, New York.

Reid, S.D., and D.G. McDonald. 1988. Effects of cadmium, copper, and low pH on ion fluxes in the rainbow trout, *Salmo gairdneri. Canadian Journal of Fisheries and Aquatic Sciences* 45:244–253.

Samanidou, V., and K. Fytianos. 1987. Partitioning of heavy metals into selective chemical fractions in sediments from rivers in northern Greece. *Science of the Total Environment* 67:279–285.

Scott, R., E. Aughey, G.S. Fell, and M.J. Quinn. 1987. Cadmium concentrations in human kidneys from the UK. *Human Toxicology* 6:111–119.

Sears, J.R., K.J. Pecci, and R.A. Cooper. 1985. Trace metal concentrations in offshore, deep-water seaweeds in the western North Atlantic. *Marine Pollution Bulletin* 16:325–328.

Sedlacek, J., T. Kallqvist, and E. Gjessing. 1983. Effect of aquatic humus on uptake and toxicity of cadmium to *Selenastrum capricornutum* Printz. *In: Aquatic and terrestrial humic materials,* eds. R.F. Christman and E.T. Gjessing, 495–516. Ann Arbor Science, Ann Arbor, MI.

Shukla, J.P., and K. Pandey. 1988. Toxicity and long term effects of a sublethal concentration of cadmium on the growth of the fingerlings of *Ophiocephalus punctatus* (Bl.). *Acta Hydrochimica Hydrobiologie* 16:537–540.

Skowronski, T. 1986. Adsorption of cadmium on green microalga *Stichococcus bacillaris. Chemosphere* 15:69–76.

Skowronski, T., and M. Przytocka-Jusiak. 1986. Cadmium removal by green alga *Stichococcus bacillaris. Chemosphere* 15:77–79.

Sprague, J.B. 1987. Effects of cadmium on freshwater fish. *In: Cadmium in the aquatic environment,* eds. J.O. Nriagu and J.B. Sprague, 139–169. Wiley, New York.

Thomas, R.L., and A. Mudroch. 1979. *Small craft harbours: sediment survey, Lakes Ontario, Erie and Lake St. Clair.* Report to Small Craft Harbours, Great Lakes Biolimnology Laboratory, Burlington, Ontario, Canada. 27 pp.

US Environmental Protection Agency. 1989. Drinking water health advisories. *Reviews of Environmental Contamination and Toxicology* 107:1–184

Uthe, J.F., and C.L. Chou. 1987. Cadmium in sea scallop (*Placopecten magellanicus*) tissues from clean and contaminated areas. *Canadian Journal of Fisheries and Aquatic Sciences* 44:91–98.

Vas, P., and J.D.M. Gordon. 1988. Trace metal concentrations in the scyliorhinid shark *Galeus melastomus* from the Rockall trough. *Marine Pollution Bulletin* 19:396–398.

V.-Balogh, K., D.S. Fernandez, and J. Salanki. 1988. Heavy metal concentrations of *Lymnaea stagnalis* L. in the environs of Lake Balaton (Hungary). *Water Research* 22:1205–1210.

Wang, W. 1986a. The effect of river water on phytotoxicity of Ba, Cd and Cr. *Environmental Pollution* 11:193–204.

Wang, W. 1986b. Toxicity tests of aquatic pollutants by using common duckweed. *Environmental Pollution* 11:1–14.

Whitehead, N.E., L. Huynh-Ngoc, and S.R. Aston. 1988. Trace metals in two north Mediterranean rivers. *Water, Air, and Soil Pollution* 42:7–18.

World Bureau of Metal Statistics. 1989. *World metals statistics yearbook.* WBMS, London.

Yeats, P.A., and J.M. Bewers. 1987. Evidence for anthropogenic modification of global transport of cadmium. *In: Cadmium in the aquatic environment,* eds. J.O. Nriagu and J.B. Sprague, 19–34. Wiley, New York.

9
Chromium

Chromium occurs in the earth's crust at an average concentration of 100 mg/kg, principally as minerals in the chromite spinel group. These minerals have the generalized formula $(Mg,Fe)O(Cr,Al,Fe)_2O_3$. Depending on the degree of substitution in the Al, Fe, Cr series, the chromites contain from 13% to 65% Cr_2O_3. Numerous chromium compounds are manufactured from these minerals, most of which contain chromium in the stable 3+ or 6+ state.

Production, Sources, and Residues

Production

World production of chromite was 560×10^3 metric tons in 1930, increasing to $4,400 \times 10^3$ metric tons in 1960 and $11,200 \times 10^3$ metric tons in 1980 (US Minerals Yearbooks, 1930–1989). Production in recent years has stayed around the $10,000 \times 10^3$ mark. The leading exporters of chromite are South Africa, Albania, the USSR, and Turkey; the major importers are Japan, Sweden, the FRG, and Italy (Table 9.1).

The primary use of metallurgical grade chromium is in the production of ferrochromium alloys, which are used as additives in stainless steel and other specialized products. Chemical grade chromite ore is first converted to sodium dichromate and then used in the manufacture of chromic acid, pigments, and leather tanning agents.Chromite refractory materials are used in iron and steel processing, nonferrous alloy production, glass making, and cement production.

Table 9.1. World's major exporters and importers of chromite.

Exporting nation	Quantity (1,000 metric tons/yr)	Importing nation	Quantity (1,000 metric tons/yr)
South Africa	803	Japan	644
Albania	584	Sweden	342
USSR	496	FRG	239
Turkey	363	Italy	192
India	150	Yugoslavia	190

Source: Noetstaller (1988).

Sources

Estimates of the total anthropogenic discharge of chromium to surface waters range from 45×10^3 to 239×10^3 metric tons per year (Table 9.2). The primary sources include domestic waste water from both central and noncentral sources, manufacturing processes involving metals, and the dumping of sewage sludge. Abuzkhar et al. (1987) showed that the average concentration of chromium in sewage sludge from Tripoli (Libya) was 27.7 mg/kg dry weight, with a range of 0.005–177.4 mg/kg. Similarly, dewatered sludge from a plant in the state of Washington contained 73.1 mg Cr/L (Nevissi et al., 1988), while dried sludge from another 25 Washington plants contained residues of 6,610 mg/kg (range 83–137,000 mg/kg) (Sung et al., 1986). Such values are far higher than those (0.08–8.40 mg/L) found in most landfill leachates (reviewed by Lema et al., 1988).

Atmospheric fallout is the seventh most important source of chromium in surface waters, contributing up to 16×10^3 metric tons of Cr per year

Table 9.2. Worldwide anthropogenic input of chromium to freshwaters.

Source	Input (thousand metric tons per year)
Manufacturing processes	
metals	15–58
chemicals	2.5–24
pulp and paper	0.01–1.5
petroleum products	0–0.2
Domestic wastewater	
central	8.1–36
noncentral	6.0–42
Discharge of sewage sludge	5.8–32
Smelting and refining	
nonferrous metals	3–20
Atmospheric deposition	2.2–16
Base metal mining and dressing	0–0.7

Source: Nriagu and Pacyna (1988).

Table 9.3. Annual input of total chromium to The Netherlands' part of the North Sea in 1980 (actual) and 1990 (projected).

Source	Input 1980 (metric tons/yr)	Input 1990 (metric tons/yr)
Total input	1,600	440–580
Atmospheric deposition	17	17
Rivers	1,000	400–540
Coastal discharges	10	2.7
Dredging sludges	250	5
Industrial wastes	270	0
Incineration at sea	0.3	0
Offshore mining	25	17

Source: Beukema et al. (1986).

(Table 9.2). Only about 41% of global atmospheric emissions are from anthropogenic sources, the remainder originating mainly from windborne soil particles and volcanic emissions (Nriagu, 1989). Chromium is not normally a major contaminant of urban precipitation, even in highly industrialized countries such as Japan (Sakai et al., 1988).

Coastal marine sources are dominated by input from rivers and, to a lesser degree, dredging sludges and dumping of industrial wastes. For example, of the total input of chromium to The Netherlands' part of the North Sea (1,600 metric tons per year), 1,520 metric tons come from these three sources annually (Table 9.3).

Residues

Total Cr is generally detected at low concentrations in freshwaters. Water from the Great Lakes of North America has an average total Cr of only 0.001 mg/L (reviewed in Canadian Water Quality Guidelines, 1987). Similarly, total Cr was not detected in 6 of 10 samples from the Mississippi River, and was present at an average concentration of 0.006 mg/L in the remaining four samples (DeLeon et al., 1986). Residues in a highly polluted creek draining into Lake Erie averaged 0.002–0.003 mg/L (US Environmental Protection Agency, 1988). Total Cr also occurs in relatively low concentrations in seawater; for example, levels in the highly polluted North Sea near The Netherlands are approximately 0.002 mg/L, declining to 0.0006 mg/L in offshore areas (Beukema et al., 1986).

Total Cr in the surficial sediments of the depositional basins of the Great Lakes ranges from 8 to 362 mg/kg dry weight, compared to background concentrations of 9–86 mg/kg (Mudroch et al., 1988). Substantially higher levels, up to 665 mg/kg, were found in the same study in the surficial sediments of embayments where the background residues were similar to those found in the depositional basins. Mohapatra (1988) reported that total Cr in the sediments of a polluted creek near Bombay (India) averaged 39.9 mg/kg dry weight (range 24.2–74.5 mg/kg). Sediment samples taken from the Belgium part of the North Sea, plus the Scheldt Estuary, gave average concentrations of 44–114 mg/kg, depending on sediment particle size (Araujo et al., 1988), whereas much higher levels (up to 1,600 mg/kg) were found in sediments offshore of Los Angeles in 1981 (Stull and Baird, 1985). Those extremely high residues were due to the presence of a nearby sewage outfall.

Chemistry

Cr^{3+} is classified as a hard acid and forms relatively strong complexes with oxygen donor ligands. These complexes, which are generally stable and kinetically inert, can often be isolated as solids. Cr^{3+} is hexacoordinate in its reactions, preferentially forming complexes with amines, $[CrAm_{6-n-m}(H_2O)_nR_m]^{3-m})^+$ where Am = NH_3 or polydentate amine such as ethylenediamine, and R is an acido ligand such as halide, nitro, or sulfate ion. The principal forms in freshwater include $CrOH^{2+}$, $Cr(OH)_2{}^+$, and $Cr(OH)_4{}^-$. Cr^{6+}, on the other hand, is water soluble, always existing in solution as a component of a complex anion. The anionic species varies with pH, and may be chromate ($CrO_4{}^{2-}$), hydroxychromate ($HCrO_4{}^-$), or dichromate ($Cr_2O_7{}^{2-}$). Dichromate is rare at circumneutral pH, and only becomes common in highly acidic waters. At pH > 6.5, most Cr^{6+} is present as the chromate ion. Eary and Ral (1987) showed that Cr^{3+} could be oxidized to Cr^{6+} by reaction with manganese dioxide. Although the rate of reaction was not appreciably affected by dissolved oxygen, slightly acidic to basic pH water limited the rate of oxidation.

Nakayama et al. (1981) showed that chromium in the Pacific Ocean and Sea of Japan existed in the following forms: inorganic Cr^{3+} (10–20%), inorganic Cr^{6+} (25–40%), and organic to Cr (45% to 65%). The concentration ratio of dissolved versus particulate chromium in the Pacific Ocean was 6:1, and in the Sea of Japan, 5.25:1 .

Adsorption of Cr^{6+} by clays, ferric hydroxide, and ferric and manganese oxides is generally a minor fate process, whereas Cr^{3+} is rapidly adsorbed, at least by clays. The rate of adsorption of Cr^{3+} increases with pH to the point where the total amount of bound Cr^{3+} exceeds that of Cr^{6+} by 30–300 times. This means that although Cr^{6+} is highly mobile in aquatic systems, Cr^{3+} is quickly immobilized in the sediments. Young et al. (1987) reported that the adsorption of Cr^{3+} was linearly dependent on soluble metal concentrations, and that subsequent desorption occurred over a period of at least 24 days. That study also showed that more than

Table 9.4. Partitioning of chromium in the sediments of the Axios and Aliakmon Rivers, and their estuaries, in Greece.

Fraction	Cr in rivers (%)	Cr in estuaries (%)
Cation-exchangeable	0.23–0.52	0.19–0.22
Carbonates	0.52–0.69	0.63–0.65
Fe–Mn hydrous oxides	3.54–8.52	2.87–9.69
Organic sulfides	32.51–66.90	30.70–69.88
Residual	28.64–57.93	26.38–58.79

Source: Samanidou and Fytianos (1987).

50% of Cr^{3+} was reversibly bound whereas the remainder was tightly bound.

Samanidou and Fytianos (1987) studied the partitioning of chromium into selective chemical fractions in the sediments of rivers in northern Greece (Table 9.4). They found that the dominant species, of those identified, were associated with organic sulfides and Fe-Mn hydrous oxides. This partitioning did not change in either the freshwater part of the rivers or their estuaries. Nriagu and Coker (1980) showed that humic and fulvic acids in the sediments of Lake Ontario concentrated chromium by a factor of 2× compared to the overlying water. The total amount of chromium bound to organic matter in that lake was 8–11%.

Volatilization, photolysis, and biotransformations do not appear to be important processes in the environmental fate of chromium.

It should be pointed out that the amount of work done on the environmental chemistry of chromium compounds is relatively small and that very little is known about organic complexation and the subsequent fate of these compounds.

Bioaccumulation

Plants

Total Cr in freshwater plants is generally low, <5 mg/kg dry weight, except near polluting sources, where residues of up to 50 mg/kg have been reported. Estabrook et al. (1985), working on Saginaw Bay of Lake Huron, found that residues in eight submersed species averaged approximately 4.4 mg/kg with a range of 0.1–18.1 mg/kg. The terrestrial species, or those with floating or emergent leaves from the same location, contained lower levels, averaging 2.7 mg/kg (range <0.1–45.2 mg/kg). Total Cr residues in marine plants are often higher than those reported for freshwater species because of the increased bioavailability of $CrCl_6$. For example, Gryzhankova et al. (1973) recorded residues of up to 140 mg/kg in 19 species from Japanese coastal waters, and Sivalingam (1978) found total Cr of up to 58 mg/kg in 20 filamentous species from Malaysia. Although these values are comparable to those found for benthic diatoms in the Calasieu Estuary, Louisiana (Ramelow et al., 1987), substantially lower levels (<2 mg/kg) were found in four deep-water seaweeds in the western north Atlantic (Sears et al., 1985).

Invertebrates

Molluscs and other invertebrate species are widely used to monitor chromium residues in both marine and freshwaters. In most cases, residues are relatively low, <5 mg/kg dry weight, except near major polluting sources, where concentrations of over 50 mg/kg have been found (Table 9.5).

Food is likely a more significant source of chromium in most invertebrate species than water. Sorption generally increases with temperature and any factor that causes an increase in the rate of metabolism. Stackhouse and Benson (1988) showed that concentrations of humic acid of up to 50 mg/L had little effect on the uptake of Cr^{6+} by the cladoceran *Daphnia pulex*. Although it is widely assumed that young animals sorb chromium at a faster rate than older specimens, V.-Balogh *et al.* (1988) observed a negative relationship for total Cr and body size in the gastropod *Lymnaea stagnalis* from Lake Balaton (Hungary). This was likely due to differing levels of exposure as the animal grew.

Table 9.5. Concentration (mg/kg dry weight) of chromium in the soft tissues of marine invertebrate species.

Species	Average (range)	Location
Bivalve molluscs, eight species[1]	1.5 (0.8–4.0)	Pacific Ocean, Fiji
Oyster, *Crassostrea virginica*[2]	1.3 (NR[a])	Lake Pontchartrain, Louisiana
Oyster, *Crassostrea brasiliana*[3]	3.4 (1.1–14.0)	Sepetiba Bay, Brazil
Mollusc, *Arctica islandica*[4]	1.9 (0.3–4.5)	Martha's Vineyard, USA
Coral, *Pocillopora damicornis*[5]	160.0 (NR)	Phuket, Thailand
Crustacean, *Callichirus laurae*[6]	3.2 (NR)	Red Sea, Jordan

[a]Not reported.
Sources: [1]Dougherty (1988), [2]Byrne and DeLeon (1986), [3]Lima et al. (1986), [4]Phillips et al. (1987), [5]Howard and Brown (1987), [6]Abu-Hilal et al. (1988).

Fish

Total Cr does not normally accumulate in fish muscle tissues, and, in most cases, residues are less than 0.5 mg/kg wet weight. Johnson (1987), for example, found that total Cr in six species of fish from 14 lakes in northern Ontario averaged 0.23 mg/kg with a range of 0.19 to 0.27 mg/kg. These fish included both predatory and bottom-feeding species. Similarly, Hornung and Ramelow (1987) showed that total Cr in 12 fish species from the eastern Mediterranean averaged 0.4 mg/kg (range 0.004–5.8 mg/kg) whereas residues in fish from the North Sea ranges from only 0.01 to 0.03 mg/kg (Hagel, 1986).

Although concentration factors relating muscle tissue residues to water residues are generally below 3, higher values have been found in the gill, liver, kidney, and other organs of several species. Brooks and Rumsey (1974), working on eight marine species from New Zealand, reported the following average residues in muscle, liver, kidney, heart, gonad, spleen, and gills: 0.02, 0.1, 0.2, 0.3, 0.2, 1.2, and 0.5 mg/kg. This means that monitoring of chromium should include analysis of a number of tissues, not just muscle.

Sriwastwa et al. (1987) exposed the freshwater fish *Mystus vittatus* to 2 mg Cr/L as potassium dichromate. Uptake, which continued over a 7-month period, was greatest in the bones and kidney, followed in decreasing order by liver, brain, and kidneys. No uptake was found in the muscle tissue. When the fish were placed in uncontaminated water, significant depuration occurred within 3 months. Saiki and May (1988), working on the San Joaquin River (California), found that total Cr in carp *Cyprinus carpio* was significantly and inversely related to dissolved oxygen level in the water. No other factor, such as temperature, pH, total alkalinity, conductivity, and turbidity, was related to total Cr residues. Johnson (1987) noted a significant correlation between total Cr in the muscle tissues of fish from lakes in northern Ontario with atmospheric deposition of chromium, primarily from anthropogenic sources.

Toxic Effects to Aquatic Organisms

Plants

Cr^{6+} is generally only moderately toxic to algae and other aquatic plants. Wang (1986a) reported that the *EC (effective concentration)*$_{50}$ for duckweed *Lemna minor* was as high as 35 mg/L, giving a maximum permissible concentration of 3.5 mg/L for standard-setting purposes. In a related study, Wang (1986b) showed that these toxic effects were not ameliorated by the presence of particulate matter, either organic and inorganic, under the test conditions. Although it is likely that ligands do affect the toxicity of both Cr^{3+} and Cr^{6+} to aquatic plants under some conditions, no definitive data are available to confirm this point.

Invertebrates

Acute toxicity of chromium to marine and freshwater invertebrates is highly variable and depends on a number of factors including water hardness and pH. Martin and Holdich (1986) found that Cr^{3+} was not acutely toxic to two crustacean species, with LC_{50}s ranging up to 937 mg/L (Table 9.6). Cr^{6+}, on the other hand, had a 96-h LC_{50} of only 0.6 mg/L. Rehwoldt et al. (1973) found that the 96-h LC_{50} of Cr^{3+} to seven species of benthic invertebrates was 3–50 mg/L, whereas the corresponding concentrations for Cr^{6+} generally fell within the range 0.1–20 mg/L. Several other studies (reviewed in Canadian Water Quality Guidelines, 1987) have reported that the LC_{50} of Cr^{6+} for the cladoceran *Daphnia magna* ranged upwards from 0.015 mg/L and averaged 0.023 mg/L.

Marine molluscs are relatively insensitive to Cr^{6+} (LC_{50}, 14–105 mg/L), as are most marine crustaceans (LC_{50}, 2–98 mg/L) (Mance, 1987). Annelids are relatively sensitive (LC_{50}, 2–8 mg/L) during short-term exposures, and, over 28 days, an LC_{50} of only 0.03 mg/L was reported (Mance, 1987). Increasing salinity causes a decrease in the toxicity of Cr^{6+} in most of the species studied to date.

Essentially nothing is known about the chronic effects of either Cr^{6+} or Cr^{3+} to either marine or freshwater invertebrates.

Fish

There are numerous reports on the acute toxicity of Cr^{6+} to a large number of freshwater fish species; the 24- to 96-h LC_{50}s range from approximately 0.015 to 0.10 mg/L depending on species and environmental conditions. Although less is known about the sensitivity of marine species, those studies conducted to date have also yielded LC_{50}s in the 0.005–0.09 mg/L range.

Both Cr^{6+} and Cr^{3+} are sorbed at the gills; this leads to relatively high residues in the gills and concomitant tissue damage, including hyperpla-

Table 9.6. Acute toxicity (LC_{50}) of chromium compounds to two crustacean species.

Chromium species	48-h LC_{50} (mg/L)		96-h LC_{50} (mg/L)	
	Average	95% confidence limits	Average	95% confidence limits
Asellus aquaticus				
Cr^{3+}	937	732–1,436	442	382–527
Crangonyx pseudogracilis				
Cr^{3+}	388	370–410	291	279–304
Cr^{6+}	2.5	2.0–2.5	0.6	0.8–0.5

Source: Martin and Holdich (1986). Water hardness, 50 mg/L; pH, 6.75; water temperature, 13°C.

sia, clubbing of lamellae, and necrosis. As gill damage proceeds, so does impairment of ability to osmoregulate and respire. Since Cr^{6+} passes rapidly through the gills, a number of other tissues such as the liver, kidney, and spleen may be affected. Other subacute toxic effects include (1) increase in blood pyruvate in the freshwater teleost *Colisa fasciatus* exposed to 48 mg Cr^{6+}/L (Nath and Kumar, 1987); (2) depletion in liver glycogen and increase in blood glucose in *C. fasciatus* exposed to 48 mg Cr^{6+}/L (Nath and Kumar, 1988); and (3) avoidance of chromium in rainbow trout at 0.028 mg Cr^{6+}/L (Anestis and Neufeld, 1986).

In most cases, the toxicity of Cr^{6+} increases with decreasing pH. It is also assumed that decreasing hardness has the same effect on toxicity, but there have been only a handful of studies that have examined this point independent of pH. Temperature generally causes an increase in chromium sensitivity. Some species, such as juvenile Pacific salmon, suffer from osmoregulatory stress when placed in saltwater, thereby increasing their susceptibility to chromium (Sugatt, 1980).

Many nations have promulgated guidelines or standards for the protection of marine and freshwater life. Examples of some of these are listed:

Canada
- 0.02 mg Cr/L for the protection of fish
- 0.002 mg Cr/L for the protection of aquatic life including zooplankton and phytoplankton

United States
- 0.05 mg Cr^{6+}/L (4-day average)
- 1.10 mg Cr^{6+}/L (1-h average)
- 0.12 mg Cr^{3+}/L (water hardness 50 mg/L; 4-day average)
- 0.21 mg Cr^{3+}/L (water hardness 100 mg/L; 4-day average)
- 0.37 mg Cr^{3+}/L (water hardness 200 mg/L; 4-day average)

European Community
- 0.015 mg Cr/L for the protection of saltwater fish and shellfish
- 0.005 mg Cr/L for the protection of salmonids (water hardness 50 mg/L)
- 0.250 mg Cr/L for the protection of salmonids (water hardness >250 mg/L)

Health Effects

Intake

Chromium is an essential trace element, forming part of the antidiabetogenic factor, which is essential in the metabolism of insulin (Korallus, 1986). The minimum daily intake is approximately 0.05 mg, through either food or water.

Intake of chromium for a nonoccupationally exposed 70-kg reference man comes to 0.03 to 0.1 mg/day from food (Carson et al., 1987). Another

0.06 mg/day comes from drinking water, assuming the daily consumption of water is 2 L with an average chromium concentration of 0.03 mg/L. Intake from inhalation is small, <0.001 mg/day in the nonoccupationally exposed population.

Chromium is eliminated in approximately equal amounts in both the feces and urine. Araki et al. (1986) reported that chromium in the urine of 19 metal workers averaged 0.008 mg/L (range 0.002–0.011 mg/L), compared to a plasma concentration of 0.005 mg/L (range 0.003–0.007 mg/L). Much lower levels, approximately 0.0005 mg/L, were noted for the general population in two regions of northern Italy (Minoia et al., 1988). In that study, chromium residues in males were significantly different from those in females, and were also related to smoking habits.

Acute Toxicity

Cr^{6+} compounds are more toxic than Cr^{3+} compounds in humans. Acute exposure to Cr^{6+} produces nausea, diarrhea, liver and kidney damage, internal hemorrhage, dermatitis, and respiratory problems (US Environmental Protection Agency, 1989; Carson et al., 1987). Cases of acute poisoning by Cr^{3+} compounds are extremely rare, reflecting their low toxicity to the human population.

Chronic Toxicity

Chronic exposure to Cr^{3+} is often associated with allergic contact dermatitis, skin ulcers, nasal membrane and septum irritation, pulmonary congestion and edema, perforated eardrums, and nephritis. No cases of chronic toxicity following ingestion of Cr^{3+} appear to have been reported.

Carcinogenicity

The International Agency for Research on Cancer (1980) has evaluated the carcinogenicity of chromium compounds as follows. There is sufficient evidence for the carcinogenicity of calcium chromate and some relatively insoluble Cr^{6+} compounds (sintered calcium chromate, lead chromate, strontium chromate, sintered chromium trioxide, and zinc chromate) in rats. There is limited evidence for the carcinogenicity of lead chromate oxide and cobalt-chromium alloy in rats. There is sufficient evidence for respiratory carcinogenicity in men occupationally exposed during chromate production.

Water-soluble Cr^{6+} chromate compounds are often mutagenic and may also be carcinogenic (Langard, 1988; Lanfranchi et al., 1988). However, the data base is not yet strong enough to support any definitive conclusion on the carcinogenicity of these compounds to humans. Based on epidemiological and animal studies, there is no pervasive evidence for the carcinogenicity of Cr^{3+} compounds, either soluble or insoluble.

Drinking Water

Residues

Chromium is often detected in finished drinking water, generally at relatively low concentrations. In a survey by the American Water Works Association (1985) of drinking water in 39 states and three territories, there were only 18 episodes of noncompliance with guidelines and/or standards. For comparison, fluoride and nitrates were in noncompliance in 907 and 369 episodes, respectively. Chromium concentration in the episodes of noncompliance averaged 0.08 mg/L. Although Azcue et al. (1988) similarly reported that chromium concentrations in drinking water from Rio de Janeiro (Brazil) were low, averaging 0.002 mg/L with a range of 0.001 to 0.003 mg/L, higher residues (up to 0.03 mg/L) were reported for the city of Aligarh (India) (Ajmal and Uddin, 1986).

Consumption Guidelines

Many nations, plus the World Health Organization, use a maximum standard/guideline of 0.05 mg Cr/L for the protection of drinking water. It is assumed that Cr^{6+} is the dominant, if not the only, form of chromium in drinking water. The US Environmental Protection Agency (1989) has developed a less stringent long-term health advisory of 0.12 mg/L. This value was derived on the basis of a study by MacKenzie et al. (1958) on the effect of chronic ingestion of Cr^{3+} and Cr^{6+} in water by rats. The only significant adverse effect noted in that study was a reduction in water consumption at the highest dose, 25 mg Cr^{6+}/L, giving a No-Observed-Adverse-Effect Level (NOAEL) of 2.41 mg/kg/day. The following procedures were then used:

1. Determination of the Reference Dose (RD)

$$\text{RD} = (2.41 \text{ mg/kg/day})/(100)(5) = 0.0048 \text{ mg/kg/day}$$

   where 100 is the uncertainty factor and 5 is an additional uncertainty factor to compensate for less-than-lifetime exposure.
2. Determination of Drinking Water Equivalent Level (DWEL)

$$\text{DWEL} = (0.0048 \text{ mg/kg/day})(70 \text{ kg})/(2 \text{ L/day}) = 0.17 \text{ mg/L}$$

   where 70 kg refers to a standard reference man.
3. Determination of the Guideline

$$\text{Guideline} = (0.17 \text{ mg/L})(71\%) = 0.12 \text{ mg/L}$$

   where 71% is the estimated relative contribution of water to the Acceptable Daily Intake of chromium.

Because there is no evidence of carcinogenicity of chromium following oral ingestion, the guideline of 0.12 mg/L does not include an estimate of cancer risk.

Treatment

Chromium can be removed from raw water by coagulation and filtration, lime softening, ion exchange, and in the case of small systems, reverse osmosis. Lime softening is particularly effective in removing Cr^{3+}. Treatment efficiencies of 85–90% have been reported using this technique, but it is of little use on Cr^{6+}, even at pH 9.5. Coagulation is effective in the removal of Cr^{6+}, provided a reducing substance such as ferrous sulfate is also used. Since Cr^{3+} occurs with cationic species and Cr^{6+} with anionic species, a cation exchanger in series with an anion exchanger is necessary for chromium removal. The extent and type of treatment technology selected depend on the characteristics of the raw water, and compatibility with existing systems.

Recommendations

Although a considerable amount of research has been conducted on chromium, there are still a number of areas in which relatively little is known. The more important deficiencies in the environmental and health areas are as follows.

1. Environmental fate of organic chromium complexes.
2. Effects of organic ligands on the toxicity of chromium compounds to marine and freshwater plant and animal species.
3. Chronic effects of Cr^{3+} and Cr^{6+} to marine and freshwater invertebrate species.
4. Carcinogenicity of Cr^{3+} and Cr^{6+} compounds to fish.
5. Carcinogenicity of water-soluble Cr^{6+} compounds to mammals.

Monitoring programs need to stress the presence of both Cr^{3+} and Cr^{6+} in surface and drinking water. Bivalve molluscs appear to be an effective means of monitoring chromium contamination in marine and freshwaters.

References

Abu-Hilal, A., M. Badran, and J. de Vaugelas. 1988. Distribution of trace elements in *Callichirus laurae* burrows and nearby sediments in the Gulf of Aqaba, Jordan (Red Sea). *Marine Environmental Research* 25:233–248.

Abuzkhar, A.A., A.S. Gibali, Y.I. Elmehrik, and R. Ahmatullah. 1987. Chemical monitoring of sewage wastes for their use in crop production. I. Solid sludge as a soil amendment. *Environmental Monitoring and Assessment* 8:127–133.

Ajmal, M., and R. Uddin. 1986. Studies on heavy metals in the ground waters of the City of Aligarh U.P. (India). *Environmental Monitoring and Assessment* 6:181–194.

American Water Works Association. 1985. An AWWA survey of inorganic contaminants in water supplies. *Journal of the American Water Works Association* 77:67–72.

Anestis, I., and R.J. Neufeld. 1986. Avoidance-preference reactions of rainbow trout (*Salmo gairdneri*) after prolonged exposure to chromium (VI). *Water Research* 20:1233–1241.

Araki, S., H. Aono, and K. Murata. 1986. Adjustment of urinary concentration and urinary volume in relation to erythrocyte and plasma concentrations: an evaluation of urinary heavy metals and organic substances. *Archives of Environmental Health* 41:171–177.

Araujo, M.F.D., P.C. Bernard, and R.E. van Grieken. 1988. Heavy metal contamination in sediments from the Belgian coast and Scheldt Estuary. *Marine Pollution Bulletin* 19:269–273.

Azcue, J.M.P., W.C. Pfeiffer, M. Fiszman, and O. Malm. 1988. Heavy metal removal by different water treatment plants in Rio de Janeiro State, Brazil. *Environmental Technology Letters* 9:429–436.

Beukema, A.A., G.P. Hekstra, and C. Venema. 1986. The Netherlands' environmental policy for the North Sea and Wadden Sea. *Environmental Monitoring and Assessment* 7:117–155.

Brooks, R.R., and D. Rumsey. 1974. Heavy metals in some New Zealand commercial sea fishes. *New Zealand Journal of Marine and Freshwater Research* 8:155–166.

Byrne, C.J., and I.R. DeLeon. 1986. Trace metal residues in biota and sediments from Lake Pontchartrain, Louisiana. *Bulletin of Environmental Contamination and Toxicology* 37:151–158.

Canadian Water Quality Guidelines. 1987. Canadian Council of Resource and Environmental Ministers, Environment Canada, Ottawa.

Carson, B.L., H.V. Ellis, and J.L. McCann. 1987. *Toxicology and biological monitoring of metals in humans*. Lewis Publishers, Chelsea, MI. 328 pp.

DeLeon, I.R., C.J. Byrne, E.A. Peuler, S.R. Antoine, J. Schaeffer, and R.C. Murphy. 1986. Trace organic and heavy metal pollutants in the Mississippi River. *Chemosphere* 15:795–805.

Dougherty, G. 1988. Heavy metal concentrations in bivalves from Fiji's coastal waters. *Marine Pollution Bulletin* 19:81–84.

Eary, L.E., and D. Ral. 1987. Kinetics of chromium (III) oxidation to chromium (VI) by reaction with manganese dioxide. *Environmental Science and Technology* 21:1187–1193.

Estabrook, G.F., D.W. Burk, D.R. Inman, P.B. Kaufman, J.R. Wells, J.D. Jones, and N. Ghosheh. 1985. Comparison of heavy metals in aquatic plants on Charity Island, Saginaw Bay, Lake Huron, U.S.A., with plants along the shoreline of Saginaw Bay. *American Journal of Botany* 72:209–216.

Gryzhankovà, L.N., G.N. Sayenko, A.V. Karyakin, and N.V. Laktionova. 1973. Concentration of some metals in the algae of the Sea of Japan. *Oceanology* 13:206–210.

Hagel, P. 1986. Monitoring of pollutants in Dutch fishery products. *Environmental Monitoring and Assessment* 7:257–262.

Hornung, H., and G.J. Ramelow. 1987. Distribution of Cd, Cr, Cu, and Zn in eastern Mediterranean fishes. *Marine Pollution Bulletin* 18:45–49.

Howard, L.S., and B.E. Brown. 1987. Metals in *Pocillopora damicornis* exposed to tin smelter effluent. *Marine Pollution Bulletin* 18:451–454.

International Agency for Research on Cancer. 1980. *Some metals and metallic compounds,* Volume 23. International Agency for Research on Cancer, Lyon, France. 438 pp.

Johnson, M.G. 1987. Trace element loadings to sediments of fourteen Ontario lakes and correlations with concentrations in fish. *Canadian Journal of Fisheries and Aquatic Sciences* 44:3–13.

Korallus, U. 1986. Chromium compounds: occupational health, toxicological and biological monitoring aspects. *Toxicological and Environmental Chemistry* 12:47–59.

Lanfranchi, G., S. Paglialunga, and A.G. Levis. 1988. Mammalian cell transformation induced by chromium (VI) compounds in the presence of nitrilotriacetic acid. *Journal of Toxicology and Environmental Health* 24:251–260.

Langard, S. 1988. Chromium carcinogenicity: a review of experimental animal data. *Science of the Total Environment* 71:341–350.

Lema, J.M., R. Mendez, and R. Blazquez. 1988. Characteristics of landfill leachates and alternatives for their treatment: a review. *Water, Air, and Soil Pollution* 40:223–250.

Lima, N.R.W., L.D. de Lacerda, W.C. Pfeiffer, and M. Fiszman. 1986. Temporal and spatial variability in Zn, Cr, Cd, and Fe concentrations in oysters tissues (*Crassostrea brasiliana* Lamarck, 1819) from Sepetiba Bay, Brazil. *Environmental Technology Letters* 7:453–460.

MacKenzie, R.D., R.U. Byerrum, C.F. Decker, C.A. Hoppert, and R.F. Langham. 1958. Chronic toxicity studies. II. Hexavalent and trivalent chromium administered in drinking water to rats. *American Medical Association Archives of Industrial Health* 18:232–234.

Mance, G. 1987. *Pollution threat of heavy metals in aquatic environments*. Elsevier, New York. 372 pp.

Martin, T.R., and D.M. Holdich. 1986. The acute lethal toxicity of heavy metals to peracarid crustaceans (with particular reference to fresh-water asellids and gammarids). *Water Research* 20:1137–1147.

Minoia, C., P. Apostoli, G. Maranelli, C. Baldi, L. Pozzoli, and E. Capodaglio. 1988. Urinary chromium levels in subjects living in two north Italy regions. *Science of the Total Environment* 71:527–531.

Mohapatra, S.P. 1988. Distribution of heavy metals in polluted creek sediment. *Environmental Monitoring and Assessment* 10:157–163.

Mudroch, A., L. Sarazin, and T. Lomas. 1988. Summary of surface and background concentrations of selected elements in the Great Lakes sediments. *Journal of Great Lakes Research* 14:241–251.

Nakayama, E., H. Tokoro, T. Kuwamoto, and T. Fujinaga. 1981. Dissolved state of chromium in seawater. *Nature* 290:768–770.

Nath, K., and N. Kumar. 1987. Toxic impact of hexavalent chromium on the blood pyruvate of *Colisa fasciatus*. *Acta Hydrochimica Hydrobiologie* 5:531–534.

Nath, K., and N. Kumar. 1988. Hexavalent chromium: toxicity and its impact on certain aspects of carbohydrate metabolism of the freshwater teleost, *Colisa fasciatus*. *Science of the Total Environment* 72:175–181.

Nevissi, A.E., F.B. DeWalle, J.F.C. Sung, K. Mayer, and R. Dalsey. 1988. Heavy metal variability of different municipal sludges as measured by atomic absorption and inductively coupled plasma emission spectroscopy. *Journal of Environmental Science and Health* A23:823–841.

Noetstaller, R. 1988. *Industrial minerals*. Technical Paper No. 76, World Bank, Washington, DC. 117 pp.

Nriagu, J.O. 1989. A global assessment of natural sources of atmospheric trace metals. *Nature* 338:47–49.

Nriagu, J.O., and R.D. Coker. 1980. Trace metals in humic and fulvic acids from Lake Ontario sediments. *Environmental Science and Technology* 4:443–446.

Nriagu, J.O., and J.M. Pacyna. 1988. Quantitative assessment of worldwide contamination of air, water and soils by trace metals. *Nature* 333:134–139.

Phillips, C.R., J.R. Payne, J.L. Lambach, G.H. Farmer, and R.P. Sims. 1987. Georges Bank monitoring program: hydrocarbons in bottom sediments and hydrocarbons and trace metals in tissues. *Marine Environmental Research* 22:33–74.

Ramelow, G.J., R.S. Maples, R.L. Thompson, C.S. Mueller, C. Webre, and J.N. Beck. 1987. Periphyton as monitors for heavy metal pollution in the Calcasieu River estuary. *Environmental Pollution* 43:247–261.

Rehwoldt, R., L. Lasko, C. Shaw, and E. Wirhowski. 1973. The acute toxicity of some heavy metal ions toward benthic organisms. *Bulletin of Environmental Contamination and Toxicology* 10:291–294.

Saiki, M.K., and T.W. May. 1988. Trace element residues in bluegills and common carp from the lower San Joaquin River, California, and its tributaries. *Science of the Total Environment* 74:199–217.

Sakai, H., T. Sasaki, and T. Saito. 1988. Heavy metal concentrations in urban snow as an indicator of air pollution. *Science of the Total Environment* 77:163–174.

Samanidou, V., and K. Fytianos. 1987. Partitioning of heavy metals into selective chemical fractions in sediments from rivers in northern Greece. *Science of the Total Environment* 67:279–285.

Sears, J.R., K.J. Pecci, and R.A. Cooper. 1985. Trace metal concentrations in offshore, deep-water seaweeds in the western north Atlantic Ocean. *Marine Pollution Bulletin* 16:325–328.

Sivalingam, P.M. 1978. Biodeposited trace metals and mineral content studies of some tropical marine algae. *Botanica Marina* 21:327–330.

Sriwastwa, V.M.S., R.S. Tripathi, D.K. Srivastava, and R. Kumar. 1987. Accumulation and depletion of chromium in *Mystus vittatus* (Bloch). *Acta Hydrochimica Hydrobiologie* 15:179–183.

Stackhouse, R.A., and W.H. Benson. 1988. The influence of humic acid on the toxicity and bioavailability of selected trace metals. *Aquatic Toxicology* 13:99–108.

Stull, J.K., and A.B. Baird. 1985. Trace metals in marine surface sediments of the Palos Verdes Shelf, 1974 to 1980. *Journal of the Water Pollution Control Federation* 57:833–840.

Sugatt, R.H. 1980. Effects of sublethal sodium dichromate exposure in freshwater on the salinity tolerance and serum osmolality of juvenile coho salmon, *Oncorhynchus kisutch*, in seawater. *Archives of Environmental Contamination and Toxicology* 9:41–52.

Sung, J.F.C., A.E. Nevissi, and F.B. Dewalle. 1986. Concentration and removal efficiency of major and trace elements in municipal wastewater. *Journal of Environmental Science and Health* 21:435–448.

US Environmental Protection Agency. 1988. *Distribution of contaminants in waters of Monroe Harbor (River Basin) Michigan and adjacent Lake Erie*. US Environmental Protection Agency, EPA/600/3–88/002, Duluth, MN. 153 pp.

US Environmental Protection Agency. 1989. Drinking water health advisories. *Reviews of Environmental Contamination and Toxicology* 107:1–184.

US Minerals Yearbooks. 1930–1989. Bureau of Mines, US Department of the Interior, Washington, DC.

V.-Balogh, K., D.S. Fernandez, and J. Salanki. 1988. Heavy metal concentrations of *Lymnaea stagnalis* L. in the environs of Lake Balaton (Hungary). *Water Research* 22:1205–1210.

Wang, W. 1986a. Toxicity tests of aquatic pollutants by using common duckweed. *Environmental Pollution* 11:1–14.

Wang, W. 1986b. The effect of river water on phytotoxicity of Ba, Cd and Cr. *Environmental Pollution* 11:193–204.

Young, T.C., J.V. DePinto, and T.W. Kipp. 1987. Adsorption and desorption of Zn, Cu, and Cr by sediments from the Raisin River (Michigan). *Journal of Great Lakes Research* 13:353–366.

10
Cobalt

Cobalt occurs in the Earth's crust at an average concentration of 25 mg/kg, principally as linnaeite (Co_3S_4), carrollite ($CuCo_2S_4$), safflorite ($CoAs_2$), skutterudite ($(Co,Fe)As_3$), and erythrite ($Co_3(AsO_4)_2{\cdot}8H_2O$). Although the concentration of cobalt in these minerals can be as high as 200 mg/kg, relatively large residues are also found in coal, uranium ores, some crude oils, and other sulfide ores. Vitamin B_{12}, essential to humans, contains cobalt.

Production, Sources, and Residues

Production

World production of refined cobalt was only 0.7×10^3 metric tons in 1930, increasing to 14×10^3 metric tons in 1960 and 31×10^3 metric tons in 1980 (US Minerals Yearbooks, 1930–1989; Cordero, 1988). Production in recent years has continued to increase, and is currently near 40×10^3 metric tons, making it the 68th most important mineral in terms of production weight. The world's leading producers (in metric tons per year) are as follows: Zaire, 10,700; USSR, 4,800 (approx.); Zambia, 4,300; Canada, 2,100; Norway, 1,600; and Finland, 1,400.

More than 75% of cobalt production is used in the manufacture of alloys, primarily where strength, resistance to high temperatures, and resistance to oxidation by hot gases are needed. Other uses include the production of cemented tungsten carbides, and steel implements used at high speed. Cobalt also finds application as a drying agent, and as an oxidizing catalyst in the petroleum industry and in the polymerization of unsaturated glycerides.

Sources

Total environmental flux of cobalt is relatively low compared to most inorganic agents. On a worldwide basis, some 3,500 $\times$ 10^5 metric tons are transported annually by rivers, compared to only 62 $\times$ 10^5 metric tons/yr through the atmosphere. Anthropogenic emissions, largely the burning of fossil fuels, account for approximately 55% of all cobalt in the air; the primary natural sources to the atmosphere are windborne soil particles and sea salt spray.

Cobalt occurs in coal and fly ash at concentrations ranging from 5 to 25 mg/kg, and is also found in coal gasification products, tar sands, and shale oils (Table 10.1). Municipal effluents may also contain relatively high residues, particularly in cities dominated by metal-working industries. For example, Abuzkhar et al. (1987) reported that solid sludge from the City of Tripoli (Libya) contained an average residue of 18.8 mg Co/kg dry weight, with a range of 0.02–61.1 mg/kg. Similarly, cobalt in municipal waste water from 25 plants in the state of Washington ranged up to 1.7 mg/L on a wet weight basis (Table 10.2). In a study of garbage incineration plants in Paris, cobalt was found in fly ash at concentrations ranging from 20 to 60 mg/kg (Gounon and Milhau, 1986).

Residues

Total Co in freshwaters is generally below 0.001 mg/L, increasing to 0.001–0.010 mg/L in highly industrialized or mining areas. For example, residues in a stream in Czechoslovakia averaged 0.001 mg/L with a range of 0.0002–0.005 mg/L (Vymazal, 1984). Although similar results (0.0006–

Table 10.1. Concentration (mg/kg) of cobalt in fossil fuels and related products.

Sample	Concentration
Shale oil, Queensland, Australia	0.08–0.25
Shale, Queensland, Australia	9
Discarded shale oil from processing plant	11
Lake sediment, coal-burning power plant, USA	<1.0–57
Tar from production well after gasification	0.85
Crude oils, USA	0.00–1.0
Asphaltenes	122

Sources: Patterson et al. (1987), Wilson et al. (1986), Smith and Carson (1979).

Table 10.2. Concentration (mg/L wet weight) of cobalt in municipal waste from treatment plants in the state of Washington.

Sample	Concentration	
	Average	Range
Raw sewage	0.005	ND[a]–0.018
Primary effluent	0.005	ND–0.015
Secondary effluent	0.003	ND–0.008
Discharge	0.004	ND–0.008
Primary sludge	0.517	0.017–1.70
Secondary sludge	0.226	0.055–0.414

[a]Not detected (<0.001 mg/L).
Source: Sung et al. (1986).

0.005 mg/L) were reported for a lake in Nigeria (Sridahr, 1986), much higher concentrations (0.008–0.015 mg/L) were found in the Hindon River in northern India (Ajmal et al., 1987). That river receives wastes from several municipalities located along its course.

Cobalt occurs in extremely low concentrations in marine waters, often below 10 ng/L. Higher levels have been recorded in the open ocean, but these are probably due to contamination with phytoplankton and other organisms that metabolize vitamin B_{12}.

Cobalt in uncontaminated freshwater sediments is generally found at 1–10 mg/kg dry weight, increasing to 25–50 mg/kg in anthropogenically contaminated areas. Residues in Hamilton Harbor, situated in a highly industrialized part of eastern Canada, averaged 6.5 mg/kg (range 2.3–11 mg/kg) in shallow water near the shore, increasing to 38 mg/kg (range 17–58 mg/kg) in deposition basins (Poulton, 1987). Hutchinson and Fitchko (1974) reported that outlet sediments of Great Lakes tributaries contained <5 mg Co/kg in the majority of samples, with only 10% of the samples exceeding 25 mg/kg. Appreciably higher residues, averaging 37 mg/kg (range 33–40 mg/kg), were found in Thane Creek, which receives industrial wastes from the city of Bombay, India (Mohapatra, 1988). Even higher levels (up to 57 mg/kg) were found in the sediments of lakes near coal-burning generating plants (Wilson et al., 1986).

Angino (1966) demonstrated that Antarctic pelagic sediments contained an average concentration of 27 mg Co/kg, with an average Co/Al and Ni/Co ratio of 0.0037 and 1.4, respectively. Residues of 12–34 mg/kg have been reported for the Indian Ocean, whereas sediments in the Baltic Sea and North Sea contain cobalt at 8 kg and 6–25 mg/kg, respectively (Duursma, 1973).

Chemistry

In freshwater, the dominant species are Co^{2+}, $CoCO_3$, $Co(OH)_3$, and CoS. Although lesser amounts of $CoSO_4$ and $CoCl^+$ may also be detected, chloride complexes dominate in seawater. Cyanocobalamin, or vitamin B_{12}, is a cobalt coordination compound found in surface waters, sediments, and other substrates such as sewage sludge. It has a unique structure, specifically the cobalt-carbon bond and the corrinoid ligand system, that includes four pyrrole nuclei joined in a ring (Figure 10.1). Numerous plants and microorganisms use vitamin B_{12} as their primary source of cobalt.

Cobalt forms chelates of moderate stability—greater than those of iron and manganese, but less than those of mercury, copper, nickel, and lead (Hutzinger, 1980).Huljev (1986) showed that humic acids from the Sava River (Yugoslavia) contained cobalt at an average concentration of 5 mg/kg; cobalt was also detected in the hydrolysis products of these same humic acids in the following concentrations (mg/kg): amino acids 7, phenols and phenolic acids 0.2, carbohydrates 8.0, polycyclic aromatics 0.4.

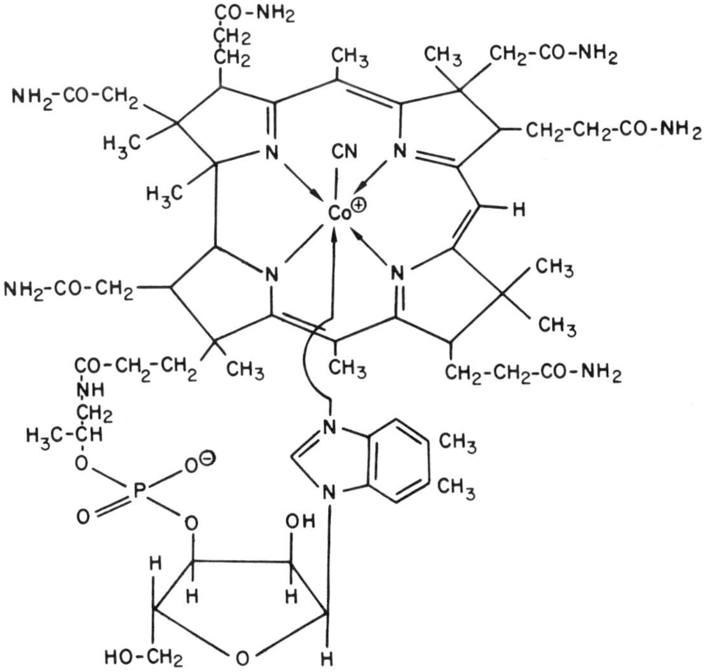

Figure 10.1. Structure of cyanocobalamin.

Theis et al. (1988) demonstrated that the concentration of dissolved metal was significantly correlated with pH and the concentration of suspended solids in the water.

Murray and Mayer (1986) similarly showed that the retention of radioactive ^{60}Co by sediments in a bay along the coast of Maine was inversely correlated with pH in seawater at salinities of 13 and 26 ‰. Studies in oxygen-deficit waters of the Santa Monica Basin (California) indicated that cobalt was scavenged by manganese oxide particles (Johnson et al., 1988). Because the subsequent reduction of manganese oxide was accelerated at low oxygen levels, cobalt near the bottom was elevated about four times above background.

Bioaccumulation

Plants

Total Co in freshwater plants is generally low, <2 mg/kg dry weight, but there are occasional reports of higher levels in the 5–25 mg/kg range. Estabrook et al. (1985), working on Saginaw Bay of Lake Huron, found that residues in eight submerged species averaged approximately 0.8 mg/kg with a range of 0.4–2.6 mg/kg. The terrestrial species, or those with emergent leaves, contained lower residues, 0.2 mg/kg (range 0.05–0.6 mg/kg). Ajmal et al. (1987) reported that total Co in plants (*Eicchornia crassipes*) from the highly polluted River Hindon (India) averaged 4.0 mg/kg, giving a concentration factor of 400.

Marine seaweeds contain low cobalt levels in most cases, falling in the 0.1–1.0 mg/kg dry weight range (reviewed by Smith and Carson, 1979). This reflects, in part, the relatively low concentration of cobalt in seawater. Concentration factors vary from 200 to 2,000, depending on species and environmental conditions.

Uptake of cobalt by plants appears to be extremely rapid. In a study by Vymazal (1984), over 90% of total Co was removed from culture water by *Cladophora glomerata* and *Oedogonium rivulare* in 1 h. It was also shown that the presence of humic acids in the media significantly reduced uptake.

Invertebrates

Total Co in freshwater and marine invertebrates is generally low, even in industrialized areas and in estuaries and other coastal waters (Table 10.3). The greatest residues are generally reported for sessile organisms, such as corals, burrowing crustaceans, and some bivalve molluscs. Concentration factors (soft tissue residue/water residue) generally range up to 300 for these species.

Table 10.3. Concentration (mg/kg dry weight) of cobalt in the soft tissues of marine invertebrate species.

Species	Average (range)	Location
Bivalve mollusc, *Donax serra*[1]	0.06 (0.01–0.19)[a]	South African coast
Gastropod mollusc, *Bullia rhodostroma*[1]	0.06 (0.01–0.19)[a]	South African coast
Barnacle, *Balanus improvisus*[2]	0.4 (0.3–0.6)	Gdansk Bay, Poland
Bivalve mollusc, *Mya arenia*[2]	2.6 (2.0–3.6)	Gdansk Bay, Poland
Bivalve mollusc, *Anadara granosa*[3]	4.1 (NR[b])	Bombay harbor
Coral, *Pocillopora damicornis*[4]	23.9 (NR)	Phuket, Thailand
Crustacean, *Callichirus laurae*[5]	9.4 (NR)	Red Sea, Jordan

[a]Wet weight.
[b]Not reported.
Sources: [1]Watling and Watling (1983), [2]Szefer (1986), [3]Patel et al. (1985), [4]Howard and Brown (1987), [5]Abu-Hilal et al. (1988).

Nakahara and Cross (1978) studied the distribution of radioactive ^{60}Co in the bivalve mollusc *Mercenaria mercenaria* after feeding algae. The initial percentage distribution of ^{60}Co was mantle 46.7, intestine 21.2, foot 19.2. After 4 h, the original dose had been distributed to the foot (18.5), intestine (12.4), body fluid (11.6), and liver (5.5); after 9 h the liver was the primary site of uptake (35.8), followed by the foot (2.4) and mantle (2.0). Other studies have shown that up to 90% of cobalt can be associated with the shell in molluscs and the exoskeleton in arthropods. The influence of temperature, pH, ligands, and salinity on uptake has not been definitively described. Competition with other metals for uptake sites is also poorly known.

Fish

Cobalt has not been identified as an environmentally significant contaminant of fish tissues, and, in fact, residues in muscle tissue are often <0.1 mg/kg wet weight. Reed et al. (1968), for example, found that the distribution of cobalt in black bullhead *Ictalurus melas* was muscle 0.030, blood 0.305, liver 0.116, kidney 1.068, and bone 0.083 mg/kg wet weight. Similarly, Korda et al. (1977) reported the following values for slimy sculpin *Cottus cognatus*: muscle 0.015, liver 0.080 mg/kg wet weight. Concentration factors (muscle residue/water residue) typically range from 10 to

1,000, but occasionally fall in the 1–5 range (reviewed by Smith and Carson, 1979). As is the case with invertebrate species, relatively little is known about the factors influencing cobalt uptake in either marine or freshwater fish.

Toxic Effects to Aquatic Organisms

Cobalt is moderately toxic to most aquatic species, more so than Cr^{3+}, Cr^{6+}, Mn^{7+}, and Mo^{6+}, but much less toxic than Al^{2+}, Cd^{2+} and Cu^{2+}. The LC_{50} for Co^{2+} typically ranges from 3 to >100 mg/L, depending on species (Table 10.4). Ewell et al. (1986) reported a 96-h LC_{50} of 48 mg/kg for the fathead minnow *Pimephales promelas*. This is a relatively tolerant species, compared to more sensitive fish such as rainbow trout *Oncorhynchus mykiss*.

A number of toxic effects have been reported following long-term exposure of fish to cobalt. These effects include decrease in muscle glycogen, hyperlacticemia, and necrosis of gill epithelial cells and concomitant decrease in oxygen uptake (Nath and Kumar, 1988). Although changes in oxygen uptake would also probably result in increased CO_2 levels in the blood and decreased blood pH, no studies appear to confirm this point.

Because cobalt generally occurs in extremely low concentrations in surface waters, most nations have not promulgated regulations aimed at protecting fish and other aquatic species.

Table 10.4. Acute toxicity (LC_{50}) of cobalt (Co^{2+}) to some aquatic organisms.

Species	96-h LC_{50} (mg/L)	Conditions
Cladoceran, *Daphnia magna*[1]	3.2	pH 7.2, total hardness 130 mg/L
Cladoceran, *Daphnia magna*[2]	2.1	pH 7.6, total hardness 240 mg/L
Flatworm, *Dugesia tigrina*[1]	25	pH 7.2, total hardness 130 mg/L
Gastropod, *Helisoma trivolvis*[1]	>100	pH 7.2, total hardness 130 mg/L
Amphipod, *Gammarus fasciatus*[1]	>100	pH 7.2, total hardness 130 mg/L
Amphipod, *Crangonyx pseudogracilis*[3]	39	pH 6.75, total hardness 50 mg/L
Amphipod, *Crangonyx pseudogracilis*[3]	167[a]	pH 6.75, total hardness 50 mg/L

[a] 48-h LC_{50}.

Sources: [1]Ewell et al. (1986), [2]Khangarot et al. (1987), [3]Martin and Holdich (1986).

Health Effects

Intake

Recommended dietary consumption of vitamin B_{12} is 0.005 mg/day, which is equivalent to 0.0002 mg Co/day. Although most diets lead to an ingestion of 0.04–0.05 mg Co/day, the consumption of molluscs and other contaminated food products may lead to appreciably greater intake (Carson et al., 1987). Anemic medical patients have been treated with cobalt salts at concentrations of 0.2 to 3.9 mg Co/kg, equivalent to a daily intake of 12 mg at the lowest dose in a 70-kg adult. Inhalation is an insignificant route of exposure, except in cases of occupational exposure.

Excretion occurs principally through the urine, with lesser amounts leaving in the feces. Alexandersson (1988) reported that concentrations in the urine of occupationally exposed workers were as high as 4,600 nmol/L, decreasing to 2,500 nmol/L after 72 h.

The total body burden of cobalt is approximately 1.5 mg in a 70-kg adult. Residues are elevated in the liver, presumably because of portal circulation. Muscle tissue, spleen, and intestine have also been reported to contain high levels, whereas relatively low concentrations are found in the kidneys, lung, and brain.

Acute Toxicity

Acute exposure to cobalt may lead to a depression in iodine uptake, anorexia, nausea, vomiting, and diarrhea (Domingo, 1989). Neurotoxicological symptoms, including headache, peripheral neuritis, and changes in reflexes, have also been reported.

Chronic Toxicity

Chronic exposure to cobalt may lead to partial or complete loss of smell, gastrointestinal problems, dilation of the heart, secondary thrombosis, increase in erythrocytes in the blood, and decrease in uptake of iodine by the thyroid.

Carcinogenicity

Cobalt powder, cobalt sulfide, and cobalt oxide can produce injection site fibrosarcoma following subcutaneous injection, and rhabdomyosarcoma in rats but not mice following intramuscular injections (reviewed by Domingo, 1989). Pure metal cobalt has also produced malignant tumors at the injection site in rats. There does not appear to be any evidence for carcinogenicity of cobalt in humans, even those exposed occupationally through the inhalation route. There is also no evidence for carcinogenicity following ingestion of cobalt in either food or water.

Drinking Water

Residues

Cobalt often goes undetected in finished drinking water, reflecting its low concentration in surface waters and the removal of particle-bound metals during the water treatment process. When present, concentrations are generally <0.005 mg/L, though there is a report of 0.019 mg/L in drinking water at a university in India (Ajmal and Uddin, 1986).

Consumption Guidelines

Because cobalt occurs in generally low concentrations, most nations, plus the World Health Organization, have not promulgated guidelines or standards for drinking water consumption. Carson et al. (1987) noted that a guideline of 1.0 mg Co/L was suggested for the protection of drinking water in the Soviet Union.

Treatment

Because cobalt generally occurs in low concentrations, there has been little need to develop advanced systems for its removal from potable water. It is assumed, however, that use of alum or related agent, plus conventional filtration, will remove particulate-bound cobalt.

Recommendations

When cobalt is found in surface waters, residues are usually small, well below those that would represent a significant environmental or health problem. In addition, the primary sources of cobalt in the environment also emit many other potentially toxic agents, often in significant quantities. This means that potential adverse effects may not be attributable to cobalt alone.

Despite its low concentration in surface waters, cobalt is potentially toxic, at least in laboratory studies. Hence multielement monitoring surveys of surface water should continue to include cobalt as part of routine analysis. Although there are no pressing research priorities involving cobalt, studies on environmental fate and chronic toxicity to multiple species may be warranted in future years. These priorities may change if consumption patterns, or environmental mobilization, change significantly, even on a regional basis.

References

Abu-Hilal, A., M. Badran, and J. de Vaugelas. 1988. Distribution of trace elements in *Callichirus laurae* burrows and nearby sediments in the Gulf of Aqaba, Jordan (Red Sea). *Marine Environmental Research* 25:233–248.

Abuzkhar, A.A., A.S. Gibali, Y.I. Elmehrik, and R. Ahmatullah. 1987. Chemical monitoring of sewage wastes for their use in crop production. 1. Solid sludge as a soil amendment. *Environmental Monitoring and Assessment* 8:127–133.

Ajmal, M., and R. Uddin. 1986. Quality of drinking water in the Aligarh Muslim University campus, Aligarh, U.P. (India) with respect to heavy metals. *Environmental Monitoring and Assessment* 6:195–205.

Ajmal, M., R. Khan, and A.U. Khan. 1987. Heavy metals in water, sediments, fish and plants of river Hindon, U.P., India. *Hydrobiologia* 148:151–157.

Alexandersson, R. 1988. Blood and urinary concentrations as estimators of cobalt exposure. *Archives of Environmental Health* 43:299–303.

Angino, E.E. 1966. Geochemistry of Antarctic pelagic sediments. *Geochimica Cosmochimica Acta* 30:939–961.

Carson, B.L., H.V. Ellis, and J.L. McCann. 1987. Toxicology and biological monitoring of metals in humans. Lewis Publishers, Chelsea, MI. 328 pp.

Cordero, R. 1988. *Metal Bulletin's prices and data 1988*. Metal Bulletin Books, Surrey, England. 375 pp.

Domingo, J.L. 1989. Cobalt in the environment and its toxicological implications. *Reviews of Environmental Contamination and Toxicology* 108:105–132.

Duursma, E.K. 1973. Specific activity of radionuclides sorbed by marine sediments in relation to the stable element composition. *In: Radioactive contamination of the marine environment,* 57–71, International Atomic Energy Agency, Vienna.

Estabrook, G.F., D.W. Burk, D.R. Inman, P.B. Kaufman, J.R. Wells, J.D. Jones, and N. Ghosheh. 1985. Comparison of heavy metals in aquatic plants on Charity Island, Saginaw Bay, Lake Huron, U.S.A., with plants along the shoreline of Saginaw Bay. *American Journal of Botany* 72:209–216.

Ewell, W.S., J.W. Gorsuch, R.O. Kringle, K.A. Robillard, and R.C. Spiegel. 1986. Simultaneous evaluation of the acute effects of chemicals on seven aquatic species. *Environmental Toxicology and Chemistry* 5:831–840.

Gounon, J., and A. Milhau. 1986. Analysis of inorganic pollutants emitted by the City of Paris garbage incineration plants. *Waste Management and Research* 4:95–104.

Howard, L.S., and B.E. Brown. 1987. Metals in *Pocillopora damicornis* exposed to tin smelter effluent. *Marine Pollution Bulletin* 18:451–454.

Huljev, D.J. 1986. Trace elements in humic acids and their hydrolysis products. *Environmental Research* 39:258–264.

Hutchinson, T.C., and J. Fitchko. 1974. Heavy metal concentrations and distributions in river mouth sediments around the Great Lakes. *Proceedings of the International Conference on the Transport of Persistent Chemicals in Aquatic Ecosystems* 1:69–77.

Hutzinger, O. 1980. *The handbook of environmental chemistry*. Volume 1. Part A. *The natural environmental and the biogeochemical cycles*. Springer-Verlag, New York. 258 pp.

Johnson, K.S., P.M. Stout, W.M. Berelson, and C.M. Sakamoto-Arnold. 1988. Cobalt and copper distributions in the waters of Santa Monica Basin, California. *Nature* 332:527–530.

Khangarot, B.S., P.K. Ray, and H. Chandra. 1987. *Daphnia magna* as a model to assess heavy metal toxicity: comparative assessment with mouse system. *Acta Hydrochimica Hydrobiologica* 15:427–432.

Korda, R.J., T.E. Henzler, P.A., Helmke, M.M. Jimenez, L.A. Haskin, and

E.M. Larsen. 1977. Trace elements in samples of fish, sediment and taconite from Lake Superior. *Journal of Great Lakes Research* 3:148–154.

Martin, T.R., and D.M. Holdich. 1986. The acute lethal toxicity of heavy metals to peracarid crustaceans (with particular reference to fresh-water asellids and gammarids). *Water Research* 20:1137–1147.

Mohapatra, S.P. 1988. Distribution of heavy metals in polluted creek sediment. *Environmental Monitoring and Assessment* 10:157–163.

Murray, S., and L.M. Mayer. 1986. Retention of Co = 60 by the sediments of Montsweag Bay, Maine. *Marine Environmental Research* 18:29–41.

Nakahara, M., and F.A. Cross. 1978. Transfer of cobalt-60 from phytoplankton to the clam (*Mercenaria mercenaria*). *Nippon Suisan Gakkaishi* 44:419–425.

Nath, K., and N. Kumar. 1988. Cobalt induced alterations in the carbohydrate metabolism of a freshwater tropical perch, *Colisa fasciatus*. *Chemosphere* 17:465–474.

Patel, B., V.S. Bangera, S. Patel, and M.C. Balani. 1985. Heavy metals in the Bombay Harbour area. *Marine Pollution Bulletin* 16:22–28.

Patterson, J.H., L.S. Dale, and J.F. Chapman. 1987. Trace element partitioning during the retorting of Julia Creek oil shale. *Environmental Science and Technology* 21:490–494.

Poulton, D.J. 1987. Trace contaminant status of Hamilton Harbour. *Journal of Great Lakes Research* 13:193–201.

Reed, J.R., N.A. Griffith, and C.H. Courtney. 1968. *Uptake and excretion of ^{60}Co by black bullheads (Ictalurus melas)*. National Technical Information Service, ORNL-4316, US Department of Commerce, Springfield, VA.

Smith, I.C., and B.L. Carson. 1979. *Trace metals in the environment*. Volume 6, *Cobalt*. Ann Arbor Science, Ann Arbor, MI. 1,202 pp.

Sridhar, M.K.C. 1986. Trace element composition of *Pistia stratiotes* L. in a polluted lake in Nigeria. *Hydrobiologia* 131:273–276.

Sung, J.F.C., A.E. Nevissi, and F.B. Dewalle. 1986. Concentration and removal efficiency of major and trace elements in municipal wastewater. *Journal of Environmental Science and Health* 21:435–448.

Szefer, P. 1986. Some metals in benthic invertebrates in Gdansk Bay. *Marine Pollution Bulletin* 17:503–507.

Theis, T.L., T.C. Young, and J.V. DePinto. 1988. Factors affecting metal partitioning during resuspension of sediments from the Detroit River. *Journal of Great Lakes Research* 14:216–226.

US Minerals Yearbooks. 1930–1989. Bureau of Mines, US Department of the Interior, Washington, DC.

Vymazal, J. 1984. Short-term uptake of heavy metals by periphyton algae. *Hydrobiologia* 119:171–179.

Watling, H.R., and R.J. Watling. 1983. Sandy beach molluscs as possible bioindicators of metal pollution. 1. Field survey. *Bulletin of Environmental Contamination and Toxicology* 31:331–338.

Wilson, B.L., R.R. Schwarzer, and N. Etonyeaku. 1986. The evaluation of heavy metals (chromium, nickel, and cobalt) in the aqueous sediment surrounding a coal burning generating plant. *Journal of Environmental Science and Health* 21:791–808.

11
Copper

Copper occurs in the earth's crust at an average concentration of approximately 50 mg/kg, principally as a sulfide, both as the simple sulfide and in numerous sulfide minerals. Although the primary copper mineral is chalcopyrite ($CuFeS_2$), metallic copper, chalcocite (CuS_2), and bornite are also economically important. Copper is essential in many enzymatic reactions in mammals.

Production, Sources, and Residues

Production

World production of copper was 1,611 $\times$ 10^3 metric tons in 1930, increasing to 4,212 x 10^3 metric tons in 1960 and 7,660 x 10^3 metric tons in 1980 (US Minerals Yearbooks, 1930–1989). Production in recent years has exceeded 8,500 x 10^3 metric tons annually. The leading producers are Chile, the USA, the USSR, Canada, and Zambia; the major consumers are the USA, the USSR, Japan, the FRG, and China (Table 11.1).

Copper's usefulness as a metal is second only to that of iron. Modern uses include electrical wiring and electroplating, the production of alloys such as bronze and brass, antifouling paint, construction, plumbing, and a host of minor applications.

Sources

Estimates of the total anthropogenic discharge of copper to surface waters range from 35 x 10^3 to 90 x 10^3 metric tons per year (Table 11.2).

Table 11.1. World's major producers and consumers of copper.

Producing nation	Quantity (1,000 metric tons/yr)	Consuming nation	Quantity (1,000 metric tons/yr)
Chile	1,356	USA	1,918
USA	1,092	USSR	1,280 (approx.)
USSR	1,020 (approx.)	Japan	1,231
Canada	724	FRG	755
Zambia	520	China	410 (approx.)

Source: Cordero (1988).

The primary sources include domestic waste water from both central and noncentral sources, manufacturing processes involving metals, steam electrical production, and the dumping of sewage sludge. Atmospheric deposition, of which approximately 56% comes from anthropogenic emissions (Nriagu, 1989), is another major source to water. Abuzkhar et al. (1987) reported that the average concentration of copper in sewage sludge from Tripoli (Libya) was 172 mg/kg dry weight, with a range of 112–360 mg/kg. Similarly high values were reported from a treatment plant in the state of Washington (Nevissi et al., 1988). Sung et al. (1986), working on wastes from 25 plants in the state of Washington, found residues as high as 294 mg/L wet weight in primary sludge (Table 11.3). These values are

Table 11.2. Worldwide anthropogenic input of copper to freshwaters.

Source	Input (thousand metric tons per year)
Domestic wastewater	
central	4.5–18
noncentral	4.2–30
Manufacturing processes	
metals	10–38
chemicals	1–18
pulp and paper	0.04–0.4
petroleum products	0–0.06
Steam electrical production	3.6–23
Dumping of sewage sludge	2.9–22
Smelting and refining	
nonferrous metals	2.4–17
Atmospheric deposition	6.0–15
Base metal mining and dressing	0.1–9
Total input	35–90

Source: Nriagu and Pacyna (1988).

Table 11.3. Concentration (mg/L wet weight) of copper in municipal waste from treatment plants in the state of Washington.

	Concentration	
Sample	Average	Range
Raw sewage	0.39	0.03–1.96
Primary effluent	0.34	0.02–1.56
Secondary effluent	0.09	ND[a]–0.35
Discharge	0.09	ND –0.55
Primary sludge	61.2	1.25–294
Secondary sludge	37.9	4.73–116

[a]Not detected (<0.001 mg/L).
Source: Sung et al. (1986).

far higher than those (0.02–0.30 mg/L) commonly reported for landfill leachates.

Gounon and Milhau (1986) reported that municipal incinerators in the city of Paris produced emissions with a copper concentration of up to 5.2 mg/m^3. Similarly high concentrations (910 mg/kg) were found in slag from two municipal incinerators in Switzerland, and the dust from electrostatic precipitators contained even higher levels at 1,100 mg/kg (Brunner and Monch, 1986).

Several industries contribute substantially to the copper burden in the environment. For example, waste from a copper mine in England contained residues of 15,400 mg Cu/kg, while lead and zinc mines produced waste with concentrations ranging from 30 to 210 mg/kg (Wong, 1986). In another study, acid mine drainage from a mining district in California contained copper up to 190 mg/L (Fillpek et al., 1987), whereas residues in soil in the Sudbury Mining district of Ontario were as high as 7 mg/kg, a result of atmospheric deposition (Taylor and Crowder, 1983).

Coastal marine waters are dominated by input from rivers and atmospheric sources, with lesser amounts coming from the dredging of sludges. Of the total copper input to The Netherlands' part of the North Sea in 1980 (1,500 metric tons per year), 900 metric tons came from rivers and 340 metric tons from the atmosphere (Table 11.4). Incineration of wastes at sea, often a politically controversial issue, produced an input of only 1.6 metric tons per year.

Residues

Total copper is detected at low concentrations, generally at <0.020 mg/L, in most freshwaters. Whitehead et al. (1988), working on a north Mediterranean river, found that copper in filtered water samples, which were

Table 11.4. Annual input of total copper to The Netherlands' part of the North Sea in 1980 (actual) and 1990 (projected).

Source	Input 1980 (metric tons per year)	Input 1990 (metric tons per year)
Total input	1,500	940–1100
Atmospheric deposition	340	340
Rivers	900	530–700
Coastal discharges	40	8
Dredging sludges	190	63
Industrial wastes	4.4	0
Incineration at sea	1.6	not known
Offshore mining	0.6	1.0

Source: Beukema et al. (1986).

also acidified, ranged from 100 to 280 ng/L; however, much higher levels, 0.1–6.6 mg/L, were found in suspended solids. DeLeon et al. (1986) reported that residues in the Mississippi River ranged from 0.002 to 0.039 mg/L; these residues were based on acidified and filtered samples. Dissolved levels in Canadian freshwaters rarely exceed 0.005 mg/L in the absence of an anthropogenic source of copper (reviewed in Canadian Water Quality Guidelines, 1987).

Dissolved Cu in the freshwater/saltwater mixing zones of estuaries are often greater than those levels in the inflowing rivers. This is generally due to desorption caused by competition with the chloride ion and, to a lesser degree, bacterial-mediated decomposition of organic material. Patel et al. (1985) reported that dissolved Cu in Bombay Harbor, which receives large amounts of industrial and municipal wastes, averaged 0.63 mg/L, with a range of 0.02 to 1.30 mg/L. Although these values are obviously high, dissolved Cu residues of up to 0.016 mg/L were found in the Mindhola Estuary (India), which also receives waste water (Zingde et al., 1988). The open waters of the North Sea contained dissolved Cu of approximately 0.0004 mg/L (Beukema et al., 1986), whereas seawater off the coast of Malaysia was reported by Seng et al. (1987) to have residues in the 0.0006–0.0008 mg/L range.

Because copper has a strong affinity for clays, iron and manganese oxides, and carbonate materials (discussed in the Chemistry section), residues are often elevated in sediments, both freshwater and marine. For example, total Cu in the sediments of Hamilton Harbor in eastern Canada ranged from 34 to 210 mg/kg dry weight (Poulton, 1987). Similarly, in their review of literature on the Great Lakes, Mudroch et al. (1988) reported the following ranges (mg/kg dry weight) for depositional basins: Ontario, 26–109; Erie, 5–207; Huron, 3–78; Michigan, 15–54; Superior, 30–173. Highly polluted parts of Chesapeake Bay (USA) contain total Cu in the

250–1500 mg/kg dry weight range (Sinex and Wright, 1988) while in San Francisco Bay, residues of 37–380 mg/kg have been reported (Luoma and Phillips, 1988). Concentration factors (sediment residue/water residue) typically exceed 100 in both marine and freshwaters.

Chemistry

Copper shows a pronounced tendency to form complexes with inorganic and organic ligands. In fact, in freshwater at circumneutral pH, most of the inorganic copper in solution is present as complexes with carbonate, nitrate, sulfate, and chloride, rather than as the hydrated divalent cupric ion. The same distribution is also found in seawaters (Luther et al., 1986). Neutral ligands such as ammonia, ethylenediamine, and pyridine also form strong 4-coordinated complexes. Being an intermediate receptor between hard and soft acids, copper forms relatively insoluble complexes with sulfides. The stability constants for some Cu^{2+} complexes are listed in Table 11.5.

In some freshwaters, more than 90% of total Cu may be bound to humic acids (Mantoura et al., 1978). These complexes are often quite stable, more so than those humic acid complexes formed with zinc and manganese (Campanella et al., 1987a,b). In seawater, no more than 10% of total Cu may be bound to humic acids (Mantoura et al., 1978). Although it is generally assumed that calcium and magnesium displace copper from humic acids, Hering and Morel (1988) showed that there were no competitive effects between copper and calcium, at least in laboratory experiments. This suggested either that copper and calcium bind to different sites in humic acids or that a binding mechanism other than discrete ligand binding must have been operative. Similarly, Johnson et al. (1988) reported that copper was scavenged principally by organic compounds in the waters of the Santa Monica Basin, California.

Copper has a strong affinity for hydrous iron and manganese oxides,

Table 11.5. Stability constants for Cu^{2+} complexes at 25°C.

Ligand	Log K	Ligand	Log K
CO_3^{2-}	6.75	NO_2^-	2.02
HPO_3^{2-}	4.6	F^-	1.2
SO_3^{2-}	4.2	NO_3^-	0.5
N_3^-	2.9	Cl^-	0.4
SO_4^{2-}	2.4	Br^-	0.3
NCS^-	2.3		

Source: Conklin and Hoffmann (1988).

Table 11.6. Partitioning of copper in the sediments of the Axios River and Estuary (Greece).

	Distribution (%)	
Fraction	River	Estuary
Cation-exchangeable	0.9	1.0
Carbonates	1.5	2.2
Fe–Mn hydrous oxides	3.1	6.0
Organic sulfides	44.7	43.0
Residual	49.8	47.8

Source: Samanidou and Fytianos (1987).

carbonate materials, clays, and organic matter in bottom sediments. In a study of one Greek river, Samanidou and Fytianos (1987) showed that almost 45% of the copper was bound to organic sulfides, and that only slightly lower levels were found in the river's estuary (Table 11.6). Tessier et al. (1980), working on two rivers in Canada, reported the following distributions: organic, 31–52%; residual, 22–41%; Fe-Mn hydrous oxides, 12–20%; carbonates, 8–14%; and exchangeable, 1.0–1.2%.

Binding to particulates such as clay results in significant downstream transport of copper. Depending on river and environmental conditions, from 10% to 95% of total Cu was transported by this mechanism. Young et al. (1987) showed that adsorption to river sediments depended linearly on soluble metal concentration to sorption concentrations of 6,000 mg/kg dry weight. Copper desorbed from the sediments, falling to only 50% of the initial concentration within 96 h. Several other studies have shown that desorption is enhanced by increasing salinity and the concomitant competition for binding sites with the chloride ion.

Bioaccumulation

Plants

Total Cu in marine and freshwater plants is typically <10 mg/kg dry weight, except near polluting sources, where residues of over 100 mg/kg have been found. Soderlund et al. (1988), working on the northern coast of the Baltic Sea and the southern Bothian Sea, reported that residues in the brown seaweed *Fuscus vesiculosus* ranged from 2.1 to 8.0 mg/kg dry weight, whereas the corresponding range for the aquatic moss *Fontinalis dalecarlica* was 14.4 to 16.3 mg/kg. Comparably low levels (1.4–3.8 mg/kg wet weight) were found in four deep-water seaweeds from the northwest Atlantic (Sears et al., 1985). However, residues in *Ascophyllum nodo-*

sum, collected from a polluted fjord in Norway, averaged 166 mg/kg dry weight, compared to control plant residues of 6 mg/kg (Haug et al., 1974). Ramelow et al. (1987) reported much higher concentrations (16–68 mg/kg dry weight) in microscopic diatoms from a bayou in Louisiana. Such values likely reflect the high surface area:volume ratio of small plants, which permit a relatively rapid uptake of copper.

Uptake of copper is rapid in most plant species studied to date. Drbal et al. (1985), working with the microscopic alga *Scenedesmus obliquus*, reported an 83% removal of copper from solution in 0.5 h. Similarly, Lee and Hardy (1987) found that the initial rapid uptake phase by the water hyacinth *Eichornia crassipes* lasted 4 hours, and was followed by a slower, near linear uptake phase extending beyond 48 h. Although uptake is pH dependent in some species, such as *Scenedesmus pannonicus* (Demon et al., 1988), some species apparently go unaffected by pH (Lee and Hardy, 1987). Increasing salinity generally reduces absorption of copper (Lustigman et al., 1985), presumably owing to competition with chloride ions for binding sites.

Invertebrates

Bivalve molluscs are widely used to monitor copper in marine and, to a lesser degree, freshwaters. Some of the highest residues on record are for oysters *Crassostrea gigas* (6,480 mg/kg wet weight) and molluscs *Nucella lapillus* (1,750 mg/kg) from the Bristol Channel and Severn Estuary in the United Kingdom (Stenner and Nickless, 1974; Boyden and Romeril, 1974). Concentrations in the 1–200 mg/kg range have been reported from a number of other marine locations (Table 11.7).

Copper generally occurs in relatively low concentrations in freshwater invertebrates. There also appears to be little or no concentration through the invertebrate food chain. For example, Anderson (1977) showed that total Cu in herbivorous, omnivorous, and carnivorous invertebrates collected from a stream in an industrial area of Wisconsin averaged 13.7, 70.9, and 30.5 mg/kg wet weight, respectively

Copper uptake generally depends on residues in the water and, in the case of benthic invertebrates, residues in the sediments. Based on laboratory experiments, increasing salinity causes a decrease in the rate of uptake, reflecting the availability of the free copper ion (Wright and Zamuda, 1987). Although there are many exceptions, concentrations generally increase with the size and age of animal.

Fish

Copper residues in fish muscle tissues are generally low, so there are only a few instances worldwide in which fisheries utilization has been limited owing to copper contamination. Ashraf and Jaffer (1988), working on six

Table 11.7. Concentration (mg/kg wet weight) of copper in the soft tissues of marine and freshwater invertebrate species.

Species	Average (range)	Location
Mussel, *Mytilus edulis*[1]	7.9 (6.8–9.8)	Eastern Scheldt, Netherlands
Mussel, *Mytilus edulis*[1]	20.5 (7.5–41.5)	Lake Verre, Netherlands
Bivalve molluscs, eight species[2]	20.2 (3.5–160.0)	Pacific Ocean, Fiji
Bivalve mollusc, *Donax serra*[3]	0.83 (0.12–1.35)[a]	South African coast
Gastropod mollusc, *Bullia rhodostroma*[3]	1.11 (0.27–4.42)[a]	South African coast
Oyster, *Crassostrea angulata*[4]	180.5 (26.4–535.1)	Huelva Estuary, Spain
Cockle, *Cardium edule*[4]	6.4 (2.3–10.9)	Huelva Estuary, Spain
Pearl oyster, *Pinctada radiata*[5]	2.0 (0.4–8.8)	Arabian Gulf, Saudi Arabia
Grass shrimp, *Palaemonetes pugio*[6]	40.0 (NR[b])	Estuaries, New Jersey
Gastropod, *Lymnaea stagnalis*[7]	20.0 (11–31)	Lake Balaton, Hungary

[a]Dry weight.
[b]Not reported.
Sources: [1]De Kock (1986), [2]Dougherty (1988), [3]Watling and Watling (1983), [4]Lopez-Artiguez et al. (1989), [5]Sadig and Alam (1989), [6]Khan et al. (1989), V.-Balogh et al. (1988).

species from the Arabian Sea, recorded muscle levels of only 0.10–0.51 mg/kg wet weight. Similarly, residues in Atlantic bonito *Sarda sarda* from the Gulf of Genona ranged from 0.4 to 5.3 mg/kg, whereas copper in three commercial species from the North Sea ranged from only 0.5 to 1.0 mg/kg (Hagel, 1986). A freshwater teleost *Heteropnuestes fossilis* from the highly polluted Hindon River (India) contained average copper levels of 4.5 mg/kg dry weight, equivalent to approximately 1.5 mg Cu/kg wet weight (Ajmal et al., 1987).

Copper is almost always elevated in the gills, liver, pancreas, and other internal organs, making analysis of such tissues useful in monitoring residues in the environment. For example, Legorburu et al. (1988), working with eels *Anguilla anguilla* from the Urola River (Spain), reported that residues in muscle tissue reached a maximum of only 2.8 mg Cu/kg dry weight, compared to 570 mg/kg for liver and 5.9 mg/kg dry weight for gills. Denton and Burdon-Jones (1986) similarly showed that residues in the livers of all 50 species of fish examined from the Great Barrier Reef were greater than those found in muscle tissue. Periodically, exceptions are reported to this general pattern. For example, in their study on dogfish *Galeus melastomus* near the British Isles, Vas and Gordon (1988) found that copper in muscle averaged 0.22 mg/kg wet weight, compared to 0.25 mg/kg for liver and 0.05 mg/kg for gills.

Copper in fish tissues is readily bound by metallothionein and related protein structures. Studies by Krezoski et al. (1988) on 11 species of freshwater fish showed the presence of large amounts of a 10,000-Da metal-binding protein containing copper, as well as zinc. These proteins were found in both the liver and kidney. Although it is widely assumed that the quantity of metallothionein increases with exposure to copper, Krezoski et al. (1988) reported that metallothionein levels were high in unexposed fish. This means that the routine determination of metallothionein may not be a useful monitoring tool for heavy metal pollution.

Toxic Effects to Aquatic Organisms

Plants

Copper is highly toxic to most species of aquatic plants, and is routinely used as an algicide and herbicide. Inhibition of growth generally occurs at concentrations <0.1 mg Cu/L, whereas chronic effects such as reduced carbon uptake may occur at concentrations in the 0.003–0.03 mg/L range. Some species, such as duckweed *Lemna minor*, are relatively insensitive, showing an EC_{50} (Effective Concentration) of 1.1 mg/L (Wang, 1986).

A considerable body of knowledge is available on the factors affecting the toxicity of copper to aquatic plants (Table 11.8). Water-soluble li-

Table 11.8. Factors influencing the toxicity of copper to marine and freshwater plants.

Species	Factor	Response
Blue-green alga, *Microcystis* sp.[1]	Nitrate–N	Improvement in growth with increasing nutrient level
Marine diatom, *Nitzschia closterium*[2]	Iron	Decrease in copper sensitivity with increasing Fe
Marine diatom, *Nitzschia closterium*[3]	Manganese	Decrease in copper sensitivity with increasing Mn
Green alga, *Chlorella vulgaris*[4]	Nitrogen and phosphorus	Decrease in growth with nutrient limitation
Green alga, *Scenedesmus quadricauda*[5]	pH	Decrease in phosphorus uptake with increasing pH (5.0–6.5)

Sources: [1]Gupta (1989), [2]Stauber and Florence (1985a), [3]Stauber and Florence (1985b), [4]Hall et al. (1989), [5]Peterson et al. (1984).

gands are widely suspected of binding copper, thereby reducing potential toxic effects. In a study on the marine diatom *Nitzschia closterium*, there was at least a 35-fold difference in the toxicity of copper complexes (Florence and Stauber, 1986). Similarly, Starodub et al. (1987) showed that the EC_{50} for the green alga *Scenedesmus quadricauda* increased from 0.10 mg/L for the free Cu ion to 0.30 mg/L for a Cu–EDTA (ethylenediaminetetraacetic acid) complex, and 0.14 mg/L for a Cu–citric acid complex. Other studies have shown that nutrient limitation increases the susceptibility of plants to copper, while the presence of manganese and iron reduces the toxic effect.

Invertebrates

Copper is also highly toxic to most invertebrate species, both marine and freshwater. LC_{50}s are typically less than 0.5 mg/L, but may range from 0.005 to >200 mg/L under certain conditions.

One of the key factors in determining the toxicity to copper is the presence of inorganic and organic ligands.For example, McLeese and Ray (1986) showed that the 144-h LC_{50} of $CdCl_2$ in the crustacean *Crangon septemspinosa* was 2.8 mg/L but increased to >30 mg/L for a Cu–EDTA complex; the corresponding values for another crustacean, *Pandalus montagui*, were 0.05 and >30 mg/L, respectively. Similarly, the 96-h LC_{50} for the oligochaete *Tubifex tubifex* increased by a factor of 150 when water hardness increased from 0.1 to 260 mg $CaCO_3$/L (Brkovic-Popovic and Popovic, 1977). In another study, neonate production in *Ceriodaphnia dubia* (a measure of chronic toxicity) was significantly impaired at 0.008 mg Cu/L at a water hardness of 94 mg/L, and at 0.010 mg Cu/L at a water hardness of 170 mg/L (Belanger *et al.*, 1989). De March (1988) reported that combinations of Cu and Cd, and Cu and Zn produced more than additive effects in the amphipod *Gammarus lacustris*.

In general, sensitivity is inversely related to the age and size of animal. In one particularly good example, the 24-h LC_{50} for the veliger larvae of the mollusc *Corbicula manilensis* was only 0.028 mg Cu/L (Harrison et al., 1984); however, the corresponding values for the juvenile larvae and the adults were 0.60 and 2.60 mg/L. Many other invertebrate species show a less dramatic response to copper as they age.

Numerous chronic effects have been reported following long-term exposure of invertebrates to copper. These effects include (1) reduction in ciliary function in gill tissue of the mollusc *Mytilus californianus* following exposure at 10 mg Cu/L (Smith, 1985); (2) increase in tissue lactic acid in the mollusc *Villorita cyprinoides* following exposure at 0.30 mg/L (Sathyanathan et al., 1988); (3) decrease in triglycerides in the gastropod *Lymnaea luteola* exposed to 2 mg Cu/L (Reddy and Rao, 1987); and (4) necrosis of epithelial tissue in the gastropod *Bulinus tropicus* exposed to 1 mg $CuSO_4$/L (Wolmarans et al., 1986).

Fish

Copper is one of the most toxic heavy metals to fish. Depending on water hardness, LC_{50}s typically range from 0.02 to 1.0 mg Cu/L in freshwater. Ionic copper (Cu^{2+}) and its ionized hydroxides ($Cu_2OH_2^{2+}$, $CuOH^+$) are most toxic, giving LC_{50}s of <0.1 mg/L in rainbow trout (Brown, 1968). An increase in water hardness to approximately 500 mg/L as $CaCO_3$ also increases the LC_{50} to 0.5 mg/L in the same species. Copper toxicity is enhanced synergistically by combinations of Cu/Al/Zn, Cu/Cd/Zn, Cu/Zn, Cu/Zn/Ni/, Cu/Zn/phenol, Cu/phenol, and Cu/H^+. Organic sequestration significantly reduces acute toxicity.

As with most xenobiotics, there is considerable interspecies variability in the sensitivity of fish to copper. Some of the more sensitive species include rainbow trout and guppies *Poecilia reticulata*. Chapman (1978) reported that newly hatched alevins of steelhead trout *Oncorhynchus mykiss* were less sensitive than the swim-up and parr stages, whereas all stages of chinook salmon *Oncorhynchus tshawytscha* were equally sensitive to copper. In bluegill *Lepomis macrochirus*, survival times for age–0 males, age–0 females, age–1 males and age–1 females were 2.8, 2.9, 4.2, and 6.3 times that of juveniles (Tsai and Chang, 1984).

Several species have the ability to adapt to potentially toxic copper levels, at least on a temporary basis. Lauren and McDonald (1987a,b) showed that the ability to adapt depended on changes in both Na^+ transport and permeability in the whole body. It was also shown that rainbow trout sequestered copper in a sulfhydryl-rich, acid-soluble protein, tentatively identified as metallothionein, present in both the gills and liver.

The chronic effects of copper have been well described and include (1) inhibition of acetylcholinesterase activity in rainbow trout exposed to 0.2 mg Cu/L (Nemcsok and Hughes, 1988); (2) cellular changes in the liver of starved roach *Rutilus rutilus* exposed to 0.08 mg Cu/L (Segner, 1987); (3) reduction in growth of developing rainbow trout following exposure to intermittent levels of copper at 0.3–0.5 mg/L (Seim et al., 1984); and (4) reduction in immune response to the air-breathing teleost *Saccobranchus fossilis* exposed to 0.06 mg Cu/L (Khangarot et al., 1988).

Many nations have promulgated guidelines and standards for the protection of freshwater and marine life. Examples of some of these are listed below:

Canada
- 0.0002 mg/L (water hardness 0 to 120 mg/L)
- 0.0003 mg/L (water hardness 121 to 180 mg/L)
- 0.0004 mg/L (water hardness >180 mg/L)

USA
- 0.00066 mg/L (water hardness 50 mg/L; 4-day average)
- 0.0012 mg/L (water hardness 100 mg/L; 4-day average)
- 0.002 mg/L (water hardness 200 mg/L; 4-day average)

European Community
0.001 mg/L (water hardness 0 to 50 mg/L)
0.006 mg/L (water hardness 51 to 100 mg/L)
0.010 mg/L (water hardness 101 to 250 mg/L)
0.028 mg/L (water hardness >250 mg/L)
0.005 mg/L for the protection of saltwater fish and shellfish

Health Effects

Intake

The typical Western diet yields a daily copper intake of 2 to 4 mg in adults and 1.5 mg in young children (Federal Register, 1985). Food is the primary source, contributing at least 75% of total daily intake, followed by consumption of water. Inhalation of airborne copper and dermal absorption through intact skin are generally negligible, even in occupational settings. Approximately 80% of copper is eliminated through the feces, the remainder going through the urine and other routes. The estimated body burden of copper in a 70-kg reference person is 72 mg, of which 65 mg is in soft tissues.

Copper is regarded as an essential element in mammalian nutrition because of its key role in many enzymatic reactions. The minimum dietary requirement in humans is 0.0001 mg/day (Federal Register, 1985). Copper deficiency may result in decreased iron absorption and iron deficiency, and has also been linked with reproductive abnormalities.

Acute Toxicity

There are only a few reported cases of acute poisoning by copper and its salts. The chief symptoms following ingestion are epigastric burning, nausea, vomiting, and diarrhea (reviewed by Carson et al., 1987). There may also be lesions in the gastrointestinal tract and induction of hemolytic anemia. Inhalation of copper produces symptoms similar to those of silicosis; allergic contact dermatitis has also been reported.

Chronic Toxicity

Chronic copper poisoning is rarely reported, except for individuals suffering from Wilson's disease. This is a congenital disorder causing accumulation of copper in liver, brain, and kidney, resulting in hemolytic anemia, neurological abnormalities, and corneal opacity.

Carcinogenicity

The International Agency for Research on Cancer has not evaluated the carcinogenic potential of copper. The US Environmental Protection

Agency ranks copper in Group D—inadequate data for humans and experimental animals. Orally administered copper compounds were not found to increase tumor incidence in several studies (Federal Register, 1985). Stettler et al. (1988) showed that copper slag from smelters induced adenocarcinomas in laboratory rats following intratracheal instillation of a single 20-mg dose.

Drinking Water

Residues

Copper is routinely detected in finished drinking water, generally at a concentration of <0.1 mg/L. In a survey of the American Water Works Association (1985) of drinking water in 39 states and 3 territories, there were 99 episodes of noncompliance with guidelines/standards. For comparison, fluoride and nitrates were in noncompliance in 907 and 369 episodes, respectively. Elevated concentrations of copper were generally associated with the distribution system, rather than contamination of source water. Ajmal and Uddin (1986a,b) reported that residues in hand pump water samples from the city of Aligarh (India) ranged from <0.004 to 0.083 mg Cu/L, compared to 0.001–0.015 mg/L for running-water samples from Aligarh University. Although extremely low concentrations (0.002–0.004 mg/L) were also found in drinking water from Rio de Janeiro, Brazil (Azcue et al., 1988), residues from a number of Canadian cities ranged from <0.010–0.550 mg/L (Gibson et al., 1987).

Leaching from copper pipes and brass fittings contributes in a substantial way to copper residues in drinking water, particularly in acidic and standing water. In one Illinois study, residues in standing water were elevated more than 50% over those in running water (Figure 11.1); additionally, the episodes of noncompliance with the Recommended Contaminant Limit (1.0 mg/L) were restricted to standing water samples. Ajmal and Uddin (1986b) similarly showed that copper in standing water from the Aligarh University (India) reached a maximum of 0.020 mg/L, compared to 0.015 mg/L for running water.

Consumption Guidelines

Many nations use a maximum standard/guideline of 1.0 mg Cu per liter for the protection of drinking water, based on the following considerations. The Lowest-Observed-Adverse-Effect Level (LOAEL) is 5.3 mg Cu/day based on human clinical studies, in which 5.3 mg was the lowest oral dose at which gastrointestinal effects were observed (Federal Register, 1985). An uncertainty factor of 2 was applied because (1) the effect was not permanent, (2) 5.3 mg is the lowest value reported in the literature to generate an effect, and is therefore conservative, and (3) copper

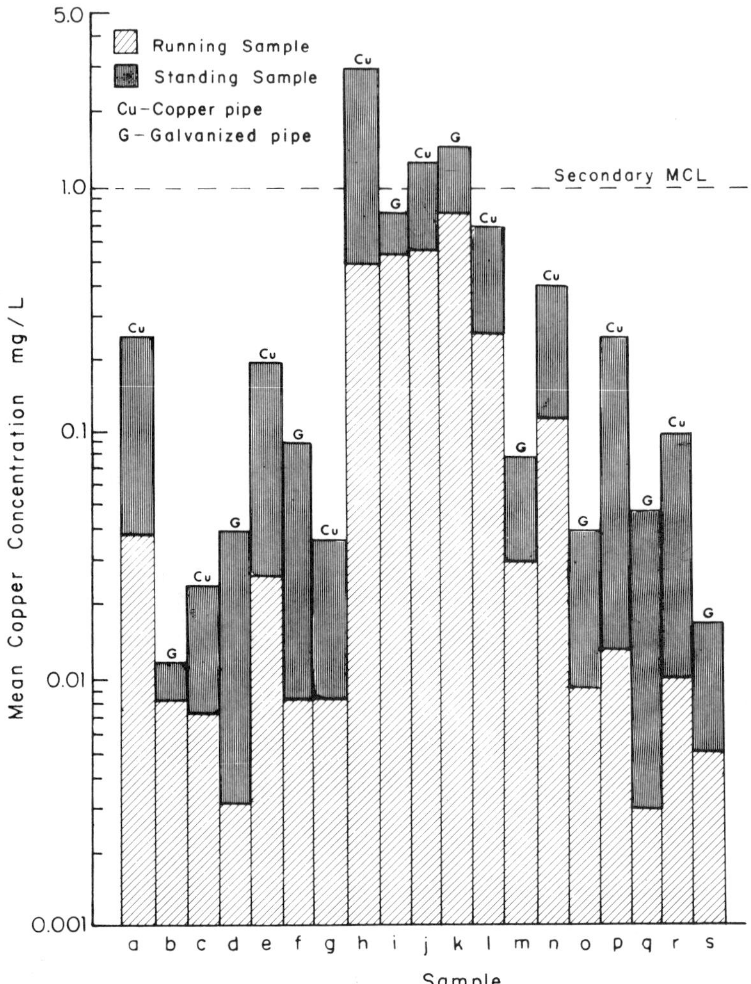

Figure 11.1. Mean copper concentrations in standing and running drinking waters from public systems in the State of Illinois. (From Schock and Neff, 1988.) Reprinted from *Journal American Water Works Association,* Vol. 80, No. 11 (November 1988), by permission. Copyright © 1988, American Water Works Association.

residues in the body are controlled by a homeostatic mechanism, thereby limiting accumulation.

The Adjusted Acceptable Daily Intake was determined for copper based on its acute toxicity, primarily in experimental animals. The same LOAEL and an uncertainty factor of 2 were used to derive the 1-day assessment, resulting in a value of 1.3 mg/L. Downward revision to 1.0 mg/L is based on taste and odor considerations (Federal Register, 1985).

Treatment

A key factor in controlling copper in drinking water is to minimize leaching from pipes and fittings. This is usually accomplished by upward pH adjustment with sodium hydroxide. Basic pH also favors the precipitation of soluble copper, which in turn facilitates removal by coagulation and filtration.

Recommendations

An enormous body of literature is available on the chemistry and the environmental and health effects of copper. Many areas (such as bioaccumulation and toxic effects to marine and freshwater organisms) have already been adequately investigated, and any further research cannot be considered high priority at this time. Copper does not threaten most drinking water supplies, and in cases of noncompliance with regulations, elevated residues are generally due to problems with the distribution system rather than treatment system or source water. Overall, therefore, the environmental and health threat of copper is well recognized and easily controlled.

Although there are no pressing research priorities involving copper, it is important to maintain multimedia monitoring programs, simply because excursions from guidelines and standards will inevitably occur and so threaten the multiple use of water. This assumes that consumption patterns and environmental mobilization of copper undergo no significant change, even on a regional basis. The relatively low research priority placed on copper may need to be changed if these latter considerations also change.

References

Abuzkhar, A.A., A.S. Gibali, Y.I. Elmerik, and R. Ahmatullah. 1987. Chemical monitoring of sewage wastes for their use in crop protection. I. solid sludge as a soil amendment. *Environmental Monitoring and Assessment* 8:127–133.

Ajmal, M., and R. Uddin. 1986a. Studies on heavy metals in the ground waters of the City of Aligarh U.P. (India). *Environmental Monitoring and Assessment* 6:181–194.

Ajmal, M., and R. Uddin. 1986b. Quality of drinking water in the Aligarh Muslim University campus, Aligarh, U.P. (India) with respect to heavy metals. *Environmental Monitoring and Assessment* 6:195–205.

Ajmal, M., R. Khan, and A.U. Khan. 1987. Heavy metals in water, sediments, fish and plants of river Hindon, U.P., India. *Hydrobiologia* 148:151–157.

American Water Works Association. 1985. An AWWA survey of inorganic contaminants in water supplies. *Journal of the American Water Works Association* 77:67–72.

Anderson, R.V. 1977. Concentration of cadmium, copper, lead, and zinc in thirty-five genera of freshwater macroinvertebrates from the Fox River, Illinois and Wisconsin. *Bulletin of Environmental Contamination and Toxicology* 18:345–349.

Ashraf, M., and M. Jaffer. 1988. Correlation between some selected trace metal concentrations in six species of fish from the Arabian Sea. *Bulletin of Environmental Contamination and Toxicology* 41:86–93.

Azcue, J.M.P., W.C. Pfeiffer, M., Fiszman, and O. Maim. 1988. Heavy metal removal by different water treatment plants in Rio de Janeiro, Brazil. *Environmental Technology Letters* 9:429–436.

Belanger, S.E., J.L. Farris, and D.S. Cherry. 1989. Effects of diet, water hardness, and population source on acute and chronic copper toxicity to *Ceriodaphnia dubia. Archives of Environmental Contamination and Toxicology* 18:601–611.

Beukema, A.A., G.P. Hekstra, and C. Venema. 1986. The Netherlands' environmental policy for the North Sea and Wadden Sea. *Environmental Monitoring and Assessment* 7:117–155.

Boyden, C.R., and M.G. Romeril. 1974. A trace metal problem in pond oyster culture. *Marine Pollution Bulletin* 5:74–78.

Brkovic-Popovic, I., and Popovic. 1977. Effects of heavy metals on survival and respiration rate of tubificid worms. I. Effects on survival. *Environmental Pollution* 13:65–72.

Brown, V.M. 1968. The calculation of the acute toxicity of mixtures of poisons to rainbow trout. *Water Research* 10:555–559.

Brunner, P.H., and H. Monch. 1986. The flux of metals through municipal solid waste incinerators. *Waste Management and Research* 4:105–119.

Campanella, L., E. Cardarelli, T. Ferri, B.M. Petronia, and A. Pupella. 1987a. Evaluation of heavy metals speciation in an urban sludge. I. batch method. *Science of the Total Environment* 61:217–228.

Campanella, L., E. Cardarelli, T. Ferri, B.M. Petronia, and A. Pupella. 1987b. Evaluation of heavy metals speciation in an urban sludge. II. column method. *Science of the Total Environment* 61:229–234.

Canadian Water Quality Guidelines. 1987. Canadian Council for Resource and Environment Ministers. Environment Canada, Ottawa.

Capelli, R., and V. Minganti. 1987. Total mercury, organic mercury, copper, manganese, selenium, and zinc in Sarda sarda from the Gulf of Genoa. *Science of the Total Environment* 63:83–99.

Carson, B.L., H.V. Ellis, and J.L. McCann. 1987. *Toxicology and biological monitoring of metals in humans*. Lewis Publishers, Chelsea, MI. 328 pp.

Chapman, G.A. 1978. Toxicities of cadmium, copper and zinc to four juvenile stages of chinook salmon and steelhead. *Transactions of the American Fisheries Society* 107:841–847.

Conklin, M.H., and M.R. Hoffmann. 1988. Metal ion-sulfur (IV) chemistry. 1. structure and thermodynamics of transient copper (II)-sulfur (IV) complexes. *Environmental Science and Technology* 22:883–891.

Cordero, R. 1988. *Metal Bulletin's prices and data 1988*. Metal Bulletin Books, Surrey, England. 375 pp.

De Kock, W.C. 1986. Monitoring bio-available marine contaminants with mussels *(Mytilus edulis L.)* in the Netherlands. *Environmental Monitoring and Assessment* 7:209–220.

De March, B.G.E. 1988. Acute toxicity of binary mixtures of five cations (Cu^{2+}, Cd^{2+}, Zn^{2+}, Mg^{2+}, and K^{+}) to the freshwater amphipod *Gammarus lacustris* (Sars): alternative descriptive models. *Canadian Journal of Fisheries and Aquatic Sciences* 45:625–633.

DeLeon, I.R., C.J. Byrne, E.A. Peuler, S.R. Antoine, J. Schaeffer, and R.C. Murphy. 1986. Trace organic and heavy metal pollutants in the Mississippi River. *Chemosphere* 15:795–805.

Demon, A., M. De Bruin, and H.T. Wolterbeek. 1988. The influence of pH on trace element uptake by an alga *(Scenedesmus pannonicus* subsp. *berlin)* and fungus *(Aureobasidium pullulans)*. *Environmental Monitoring and Assessment* 10:165–173.

Denton, G.R.W., and C. Burdon-Jones. 1986. Trace metals in fish from the Great Barrier Reef. *Marine Pollution Bulletin* 17:201–209.

Dougherty, G. 1988. Heavy metal concentrations in bivalves from Fiji's coastal waters. *Marine Pollution Bulletin* 19:81–84.

Drbal, K., K. Veber, and J. Zahradnik. 1985. Toxicity and accumulation of copper and cadmium in the alga *Scenedesmus obliquus* LH. *Bulletin of Environmental Contamination and Toxicology* 34:904–908.

Federal Register. 1985. National primary drinking water regulations. *Federal Register* 50:46936–47022.

Fillpek, L.H., D.K. Nordstrom, and W.H. Ficklin. 1987. Interaction of acid mine drainage with waters and sediments of West Squaw Creek in the West Shasta mining district, California. *Environmental Science and Technology* 21: 388–396.

Florence, T.M., and J.L. Stauber. 1986. Toxicity of copper complexes to the marine diatom *Nitzschia closterium*. *Aquatic Toxicology* 8:11–26.

Gibson, R.S., P.S. Vanderkooy, C.E. McLennan, and N.M. Mercer. 1987. Contribution of tap water to mineral intakes of Canadian preschool children. *Archives of Environmental Health* 42:165–169.

Gounon, J., and A. Milhau. 1986. Analysis of inorganic pollutants emitted by the City of Paris garbage incineration plants. *Waste Management and Research* 4:95–104.

Gupta, S.L. 1989. Interactive effects of nitrogen and copper on growth of cyanobacterium *Microcystis*. *Bulletin of Environmental Contamination and Toxicology* 42:270–275.

Hagel, P. 1986. Monitoring of pollutants in Dutch fishery products. *Environmental Monitoring and Assessment* 7:257–262.

Hall, J., F.P. Healey, and G.G.C. Robinson. 1989. The interaction of chronic toxicity with nutrient limitation in chemostat cultures of *Chlorella*. *Aquatic Toxicology* 14:15–26.

Harrison, F.L., J.P. Knezovich, and D.W. Rice. 1984. The toxicity of copper to the adult and early life stages of the freshwater clam, *Corbicula manilensis*. *Archives of Environmental Contamination and Toxicology* 13:85–92.

Haug, A., S. Melsom, and S. Omang. 1974. Estimation of heavy metal pollution in two Norwegian fjord areas by analysis of the brown alga *Ascophyllum nodosum*. *Environmental Pollution* 7:179–192.

Hering, J.G., and F.M.M. Morel. 1988. Humic acid complexation of calcium and copper. *Environmental Science and Technology* 22:1234–1237.

Johnson, K.S., P.M. Stout, W.M. Berelson, and C.M. Sakamoto-Arnold. 1988.

Cobalt and copper distributions in the waters of Santa Monica Basin, California. *Nature* 332:527–530.

Khan, A.T., J.S. Weis, and L. D'Andrea. 1989. Bioaccumulation of four heavy metals in two populations of grass shrimp, *Palaemonetes pugio*. *Bulletin of Environmental Contamination and Toxicology* 42:339–343.

Khangarot, B.S., P.K. Ray, and K.P. Singh. 1988. Influence of copper treatment on the immune response in an air-breathing teleost, *Saccobranchus fossilis*. *Bulletin of Environmental Contamination and Toxicology* 41:222–226.

Krezoski, S., J. Laib, P. Onana, T. Hartmann, P. Chen, C.F. Shaw, and D.H. Petering. 1988. Presence of Zn, Cu-binding protein in liver of freshwater fishes in the absence of elevated exogenous metal: relevance to toxic metal exposure. *Marine Environmental Research* 24:147–150.

Lauren, D.J., and D.G. McDonald. 1987a. Acclimation to copper by rainbow trout, *Salmo gairdneri:* physiology. *Canadian Journal of Fisheries and Aquatic Sciences* 44:99–104.

Lauren, D.J., and D.G. McDonald. 1987b. Acclimation to copper by rainbow trout, *Salmo gairdneri:* biochemistry. *Canadian Journal of Fisheries and Aquatic Sciences* 44:105–111.

Lee, T.A., and J.K. Hardy. 1987. Copper uptake by the water hyacinth. *Journal of Environmental Science and Health* 22:141–160.

Legorburu, I., L. Canton, E. Millan, and A. Casado. 1988. Trace metal levels in fish from Urola River (Spain), Anguillidae, Mugillidae and Salmonidae. *Environmental Technology Letters* 9:1373–1378.

Lema, J.M., R. Mendez, and R. Blazquez. 1988. Characteristics of landfill leachates and alternatives for their treatment: a review. *Water, Air, and Soil Pollution* 40:223–250.

Lopez-Artiguez, M., M.L. Soria, and M. Repetto. 1989. Heavy metals in bivalve molluscs in the Huelva Estuary. *Bulletin of Environmental Contamination and Toxicology* 42:634–642.

Luoma, S.N., and D.J.H. Phillips. 1988. Distribution, variability, and impacts of trace elements in San Francisco Bay. *Marine Pollution Bulletin* 19:413–425.

Lustigman, B., J. Korky, A. Zabady, and J.M. McCormick. 1985. Absorption of Cu^{++} by long-term cultures of *Dunaliella salina, D. tertiolecta,* and *D. viridis*. *Bulletin of Environmental Contamination and Toxicology* 35:362–367.

Luther, G.W., Z. Wilk, R.A. Ryans, and A.L. Meyerson. 1986. On the speciation of metals in the water column of a polluted estuary. *Marine Pollution Bulletin* 17:535–542.

Mantoura, R.F.C., A. Dickson, and J.P. Riley. 1978. The complexation of metals with humic materials in natural waters. *Estuarine and Coastal Marine Science* 6:387–408.

McLeese, D.W., and S. Ray. 1986. Toxicity of $CdCl_2$, CdEDTA, $CuCl_2$, and CuEDTA to marine invertebrates. *Bulletin of Environmental Contamination and Toxicology* 36:749–755.

Mudroch, A., L. Sarazin, and T. Lomas. 1988. Summary of surface and background concentrations of selected elements in the Great Lakes sediments. *Journal of Great Lakes Research* 14:241–251.

Nemcsok, J.G., and G.M. Hughes. 1988. The effect of copper sulphate on some biochemical parameters of rainbow trout. *Environmental Pollution* 49:77–85.

Nevissi, A.E., F.B. DeWalle, J.F.C. Sung, K. Mayer, and R. Dalsey. 1988.

Heavy metal variability of different municipal sludges as measured by atomic absorption and inductively coupled plasma emission spectroscopy. *Journal of Environmental Science and Health* 23:823–841.

Nriagu, J.O. 1989. A global assessment of natural sources of atmospheric trace metals. *Nature* 338:47–49.

Nriagu, J.O., and J.M. Pacyna. 1988. Quantitative assessment of worldwide contamination of air, water and soils by trace metals. *Nature* 333:134–139.

Patel, B., V.S. Bangera, S. Patel, and M.C. Balani. 1985. Heavy metals in the Bombay Harbour area. *Marine Pollution Bulletin* 16:22–28.

Peterson, H.G., F.P. Healey, and R. Wagemann. 1984. Metal toxicity to algae: a highly pH dependent variable. *Canadian Journal of Fisheries and Aquatic Sciences* 41:974–979.

Poulton, D.J. 1987. Trace contaminant status on Hamilton Harbour. *Journal of Great Lakes Research* 13:193–201.

Ramelow, G.J., R.S. Maples, R.L. Thompson, C.S. Mueller, C. Webre, and J.N. Beck. 1987. Periphyton as monitors of heavy metal pollution in the Calcasieu River Estuary. *Environmental Pollution* 43:247–261.

Reddy, N.M., and P.V. Rao. 1987. Copper toxicity to the fresh water snail, *Lymnaea luteola*. *Bulletin of Environmental Contamination and Toxicology* 39:50–55.

Sadig, M., and I. Alam. 1989. Metal concentrations in pearl oyster, *Pinctada radiata,* collected from Saudi Arabian coast of the Arabian Gulf. *Bulletin of Environmental Contamination and Toxicology* 42:111–118.

Samanidou, V., and K. Fytianos. 1987. Partitioning of heavy metals into selective chemical fractions in sediments from rivers in northern Greece. *Science of the Total Environment* 67:279–285.

Sathyanathan, B., S.M. Nair, J. Chacko, and P.N.K. Namoisan. 1988. Sublethal effects of copper and mercury on some biochemical constituents of the estuarine clam *Villorita cyprinoides* var. *cochinensis* (Hanley). *Bulletin of Environmental Contamination and Toxicology* 40:510–516.

Schock, M.R., and C.H. Neff. 1988. Trace metal contamination from brass fittings. *Journal of the American Water Works Association* 80:47–56.

Sears, J.R., K.J. Pecci, and R.A. Cooper. 1985. Trace metal concentrations in offshore, deep-water seaweeds in the western North Atlantic Ocean. *Marine Pollution Bulletin* 16:325–328.

Segner, H. 1987. Response of fed and starved roach, *Rutilus rutilus,* to sublethal copper contamination. *Journal of Fish Biology* 30:423–437.

Seim, W.K., L.R. Curtis, S.W. Glenn, and G.A. Chapman. 1984. Growth and survival of developing steelhead trout *(Salmo gairdneri)* continuously or intermittently exposed to copper. *Canadian Journal of Fisheries and Aquatic Sciences* 41:433–438.

Seng, C., P. Lim, and T. Ang. 1987. Heavy metal concentrations in coastal seawater and sediments off Prai industrial estate, Penang, Malaysia. *Marine Pollution Bulletin* 18:611–612.

Sinex, S.A., and D.A. Wright. 1988. Distribution of trace metals in the sediments and biota of Chesapeake Bay. *Marine Pollution Bulletin* 19:425–431.

Smith, J.R. 1985. Copper exposure and ciliary function in gill tissue of *Mytilus californianus*. *Bulletin of Environmental Contamination and Toxicology* 35:556–563.

Soderlund, S., A. Forsberg, and M. Pedersen. 1988. Concentrations of cadmium and other metals in *Fucus vesiculosus* L. and *Fontinalis dalecarlica* Br. Eur. from the northern Baltic Sea and the southern Bothnian Sea. *Environmental Pollution* 51:197–212.

Starodub, M.E., P.T.S. Wong, C.I. Mayfield, and Y.K. Chau. 1987. Influence of complexation and pH on individual and combined heavy metal toxicity to a freshwater green alga. *Canadian Journal of Fisheries and Aquatic Sciences* 44:1173–1180.

Stauber, J.L., and T.M. Florence. 1985a. The influence of iron on copper toxicity to the marine diatom, *Nitzschia closterium* (Ehrenberg) W. Smith. *Aquatic Toxicology* 6:297–305.

Stauber, J.L., and T.M. Florence. 1985b. Interactions of copper and manganese: a mechanism by which manganese alleviates copper toxicity to the marine diatom, *Nitzschia closterium* (Ehrenberg) W. Smith. *Aquatic Toxicology* 7:241–254.

Stenner, R.D., and G. Nickless. 1974. Absorption of cadmium, copper, and zinc by dog whelks in the Bristol Channel. *Nature* 247:198–199.

Stettler, L.E., J.E. Proctor, S.F. Platek, R.J. Carolan, R.J. Smith, and H.M. Donaldson. 1988. Fibrogenicity and carcinogenic potential of smelter slags used as abrasive blasting substitutes *Journal of Toxicology and Environmental Health* 25:35–56.

Sung, J.F.C., A.E. Nevissi, and F.B. DeWalle. 1986. Concentration and removal efficiency of major and trace elements in municipal wastewater. *Journal of Environmental Science and Health* 21:435–448.

Taylor, G.J., and A.A. Crowder. 1983. Accumulation of atmospherically deposited metals in wetland soils of Sudbury, Ontario. *Water, Air, and Soil Pollution* 19:29–42.

Tessier, A., P.G.C. Campbell, and M. Bisson. 1980. Trace metal speciation in the Yamaska and St. Francois Rivers (Quebec). *Canadian Journal of Earth Sciences* 17:90–105.

Tsai, C., and K. Chang. 1984. Intraspecific variation in copper susceptibility of the bluegill sunfish. *Archives of Environmental Contamination and Toxicology* 13:93–99.

US Minerals Yearbooks. 1930–1989. Bureau of Mines, US Department of the Interior, Washington, DC.

V.-Balogh, K., D.S. Fernandez, and J. Salanki. 1988. Heavy metal concentrations of *Lymnaea stagnalis* L. in the environs of Lake Balaton (Hungary). *Water Research* 22:1205–1210.

Vas, P., and J.D.M. Gordon. 1988. Trace metal concentrations in the scyliorhinid shark *Galeus melastomus* from the Rockall Trough. *Marine Pollution Bulletin* 19:396–397.

Wang, W. 1986. Toxicity tests of aquatic pollutants by using common duckweed. *Environmental Pollution* 11:1–14.

Watling, H.R., and R.J. Watling. 1983. Sandy beach molluscs as possible bioindicators of metal pollution. 1. Field survey. *Bulletin of Environmental Contamination and Toxicology* 31:331–338.

Whitehead, N.E., L. Huynh-Ngoc, and S.R. Aston. 1988. Trace metals in two north Mediterranean rivers. *Water, Air, and Soil Pollution* 42:7–18.

Wolmarans, C.T., W.J. van Aardt, and J. Coetzee. 1986. Histopathological effects of copper on selected epithelial tissues of snails. *Bulletin of Environmental Contamination and Toxicology* 36:906–911.

Wong, M.H. 1986. Reclamation of wastes contaminated by copper, lead, and zinc. *Environmental Management* 10:707–713.

Wright, D.A., and C.D. Zamuda. 1987. Copper accumulation by two bivalve molluscs: salinity effect is independent of cupric ion activity. *Marine Environmental Research* 23:1–14.

Young, T.C., J.V. DePinto, and T.W. Kipp. 1987. Adsorption and desorption of Zn, Cu, and Cr by sediments from the Raisin River (Michigan). *Journal of Great Lakes Research* 13:353–336.

Zingde, M.D., M.A. Rokade, and A.V. Mandalia. 1988. Heavy metals in Mindhola River estuary, India. *Marine Pollution Bulletin* 19:538–540.

12
Cyanides

Cyanides are a group of inorganic and organic compounds that contain the cyano (CN) as an integral part of their structure. They exist in numerous forms, many with remarkably different environmental fate and toxicity. Some cyanides are highly reactive and exist only rarely in nature whereas others are moderately persistent and form inorganic complexes with metals.

Production, Sources, and Residues

Production

Although cyanides find commercial application, they are also produced as by-products of other industrial processes, such as fertilizer production and gold mining. Consequently, it is difficult to determine both worldwide and national production rates. In addition, cyanides are generated in many different forms, many of which are not monitored prior to release to the environment, and are also produced by natural degradation processes in the aquatic environment. The following data, given only as an indication of the large amounts of cyanide used in industrial processes, are not indicative of total production: (1) importation of cyanide compounds into Canada 7,500 metric tons/yr (Canadian Water Quality Guidelines, 1987); (2) cyanide production in the United States 317,000 metric tons/yr (US Environmental Protection Agency, 1989).

Cyanides are used in the production of acrylonitrile, adiponitrile, and methylmethacrylate. They also find application in rodenticides, silver and metal polishes, fumigation solutions, and photographic products. Sodium cyanide is used in the extraction of gold and silver from low-grade ores, in steel production, and in electroplating.

Sources

Sodium cyanide is used at a concentration of approximately 250 mg/L for gold and silver leaching (US Environmental Protection Agency, 1988). Since pH is maintained at 9–11 during the process, free cyanide dominates the leaching solution. Although the waste from most mines is treated, relatively high cyanide residues are still detected in waste streams. For example, in a study of six Canadian gold mines, residues ranged from 0.3 to 26.5 mg/L (reviewed in Canadian Water Quality Guidelines, 1987).

Total cyanide concentrations ranging from 30 to 60 mg/L have been found in steel mill effluents, from 2 to 4 mg/L in oil refinery effluents, and at an average of 3 mg/L in electroplating wastes (Canadian Water Quality Guidelines, 1987). Cyanides can be readily removed from these wastes using a number of processes such as air stripping, oxidation with ozone, and photozone destruction (Hazardous Waste Consultant, 1987; Pearson and Bowers, 1988).

Residues

Because cyanides are highly susceptible to environmental degradation, residues are often extremely low in surface waters. In one study of Canadian rivers, total cyanide was generally <0.01 mg/L and was frequently not detected (Table 12.1). Much higher residues, >100 mg/L, can be found downstream of certain waste producers, such as fertilizer plants, particularly during periods of low flow. Several cyanides also accumulate in bottom sediments, but residues are generally low, <1 mg/kg, even near polluting sources.

Table 12.1. Concentration (mg/L) of total cyanide in Canadian surface waters.

Region	Number of samples	Concentration range
Pacific	147	<0.0005–0.097
Western	33	<0.002
Central	134	<0.002–0.370

Source: Naquadat (1985).

Chemistry

Cyanides occur in water as hydrocyanic acid (HCN), the cyanide ion (CN^-), simple cyanides, and metallocyanide complexes, and in combination with various organic compounds referred to as nitriles. The term "free cyanide" refers to the combination of HCN and CN^-. The HCN component dominates when pH is <8 and temperature is <25°C (Figure 12.1). Simple cyanides such as sodium cyanide and potassium cyanide readily dissociate and hydrolyze to form CN^- and HCN.

Iron, cobalt, and gold form strong complexes with cyanide to form $Fe(CN)_6^{4-}$, $Co(CN)_6^{4-}$, and $Au(CN)_2^-$, respectively (US Environmental Protection Agency, 1988). Moderately strong complexes include those of copper, nickel, mercury, and silver—$Cu(CN)_2^-$, $Ni(CN)_3^{2-}$, $Hg(CN)_4^{2-}$, and $Ag(CN)_2^{2-}$, respectively, whereas zinc and cadmium form relatively weak complexes of $Zn(CN)_4^{2-}$ and $Cd(CN)_3^-$. As these complexes dissociate, they may form other complex ions and free cyanide.

Photolysis is an important fate process for several compounds, including ferrocyanide and ferricyanide, but not for hydrogen cyanide. This latter compound is highly volatile (vapor pressure 360 torr), and is rapidly removed from surface waters by that process. The majority of metallic complexes, while susceptible to photolysis, are more resistant to volatilization.

Many cyanides are presumed to be rapidly sorbed by sediments, clays, and biological substrates. In a study of gold and silver mines in the United States, free cyanide was detected in tailings at concentrations of 8–200 mg/kg in samples taken 3 months after the use of cyanide had terminated

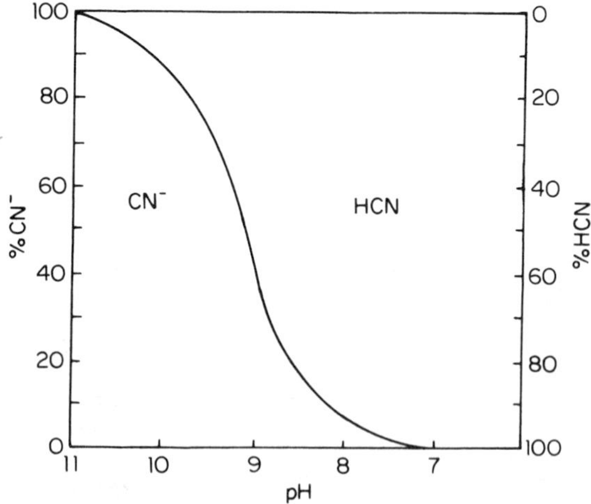

Figure 12.1. Effect of pH on dissociation of hydrogen cyanide.

at the mines (US Environmental Protection Agency, 1988). It is similarly assumed that metallic cyanide complexes also concentrate in bottom sediments.

Hydrogen cyanide, metallic complexes, and nitriles are subject to significant biological transformation during anaerobic and aerobic sewage treatment. The processes are slower in surface waters, but are mediated by bacteria and rate-limited by nutrient levels and dissolved gases. The typical metabolic pathways for degradation of cyanide compounds include (1) production of thiocyanate, (2) reaction with hydroxocobalamin to form cyanocobalamin, (3) oxidation to carbon dioxide and the formate ion, and (4) combination with amino acids.

Hardy and Knight (1967) showed that, under reducing conditions, the bacteria *Clostridium pasteurnianum* and *Azotobacter vinelandii* transformed HCN to CH_4 and NH_3. Similarly, DiGeronimo and Antoine (1976) reported that the bacterium *Nocardia rhodochrous* could metabolize succinonitrile, butyronitrile, hydroacrylonitrile, acetonitrile, and propionitrile. This process occurred through a two-step, enzymatically mediated hydrolysis with an amide as the intermediate product and ammonia and carboxylic acid as the end products. Other bacteria presumably produce other intermediate and final products.

It should be emphasized that the literature on the environmental fate and speciation of cyanides is not adequate to effectively manage the multiple use of surface waters. At present, most monitoring programs determine only free cyanide or total cyanide, and this obviously does not address the fate of the different compounds.

Bioaccumulation

Many cyanide compounds, including HCN, are rapidly metabolized in biological systems, thereby limiting the potential for bioaccumulation of these compounds. Pennington (1982), for example, did not detect cyanide in four species of fish from a Mississippi lake. Although other studies (Holden and Marsden, 1964; US Environmental Protection Agency, 1985) have similarly shown little or no uptake of HCN in several fish species, metallic cyanide complexes are bioaccumulated under some conditions. Broderius (1973), for example, detected silver cyanide at concentrations of up to 168 mg/kg wet weight in tissues of bluegill *Lepomis macrochirus*, whereas copper cyanide was found at a maximum residue of 304 mg/kg wet weight. The greatest residues were found in the liver, presumably reflecting the presence of metal-binding proteins (metallothionein).

There are a number of records of cyanide in aquatic plants. Apparently, many plants synthesize and accumulate cyanogenic glycosides. When the plant is prepared for analysis, hydrolytic enzymes are released, degrading the cyanogenic glycosides to cyanide.

Toxic Effects to Aquatic Organisms

Cyanide toxicity in animals is essentially an inhibition of oxygen metabolism, rendering tissues incapable of exchanging oxygen. The cytochrome oxidase system, which functions in electron transfer from reduced metabolites to molecular oxygen, is generally inhibited. Cyanide specifically combines with Fe^{3+} in the ferric iron–porphyrin molecule, responsible for the catalytic action of cytochrome oxidase. Other enzymes containing a metal porphyrin molecule are also strongly inhibited by cyanide.

The toxicity of cyanides to marine and freshwater organisms is due mainly to the presence of HCN and, to a lesser degree, CN^-. These two compounds form the free cyanide complex. The HCN component is generally derived from the dissociation, photodecomposition, and hydrolysis of simple cyanides. Most metallic cyanide complexes are not acutely toxic to aquatic species.

Invertebrates

The acute toxicity of free cyanide (expressed as CN) to invertebrates is highly variable, depending to a large degree on species and the method of exposure. LC_{50}s range from 0.08 to 2.49 mg/L in freshwater species, and 0.004 to >10.0 mg/L in saltwater species (reviewed by US Environmental Protection Agency, 1985). Similarly, Kononen (1988), working with the extremely toxic cyanogen chloride, reported that the 48-h LC_{50} for *Daphnia magna* neonates was 0.029 mg/L, compared to the 24-h LC_{50} of 0.040 mg/L. The chronic toxicity of free cyanide has not been intensively investigated.

Fish

The acute toxicity of free cyanide (expressed as CN) to salmonids is relatively constant, with LC_{50}s ranging from 0.03 to 0.16 mg/L depending on species and test conditions (US Environmental Protection Agency, 1985). Fathead minnow *Pimephales promelas* is more tolerant (LC_{50} 0.08–0.35 mg/L), as are bluegill *Lepomis macrochirus* and yellow perch *Perca flavescens* (Solbe et al., 1985; US Environmental Protection Agency, 1985). The LC_{50} for the sheepshead minnow *Cyprinodon variegatus*, a marine species, is 0.30 mg/L, whereas the corresponding concentrations for winter flounder *Pseudopleuronectes americanus* and Atlantic silverside *Menidia menidia* are 0.37 and 0.06 mg/L, respectively (Suter and Rosen, 1988; US Environmental Protection Agency, 1985). Chronic effects of cyanide intoxication in fish have been adequately described and include (1) reduction in serum calcium levels but no effect upon serum phosphoprotein phosphorus levels in rainbow trout *Oncorhynchus mykiss* exposed to 0.01 mg HCN/L (Da Costa and Ruby, 1984); (2) reduction in sperm viability in rainbow trout exposed to potassium cyanide at 0.001

mg CN/L (Billard and Roubaud, 1985); (3) disruption of nucleic acid and protein synthesis in fathead minnow *Pimephales promelas* exposed to 0.05 mg HCN/L (Barron and Adelman, 1984); (4) reduced uptake of vitellogenin at the ovarian level during late vitellogenesis in Atlantic salmon *Salmo salar* (Ruby et al., 1987); (5) reduced electric organ discharge in the electric fish *Gnathonemus petersi* exposed to 0.1 mg CN^-/L (Geller, 1984); and (6) rapid mobilization of liver glycogen in rainbow trout exposed to 0.03 mg HCN/L (Raymond et al., 1986.)

The toxicity (both acute and chronic) of cyanides is not substantially influenced by alkalinity, hardness, or pH below about 8.3. Low dissolved oxygen levels generally have a synergistic effect on toxicity, as does a number of heavy metals including copper, lead, and zinc. High temperatures also increase toxicity, presumably owing to the increase in respiration and concomitant uptake of cyanide. Although diet appears to have little effect on the sensitivity to fish, lack of exercise greatly increases susceptibility to cyanide (Marking et al., 1984; McGeachy and Leduc, 1988).

Several nations have formulated guidelines for the protection of fisheries resources, as shown in the following examples:

Canada
- 0.005 mg CN/L of free cyanide

USA
- 0.005 mg CN/L of free cyanide (4-day average once every 3 years)
- 0.022 mg CN/L of free cyanide (1-h average once every 3 years)

European Community
- 0.005 mg CN/L.

Health Effects

Intake

Most food contains traces of cyanides, particularly foods of plant origin. Because cyanides generally decompose on heating, cooked food contains lower residues than uncooked food. The acceptable daily intake of cyanide via ingestion of fumigated foods is 0.5 mg/kg body weight. Exposure by the inhalation and dermal routes is not usually significant, but, in cyanide-using jobs such as HCN fumigation, metal polishing, and electroplating, both routes may lead to significant exposure. The inhalation of HCN from burning polyacrylonitrile is probably the most important fire narcotic other than carbon monoxide (Purser et al., 1984).

Acute Toxicity

Fatal oral doses of cyanide compounds range from 50 to 200 mg, equal to 0.7 to 2.9 mg CN^-/kg. Symptoms of acute poisoning include hyperventilation, gasping, vomiting, convulsions, irregular and rapid heart rate, and vascular collapse.

Chronic Toxicity

An oral dose of 2.9 to 4.7 mg/day induces little or no effect in most people. Symptoms of chronic poisoning from occupational settings include headache, dizziness, and thyroid enlargement. This last disorder has also been observed in certain peoples living in tropical Africa, where the staple diet includes cassava. Cassava contains cyanogenic glycoside and linamarin, which release cyanide on metabolism or acid hydrolysis (Osuntokun et al., 1969).

Cyanides are metabolized by rhodanese (an intramitochondrial enzyme), which catalyzes the transfer of sulfur to cyanide to form thiocyanate, which is less toxic than cyanide (Marrs and Bright, 1988; US Environmental Protection Agency, 1989). Rhodanese is found primarily in the liver, but lesser amounts are distributed throughout the body. Other detoxification pathways are (1) reaction with cystine to form 2-imino–4-thiozolidene carboxylic acid, and (2) reaction with hydroxycobalamine to form cyanocobalamine (vitamin B_{12}). The rate of detoxification depends on the plasma cyanide concentration (Bright and Marrs, 1988). Excretion of all of the preceding metabolites occurs principally through the urine, with much smaller amounts leaving via expired air.

Carcinogenicity

The International Agency for Research on Cancer has not evaluated the carcinogenic potential of cyanides for any route of exposure. The information available to date indicates that there is inadequate or no evidence for carcinogenicity in animals or humans.

Drinking Water

Residues

Cyanides are rarely detected in finished drinking water. In a study of drinking waters in the United States, residues averaged <0.0001 mg CN/L, and the highest level was only 0.008 mg CN/L (US Environmental Protection Agency, 1989). A survey of Canadian drinking waters yielded residues below 0.10 mg/L (Canadian Water Quality Guidelines, 1987), and a report by the World Health Organization (1984) indicated that concentrations are generally well below that maximum acceptable guidelines (0.1–0.2 mg/L).

Consumption Guidelines

The World Health Organization (1984) recommends a drinking water guideline of 0.1 mg CN/L. In Canada the guideline is 0.2 mg/L, and in the USA, it is 0.15 mg/L. The basis of the derivation of these values uses the following data:

1. The No-Observed-Adverse-Effect Level (NOAEL) for cyanide in experimental animals is 10.8 mg/kg/day (Howard and Hanzal, 1955).
2. Determination of the Reference Dose (RD)

   $$\text{RD} = (10.8 \text{ mg/kg/day})/(100)(5) = 0.022 \text{ mg/kg/day}$$

   where 100 is the uncertainty factor based on an animal study, and 5 is an additional uncertainty factor to allow for potentially greater absorption of cyanide from water than from diet.
3. Determination of the Drinking Water Equivalent Level (DWEL)

   $$\text{DWEL} = (0.022 \text{ mg/kg/day})(70 \text{ kg})/(2 \text{ L/day}) = 0.77 \text{ mg/L}$$

4. Determination of Guideline

   $$\text{Guideline} = (0.77 \text{ mg/L})(20\%) = 0.154 \text{ mg/L}$$

   where 20% is the estimated relative source contribution from water.

The difference in guidelines among nations and the World Health Organization is simply due to the use of different safety factors.

Treatment

Cyanides can be effectively removed from drinking waters by oxidation with chlorine or ozone, ion exchange, or reverse osmosis. During the chlorination process, chlorine gas or the hypochlorite ion reacts with free cyanide to form cyanate or carbon dioxide. Alkaline conditions enhance the rate of reaction and prevent the formation of cyanogen chloride. Ozonation, also highly effective in cyanide removal, does not result in the production of cyanogen chloride, and minimizes the production of haloforms. Although reverse osmosis and ion exchange are effective treatment methods, they are only employed when other contaminants also need to be removed from water.

Recommendations

The management of cyanides in surface waters has always been fraught with contrasts: (1) some cyanides are highly toxic whereas others are relatively innocuous; (2) some cyanides are moderately persistent and may bioaccumulate, whereas others rarely persist in the environment or in tissues; and (3) relatively high levels of cyanide may be discharged to surface waters even though control technology is available to greatly reduce such discharges. Many administrators and bureaucrats see no pressing need to tightly control discharges and permit excursions from license limits in the belief that such agents have no lasting environmental impact. Although this assumption may accurately reflect the current state of knowledge on cyanides, the discharge of relatively high cyanide levels continues to unnecessarily detract from present-day environmental quality.

The following recommendations, along with the more scientific deficiencies, reflect these management considerations:

1. Implementation of best available technology in the control of wastewater discharges.
2. Environmental fate and speciation of cyanides in surface waters.
3. Monitoring programs that consider metallic cyanide complexes in surface waters receiving such discharges.
4. Routine analysis of fish tissues and other aquatic food products from surface waters receiving metallic cyanide complexes.
5. Chronic toxicity of cyanides to a range of invertebrate species.

References

Barron, M.G., and I.R. Adelman. 1984. Nucleic acid, protein content, and growth of larval fish sublethally exposed to various toxicants. *Canadian Journal of Fisheries and Aquatic Sciences* 41:141–150.

Billard, R., and P. Roubaud. 1985. The effect of metals and cyanide on fertilization in rainbow trout *(Salmo gairdneri). Water Research* 19:209–214.

Bright, J.E., and T.C. Marrs. 1988. Pharmacokinetics of intravenous potassium cyanide. *Human Toxicology* 7:183–186.

Broderius, S.J. 1973. Determination of molecular hydrocyanic acid in water, and studies of the chemistry and toxicity to fish of metal-cyanide complexes. Ph.D. thesis, Oregon State University, Corvallis, 287 pp.

Canadian Water Quality Guidelines. 1987. Canadian Council for Resource and Environment Ministers. Environment Canada, Ottawa.

Da Costa, H., and S.M. Ruby. 1984. The effect of sublethal cyanide on vitellogenic parameters in rainbow trout *Salmo gairdneri. Archives of Environmental Contamination and Toxicology* 13:101–104.

DiGeronimo, M.J., and A.D. Antoine. 1976. Metabolism of acetonitrile and propionitrile by *Nocardia rhodochrous. Applied Environmental Microbiology* 31:900–906.

Geller, W. 1984. A toxicity warning monitor using the weakly electric fish, *Gnathonemus petersi. Water Research* 18:1285–1290.

Hardy, R.W.F., and E. Knight. 1967. ATP-dependent reduction of azide and HCN by N_2-fixing enzymes *Azotobacter vinelandii* and *Clostridium pasteurianum. Biochimica et Biophysica Acta* 139:69–90.

Hazardous Waste Consultant. 1987. Using photozone to destroy cyanide wastes from electroplating processes. *Hazardous Waste Consultant* 5:8–10.

Holden, A.V., and K. Marsden. 1964. *Cyanide in salmon and brown trout*. Freshwater and Salmon Fisheries Research, Department of Agriculture and Fisheries for Scotland, Report Number 33, Edinburgh, Scotland.

Howard, J.W., and R.F. Hanzal. 1955. Chronic toxicity for rats of food treated with hydrogen cyanide. *Journal of Agriculture and Food Chemistry* 3:325–329.

Kononen, D.W. 1988. Acute toxicity of cyanogen chloride to *Daphnia magna*. *Bulletin of Environmental Contamination and Toxicology* 41:371–377.

Marking, L.L., T.D. Bills, and J.R. Crowther. 1984. Effects of five diets on sensitivity of rainbow trout to eleven chemicals. *Progressive Fish-Culturist* 46:1–5.

Marrs, T.C., and J.E. Bright. 1987. Effect on blood and plasma cyanide levels and on methaemoglobin levels of cyanide administered with and without previous protection using PAPP. *Human Toxicology* 6:139–145.

McGeachy, S.M., and G. Leduc. 1988. The influence of season and exercise on the lethal toxicity of cyanide to rainbow trout *(Salmo gairdneri)*. *Archives of Environmental Contamination and Toxicology* 17:313–318.

Naquadat. 1985. *National water quality data bank*. Environment Canada, Ottawa.

Osuntokun, B.O., G.L. Monekosso, and J. Wilson. 1969. Relationship of a degenerative tropical neuropathy to diet, report of a field study. *British Medical Journal* 1:547–550.

Pearson, D.E., and A.R. Bowers. 1988. An air stripping method for treatment of electroplating solutions. *Hazardous Waste and Hazardous Materials* 5:85–91.

Pennington, C.H. 1982. Contaminant levels in fishes from Brown's Lake, Mississippi. *Journal of the Mississippi Academy of Science* 27:139–147.

Purser, D.A., P. Grimshaw, and K.R. Berrill. 1984. Intoxication by cyanide in fires: a study in monkeys using polyacrylonitrile. *Archives of Environmental Health* 39:394–400.

Raymond, P., G. Leduc, and J.A. Kornblatt. 1986. Investigation sur la toxicodynamique du cyanure et sur sa biotransformation chez la truite arc-en-ciel *(Salmo gairdneri)*. *Canadian Journal of Fisheries and Aquatic Sciences* 43:2017–2024.

Ruby, S.M., D.R. Idler, and Y.P. So. 1987. Changes in plasma, liver, and ovary vitellogenin in landlocked Atlantic salmon following exposure to sublethal cyanide. *Archives of Environmental Contamination and Toxicology* 16:507–510.

Solbe, J.F. de L.G., V.A. Cooper, C.A. Willis, and M.J. Mallett. 1985. Effects of pollutants in fresh waters on European non-salmonid fish. I. Non-metals. *Journal of Fish Biology 27*(Suppl. A):197–207.

Suter, G.W., and A.E. Rosen. 1988. Comparative toxicology and risk assessment of marine fishes and crustaceans. *Environmental Science and Technology* 22:548–556.

US Environmental Protection Agency. 1985. *Ambient water quality criteria for cyanide—1984*. US Environmental Protection Agency, EPA 440/5–84–028, Washington, DC. 59 pp.

US Environmental Protection Agency. 1988. Gold/silver heap leaching and management practices that minimize the potential for cyanide releases. US Environmental Protection Agency, EPA/600/2–88/002, Cincinnati, OH. 103 pp.

US Environmental Protection Agency. 1989. Drinking water health advisories. *Reviews of Environmental Contamination and Toxicology* 107:1–184.

World Health Organization. 1984. *Guidelines for drinking-water quality*. World Health Organization, Geneva.

13
Iron

With an average abundance of 5% by weight, iron is the fourth most abundant element in the earth's crust. The principal ores include magnetite (Fe_3O_4), siderite ($FeCO_3$), limonite [FeO(OH)], and hematite (Fe_2O_3). Iron is an essential trace element, required by both animals and plants. In some waters it may limit, either directly or indirectly, the growth of algae; it is also essential to oxygen transport in the blood of all vertebrates and some invertebrates. Although iron is of little direct toxicologic significance, it often controls the concentration of other elements, including toxic heavy metals, in surface waters.

Production, Sources, and Residues

Production

World production of iron (used for the production of steel) was 80.2×10^6 metric tons in 1930, increasing to 346×10^6 metric tons in 1960 and 714×10^6 metric tons in 1980 (US Minerals Yearbooks, 1930–1989). Production in recent years has continued to increase, and is currently above 715×10^6 metric tons per year. The world's leading producers of iron ore are the USSR, Brazil, Australia, China and India; the major consumers are the USSR, Japan, China, the USA and the FRG (Table 13.1). The major producers and consumers of crude steel (a significant source of anthropogenically derived iron) are the USSR, the USA and Japan (Table 13.2).

Table 13.1. World's major producers and consumers of iron.

Producing nation	Quantity (10^6 metric tons/yr)	Producing nation	Quantity (10^6 metric tons/yr)
USSR	249.0	USSR	201.1
Brazil	132.0	Japan	130.5
Australia	90.0	China	94.0
China	90.0	USA	73.7
India	47.8	FRG	50.3

Sources: US Minerals Yearbook (1988), Cordero (1988).

Sources

Total environmental flux of iron is enormous. Approximately 9.9×10^8 metric tons are transported annually by rivers, and 0.5×10^8 by atmospheric rainout (Westall and Stumm, 1980). The river transport value is second only to aluminum, whereas the rainout value is greater than any other metal. Because iron is so common in the earth's crust, natural erosion accounts for the majority of these transport figures. Westall and Stumm (1980) estimated that 28% of the iron in the atmosphere originated from anthropogenic emissions, the remainder coming from continental dust flux (49%), volcanic dust flux (22%), and other minor sources.

Iron is routinely detected in municipal effluents, particularly in cities where iron and steel are manufactured. Sung et al. (1986), working on waste water from 25 plants in the state of Washington, found iron residues of over 900 mg/L in primary sludge and 4 mg/L in primary effluent (Table 13.3), whereas dried sludge from Rome (Italy) contained iron at 15,000 mg/kg (Campanella et al., 1987). Abuzkhar et al. (1987), studying the suitability of sewage sludge as a soil amendment, found that iron averaged 8100 mg/kg dry weight, with a maximum of 16,175 mg/kg. Similarly, residues in slag, electrostatic precipitators, and flue gas from municipal incinerators in Switzerland averaged 230,000, 30,000, and 1,800 mg/kg dry weight, respectively (Brunner and Monch, 1986).

Table 13.2. World's major producers and consumers of crude steel.

Producing nation	Quantity (10^6 metric tons/yr)	Producing nation	Quantity (10^6 metric tons/yr)
USSR	161.0	USSR	159.9
Japan	98.3	USA	103.8
USA	73.0	Japan	76.6
China	52.1	China	54.4
FRG	37.1	FRG	31.9

Sources: US Minerals Yearbook (1988), Cordero (1988).

Table 13.3. Concentration (mg/L wet weight) of iron in municipal waste from treatment plants in the state of Washington.

Sample	Concentration	
	Average	Range
Raw sewage	4.4	1.6–16.9
Primary effluent	4.4	0.8–17.7
Secondary effluent	1.1	0.1–4.3
Discharge	1.1	0.2–5.4
Primary sludge	909	61–2,870
Secondary sludge	310	72–293

Source: Sung et al. (1986).

Iron is often extremely high near base and precious metal mines. For example, a lead–zinc mine in Wales produced residues of up to 5.6 mg/L in the waste stream (Jones, 1986), whereas water from diamond drill holes from a lead–zinc mine in Idaho carried iron at 6,380 mg/L (Wai et al., 1980). Similarly, Harland and Brown (1989) found residues of up to 29.0 mg/L in the effluent of an ore dressing plant in Thailand. The concentration of iron in wetland soils near the nickel mines of Sudbury (Canada) increased significantly near the emission sources, reaching 50,000 mg/kg (Taylor and Crowder, 1983).

Residues

Iron residues in surface waters are extremely variable, reflecting differences in underlying bedrock, erosion, and, to a lesser degree, industrial and municipal discharges (Table 13.4). Concentrations of over 50 mg/L, and as low as 0.004 mg/L, have been reported for some rivers. The corres-

Table 13.4. Concentration (mg/L) of total iron in surface waters.

Location	Average (range)	Number of samples
Pacific coast rivers, Canada[1]	NR[a] (<0.03–54.0)	641
Prairie rivers, Canada[1]	NR (<0.02–14.0)	2,653
Atlantic coast rivers, Canada[1]	NR (0.004–90.0)	7,167
Awba Stream, Nigeria[2]	NR (0.90–3.00)	NR
Kandy Canal, Sri Lanka[3]	2.8 (0.2–8.5)	51
River Hindon, India[4]	0.3 (0.05–0.50)	9
Four rivers, Thailand[5]	NR (0.01–0.64)	NR

[a]Not reported.

Sources: [1]Naquadat (1985), [2]Sridhar (1986), [3]Dissanayake et al. (1987), [4]Ajmal et al. (1987), [5]Hungspreugs *et al.* (1989).

ponding variation in sediment residues is smaller in many aquatic systems, particularly in the deep depositional basins of lakes. For example, Mudroch et al. (1988) reported the following iron concentrations (mg/kg dry weight) for the Great Lakes: Ontario, 1.9–9.6; Erie, 0.9–7.8; Huron, 0.5–5.1; Superior, 3.2–5.9; Michigan (no data). Similarly, the discharge of municipal effluents into the ocean off Los Angeles resulted in surficial sediment residues of about 3.5% dry weight (Stull and Baird, 1985), while concentrations in North Sea sediments off the Belgium coast ranged from 0.9–1.5% (Araujo et al., 1988). Similarly high concentrations (0.7–1.8%) were reported for offshore areas of the Arabian Gulf (Samhan et al., 1987), the Ganges Estuary (1.2–4.6%) (Subramanian, 1988), and Bombay Harbor (6.2–7.6%) (Mohapatra, 1988).

Chemistry

The primary oxidation states of iron in water are Fe^{2+} (ferrous) and Fe^{3+} (ferric). In most surface water, Fe^{3+} predominates and, when combined with its salts, is practically insoluble, at least in aerobic waters. Fe^{2+}, on the other hand, is soluble and dominates under anaerobic conditions.

The oxidation–reduction cycle is important in controlling the fate of iron in most surface waters. The cycle varies seasonally, particularly in lakes that develop an anoxic hypolimnion during the summer. Oxygen concentrations at the water–sediment interface often approach zero. This causes the reduction of Fe^{3+} to soluble Fe^{2+}, which is then transported upward in the water column. The oxygenated water results in reoxidation to the insoluble Fe^{2+}, which settles to the bottom to repeat the cycle. In seawater, reaction with hydrogen peroxide is the dominant oxidation pathway for Fe^{2+} (Moffett and Zika, 1987).

The oxidation–reduction cycle also controls the fate of manganese through the same processes (more or less) as those of iron. However, Fe^{2+} is oxidized to particulate Fe^{3+} much more rapidly than the corresponding species of manganese. In addition, Fe^{3+} can be reduced at a lower redox potential than Mn^{4+}. As a result, soluble Fe is supplied by reduction in the sediment whereas soluble Mn may be supplied from the anoxic water column. The seasonal concentration of iron and manganese in surface waters is often different, a result of these two factors.

Many elements, plus phosphates, are scavenged by iron through adsorption onto particles. When Fe^{3+} is formed as part of the reduction-oxidation cycle in the presence of phosphate, a basic iron phosphate—$Fe_2(OH)_3PO_4$—is formed with a Fe:P stoichiometry of 2:1 (Armstrong et al., 1987). The reaction is represented as a Fe^{2+} oxidation-basic iron phosphate precipitation:

$$2Fe^{2+} + \tfrac{1}{2}O_2 + H_2O_4^- \longrightarrow Fe_2(OH)_3PO_4(s) + 3H^+$$

Another possible reaction occurs when Fe^{3+} is formed and hydrolyzed prior to interaction with phosphate; in this scenario, adsorption of phosphate by $Fe(OH)_3(s)$ results in an Fe:P ratio >5:1.

$Fe_3{}^+$ reduction in the surficial sediments solubilizes iron-bound phosphate as follows:

$$Fe_2(OH)_3PO_4(s) + \tfrac{1}{2}CH_2O + 3H^+ \longrightarrow 2Fe^{2+} + H_2PO_4{}^- + \tfrac{1}{2}CO_2 + \tfrac{5}{2}H_2O$$

Similarly, solubilization of adsorbed phosphate occurs with the reduction of $Fe(OH)_3(s)$.

Iron readily complexes with sulfates in the sediments of many surface waters. This is of considerable toxicologic significance because other elements are likely to be adsorbed by the resulting complex. The resulting stability constants are relatively high (Table 13.5). In a study of the Axios River (Greece), Samanidou and Fytianos (1987) showed that iron was predominately associated with organic sulfides in both estuarine and freshwater reaches (Table 13.6). Another investigation (Moore et al., 1988) showed that diagenic sulfides were an important sink for trace metals in reduced, sulfidic sediments, and that during reduction, oxyhydroxides of iron (and manganese) dissolved, permitting arsenic, copper and zinc sulfides to precipitate.

The proportion of heavy metals bound by Fe–Mn hydrous oxides is highly variable and depends on water depth and redox reactions with the sediments. For example, Samanidou and Fytianos (1987) showed that 23.8% of lead in the estuary of the Axios River (Greece) was partitioned in the Fe–Mn hydrous oxides, compared to 27.1%, 6.0%, 9.7%, and 22.2% for zinc, copper, chromium, and cadmium, respectively. The following percentages were noted for the Aliakmon River on the Greek/Yugoslavian border (Samanidou and Fytianos, 1987): lead (16.1), zinc (13.4), copper (5.0), chromium (3.5), and cadmium (42.8). Although other studies (reviewed by Cutter, 1989) showed that <10% of sediment-bound selenium was usually associated with Fe–Mn oxides, Dissanayake et al. (1987) noted that there was a strong positive correlation between vanadium and Fe^{2+} residues in a canal in Sri Lanka.

Table 13.5. Stability constants for Fe^{3+} complexes.

Ligand	Log K	Ligand	Log K
$HPO_4{}^{2-}$	4.92	$SO_3{}^{2-}$	6.6
$N_3{}^-$	4.85	$SO_4{}^{2-}$	4.04
NCS^-	3.02	F^-	6.0
$NO_3{}^-$	1.0	Cl^-	1.48
Br^-	0.06		

Source: Conklin and Hoffmann (1988).

Table 13.6. Partitioning (%) of iron in the estuary and freshwater parts of the Axios River (Greece).

Fraction	Estuary	Freshwater
Cation-exchangeable	0.19	0.14
Carbonates	0.96	1.60
Fe–Mn hydrous oxides	18.31	18.81
Organic sulfides	37.28	37.06
Residual	43.26	42.39

Source: Samanidou and Fytianos (1987).

Because iron plays such an important role in the fate of trace metals and nutrients, a breach in the iron redox cycle may ultimately lead to the mobilization of toxic agents in the environment. One commonly encountered scenario is the redistribution of anoxic, sulfidic sediments during dredging; another is the artificial destratification of lakes to enhance oxygenation of the water column in support of fisheries. The potential remoblization of these toxic agents will often follow a site-specific response. This means that corresponding site-specific studies will be needed to determine the exact nature of the iron redox cycle.

Bioaccumulation

Plants and Invertebrates

Because iron is found in such high concentrations in the environment, it is also plentiful in freshwater and marine plants (Table 13.7). The rate of uptake is rapid in most species, often reaching a plateau within 2 h in laboratory experiments (Vymazal, 1984). Residues in invertebrates species are also high at many locations, but are of little or no toxicologic significance (Table 13.8). In molluscs, maximum concentrations are generally found in the gills, mantle, and visceral mass (Tessier et al., 1984). Residues vary seasonally in response to the mobilization of iron in the

Table 13.7. Concentration (mg/kg dry weight) of total Fe in aquatic plants.

Species	Average (range)	Location
Eicchornia crassipes[1]	3,012 (1,762–8,437)	Hindon River, India
Pistia stratiotes[2]	33,000 (19,400–51,200)[a]	Awba Lake, Nigeria
Fucus vesiculosus[3]	148 (48–522)	Baltic Sea, Sweden
Fontinalis squamosa[4]	32,044 (17,000–58,200)	River Mawddach, Wales

[a]Ash dry weight.
Sources: [1]Ajmal et al. (1987), [2]Sridhar (1986), [3]Soderlund et al. (1988), [4]Mason and Macdonald (1988).

Table 13.8. Concentration (mg/kg wet weight) of total Fe in the soft tissues of marine and freshwater invertebrates.

Species	Average (range)	Location
Oyster, *Crassostrea brasiliana*[1]	206 (150–314)[a]	Sepetiba Bay, Brazil
Bivalve molluscs, 5 species[2]	390 (250–700)	Pacific Ocean, Fiji
Pearl oyster, *Pinctada radiata*[3]	30 (14–86)	Arabian Gulf
Bivalve mollusc, *Donax serra*[4]	153 (21–453)	South African coast
Gastropod, *Lymnaea stagnalis*[5]	900 (520–>1,500)[a]	Lake Balaton, Hungary
Coral, *Pocillopora damicornis*[6]	25 (15–40)[a]	Thailand coast

[a]Dry weight.
Sources: [1]Lima et al. (1986), [2]Dougherty (1988), [3]Sadig and Alam (1989), [4]Watling and Watling (1983), [5]V.-Balogh et al. (1988), [6]Howard and Brown (1987).

environment and in response to anthropogenically derived inputs. For example, Howard and Brown (1987) found that the concentration of iron in the soft tissues of the coral *Pocillopora damicornis* was approximately 40 mg/kg dry weight near a tin smelter but only 17 mg/kg in control areas.

Fish

Iron is rarely a toxicologically significant contaminant of fish tissues. Residues in muscle typically range from 1 to 150 mg/kg wet weight and are therefore among the highest of any metal (Villarreal-Trevino et al., 1986; Ashraf and Jaffar, 1988; Legorburu et al., 1988). Among the different tissues, concentrations are typically greatest in the gills, liver, heart, and skin (Legorburu et al., 1988; Vas and Gordon, 1988).

Iron in plant and animal tissues does not need to be determined during the course of most monitoring programs. It is of little or no toxicological significance and is not normally implicated in imparting off-flavors to tissues. Such analysis should only be conducted near known anthropogenic sources of iron.

Toxic Effects to Aquatic Organisms

Plants

Fe^{3+} is moderately toxic to many species of aquatic plants. Wang (1986) found that the EC (Effective Concentration)$_{50}$ for duckweed *Lemna minor* was 3.7 mg/L, with a Maximum Permissible Concentration of 0.37 mg/L. This made iron more toxic than chromium, manganese, barium, and lead but less toxic than selenium, copper, nickel, and cadmium. Iron appreciably reduces the toxicity to aquatic plants of other heavy metals, including copper, lead, and zinc. In one study, Stauber and Florence (1985) exposed the marine diatom *Nitzschia closterium* to combinations of copper (0.020 mg/L) and iron for 3 days. Cell density at the end of the experiments was 9.00×10^4/mL at an iron concentration of 0.790 mg/L and only 2.75×10^4 cells/mL when iron was 0.0079 mg/L. This reduction in toxicity is likely due to adsorption of copper onto colloidal ferric hydroxide, and also possibly competition with copper for binding sites on the cell wall.

Invertebrates

Fe^{2+} and Fe^{3+} are only moderately toxic to most invertebrate species. Martin and Holdich (1986), for example, showed that the LC_{50}s of both agents in the isopod *Asellus aquaticus* and the amphipod *Crangonyx pseudogracilis* ranged from 95 to 160 mg/L (Table 13.9). Another study on *Asellus aquaticus* (Maltby et al., 1987) indicated that sensitivity to Fe^{2+} was enhanced, possibly synergistically, by low pH. Harland and Brown (1989) found that the coral *Porites lutea* lost zooxanthellae (symbiotic algae) when exposed to total Fe of 0.01 mg/L. Although this is not an acute response, the coral was obviously stressed by such concentrations. The same study showed that, under natural field conditions, coral was apparently able to develop some resistance to the toxic effects of iron. Warnick and Bell (1969) reported that the toxicity of total Fe to aquatic insects was highly variable, with LC_{50}s ranging from 0.3 to 16 mg/L.

Iron precipitates are periodically deposited on the bottom of lakes and rivers. These agents, particularly $Fe(OH)_3$ and Fe_2O_3, then form gels and flocs that can suffocate benthic organisms and any planktonic species with gills. This effect may occur near industries with poor waste treatment facilities.

Table 13.9. Acute toxicity (LC_{50}) of iron to two crustacean species.

	48-h LC_{50} (mg/L)		96-h LC_{50} (mg/L)	
Species	Average	95% Confidence limits	Average	95% Confidence limits
Asellus aquaticus				
Fe^{3+}	183	164–205	124	108–144
Crangonyx pseudogracilis				
Fe^{2+}	143	114–211	95	82–117
Fe^{3+}	160	139–184	120	102–146

Water hardness, 50 mg/L; pH, 6.75; water temperature, 13°C.
Source: Martin and Holdich (1986).

Fish

The lethal concentration (LC_{50}) of total Fe to fish generally ranges from 0.3 to >10 mg/L, depending on species and test conditions. Smothering effects of $Fe(OH)_3$ on fish gills have also been reported for a number of species. Although chronic toxicity is generally reported at concentrations >0.1 mg/L, Billard and Roubaud (1985) found that sperm viability was reduced in rainbow trout *Oncorhynchus mykiss* at concentrations as low as 0.005 mg/L. Iron often causes a reduction in the toxicity of other heavy metals, including aluminum, copper, lead, and zinc (Reader et al., 1989; Hutchinson and Sprague, 1986). This is probably the result of competition between iron and other metals for binding sites and/or partial sequestration of metals by iron colloids.

Guidelines for the protection of aquatic life range from 0.3 to 1.0 mg Fe/L in many nations.

Health Effects

Intake

Iron is found in most foods. The typical Western diet yields approximately 16 mg/day in men and 12 mg/day in women. Drinking water contributes a relatively small fraction (<1 mg/day) to this total, as discussed in the following section. Inhalation of iron is negligible (<0.05 mg/day) in the nonoccupationally exposed population. The total body burden in a 70-kg reference man is approximately 4.5 g, of which 73% is tied up in hemoglobin, 24% in storage in ferritin and hemosiderin, and 3% in myoglobin and other tissues (reviewed by Carson et al., 1987).

Iron is an essential component of several cofactors including hemoglobin and the cytochromes. The recommended daily intake is 10 mg for men and 18 mg for women. Iron also has the potential to reduce the toxic effects of other heavy metals, at least in experimental animals. Sullivan and Ruemmler (1987), for example, found that an excess of ferric iron reduced the retention of lead in rats from 53% of the gavaged dose to only 3%. Similarly, Gruden and Munic (1987) showed that iron reduced the uptake of cadmium in the rat. On the other hand, some toxic metals significantly alter the distribution of iron within the body. For example, Sugawara et al. (1988) reported that an excess of cadmium inhibited the formation of ferritin and related compounds in the rat, whereas Saxena et al. (1986) found that exposure of the rat to copper and zinc increased iron residues in the central nervous system.

Acute Toxicity

Acute exposure to iron is characterized by vomiting, gastrointestinal bleeding, pneumonitis, convulsions, coma, and jaundice. If the patient survives these effects, recovery is generally rapid, even though gastrointestinal irritation and hemorrhage may continue.

Chronic Toxicity

Chronic effects of iron exposure are seldom reported. In some cases, a generalized increase in iron content (known as hemosiderosis) occurs; in other instances, a specific deposition may lead to localized fibrosis.

Although the condition is relatively benign, some studies have indicated that it is accompanied by abnormal glucose metabolism and increased heart disease. Chronic inhalation of iron produces a benign, nonfibrotic pneumoconiosis in the lungs.

Carcinogenicity

There is little or no evidence that iron compounds are carcinogenic following exposure though the oral and inhalation routes.

Drinking Water

Residues

Iron is almost always detected in finished drinking water, simply because of its abundance in the earth's crust. In a survey of the American Water Works Association (1985) of drinking water in 39 states and three territories, there were 2,200 episodes of noncompliance with the Maximum Contaminant Level of 0.3 mg/L. For comparison, fluoride and nitrates were in noncompliance in 907 and 369 episodes, respectively. Ajmal and Uddin (1986a,b) reported that residues in hand pump water samples from the city of Aligarh (India) ranged from 0.02–0.46 mg/L compared to 0.01–0.04 mg/L for treated municipal water. Khoe and Waite (1989) found extremely low residues (0.001 mg/L) in potable water from a treatment plant in Australia, a result of advanced flocculation procedures with fulvic acid and alum, and filtration. Water treatment plants in Rio de Janeiro (Brazil) produced water with a dissolved Fe content of 0.04 mg/L and a suspended Fe content of 0.20 mg/L (Azcue et al., 1988)

Consumption Guidelines

The primary concern about iron in drinking water is its objectionable taste. The taste of iron can be readily detected at 1.8 mg/L in drinking water and at 3.5 mg/L in distilled water. High concentrations also lead to staining of laundry and plumbing, and massive growths of bacteria within water systems. The drinking water guideline/standard of 0.3 mg Fe/L (used by many nations) is based on these aesthetic considerations rather than health concerns. Even at 0.3 mg/L, the intake of iron from drinking water amounts to 0.6 mg/day, far below the intake from food.

Treatment

Iron is readily removed from water (particularly groundwater) by flocculation and filtration. The first step may be mediated by the use of one or more of the following agents or processes: alum, fulvic acid, ozone, pH adjustment, aeration, chemical oxidation. Various granular media are used for filtration. The exact process depends on the quality of the source water, existing water treatment facilities, and presence of other harmful agents.

Recommendations

The main interest in iron lies in its interaction with other agents, including toxic metals and nutrients. Iron itself is only moderately toxic (at most) to the majority of freshwater and marine species. Although it may impart an off-flavor to drinking water, iron has little or no toxicological significance in most water supplies.

The No. 1 priority for research on iron appears to be its effect on the fate of more toxic agents and nutrients. Any breach in the iron redox cycle may mobilize these agents, leading to significant environmental effects. An additional related area is the sequestration of toxic metals by iron and its related compounds. Many toxic metals are readily sorbed by hydroxides, thereby reducing availability for biological uptake.

The final area of continuing interest is the antagonistic effect iron has on the toxicity of many heavy metals. Although the effect of such interactions on acute toxicity has been adequately described in a number of species, relatively little is known about the corresponding effects on chronic toxicity.

The following recommendations reflect these considerations:

1. Environmental fate of sequestered and remobilized metals in the iron redox cycle.
2. Sorption of toxic metals by iron hydroxides/oxides, and resulting impact on biological uptake.
3. Iron/toxic metal interactions and concomitant effect on chronic toxicity to marine and freshwater species.

References

Abuzkhar, A.A., A.S. Gibali, Y.I. Elmehrik, and R. Ahmatullah. 1987. Chemical monitoring of sewage wastes for their use in crop production. *Environmental Monitoring and Assessment* 8:127–133.

Ajmal, M., R. Khan, and A.U. Khan. 1987. Heavy metals in water, sediments, fish and plants of river Hindon, U.P. *Hydrobiologia* 148:151–157.

Ajmal, M., and R. Uddin. 1986a. Studies on heavy metals in the ground waters of the City of Aligarh U.P. (India). *Environmental Monitoring and Assessment* 6:181–194.

Ajmal, M., and R. Uddin. 1986b. Quality of drinking water in the Aligarh Muslim University campus, Aligarh, U.P. (India) with respect to heavy metals. *Environmental Monitoring and Assessment* 6:195–205.

American Water Works Association. 1985. An AWWA survey of inorganic contaminants in water supplies. *Journal of the American Water Works Association* 77:67–72.

Araujo, M.F.D., P.C. Bernard, and R.E. van Grieken. 1988. Heavy metal contamination in sediments from the Belgian coast and Scheldt estuary. *Marine Pollution Bulletin* 19:269–273.

Armstrong, D.E., J.P. Hurley, D.L. Swackhamer, and M.M. Shafer. 1987. Cycles of nutrient elements, hydrophobic organic compounds, and metals in Crystal Lake. *In: Sources and fates of aquatic pollutants,* eds. R.A. Hites, and S.J. Eisenreich, 491–518. American Chemical Society, Washington, DC.

Ashraf, M., and M. Jaffar. 1988. Correlation between some selected trace metal concentrations in six species of fish from the Arabian Sea. *Bulletin of Environmental Contamination and Toxicology* 41:86–93.

Azcue, J.M.P., W.C. Pfeiffer, M. Fiszman, and O. Malm. 1988. Heavy metal removal by different water treatment plants, in Rio de Janeiro State, Brazil. *Environmental Technology Letters* 9:429–436.

Billard, R., and P. Roubaud. 1985. The effect of metals and cyanide on fertilization in rainbow trout. *Water Research* 19:209–214.

Brunner, P.H., and H. Monch. 1986. The flux of metals through municipal solid waste incinerators. *Waste Management and Research* 4:105–119.

Campanella, L., E. Cardarelli, T. Ferri, B.M. Petronio, and A. Pupella. 1987. Evaluation of heavy metals speciation in an urban sludge. I. batch method. *Science of the Total Environment* 61:217–228.

Carson, B.L., H.V. Ellis, and J.L. McCann. 1987. *Toxicology and biological monitoring of metals in humans.* Lewis Publishers, Chelsea, MI. 328 pp.

Conklin, M.H., and M.R. Hoffmann. 1988. Metal ion-sulfur (IV) chemistry. 3. Thermodynamics and kinetics of transient iron (III)-sulfur (IV) complexes. *Environmental Science and Technology* 22:899–907.

Cordero, R. 1988. *Metal Bulletin's prices and data 1988.* Metal Bulletin Books, Surrey, England. 375 pp. Cutter, G.A. 1989. Freshwater systems. In: Occurrence and distribution of selenium, ed. M. Ihnat, 243–262. CRC Press, Boca Raton, Florida.

Cutler, G.A. 1989. Freshwater systems. *In: Occurrence and distribution of selenium,* ed. M. Ihnat, 243–262. CRC Press, Boca Rotan, Florida.

Dissanayake, C.B., J.M. Niwas, and S.V.R. Weerasooriya. 1987. Heavy metal pollution of the mid-canal of Kandy: an environmental case study from Sri Lanka. *Environmental Research* 42:24–35.

Dougherty, G. 1988. Heavy metal concentrations in bivalves from Fiji's coastal waters. *Marine Pollution Bulletin* 19:81–84.

Gruden, N., and S. Munic. 1987. Effect of iron upon cadmium-manganese and cadmium-iron interaction. *Bulletin of Environmental Contamination and Toxicology* 38:969–974.

Harland, A.D., and B.E. Brown. 1989. Metal tolerance in the scleratinian coral *Porites lutea. Marine Pollution Bulletin* 20:353–357.

Howard, L.S., and B.E. Brown. 1987. Metals in *Pocillopora damicornis* exposed to tin smelter effluent. *Marine Pollution Bulletin* 18:451–454.

Hungspreugs, M., W. Utoomprurkporn, S. Dharmvanij, and P. Sompongchaiyakul. 1989. The present status of the aquatic environment of Thailand. *Marine Pollution Bulletin* 20:327–332.

Hutchinson, N.J., and J.B. Sprague. 1986. Toxicity of trace metal mixtures to American flagfish *(Jordanella floridae)* in soft, acidic water and implications for cultural acidification. *Canadian Journal of Fisheries and Aquatic Sciences* 43:647–655.

Jones, K.C. 1986. The distribution and partitioning of silver and other heavy metals in sediments associated with an acid mine drainage. *Environmental Pollution* 12:249–263.

Khoe, G.H., and T.D. Waite. 1989. Manganese and iron related problems in Australian water supplies. *Environmental Technology Letters* 10:479–490.

Legorburu, I., L. Canton, E. Millan, and A. Casado. 1988. Trace metal levels in fish from Urola River (Spain). *Environmental Technology Letters* 9:1373–1378.

Lima, N.R.W., L.D. de Lacerda, W.C. Pfeiffer, and M. Fiszman. 1986. Temporal and spatial variability in Zn, Cr, Cd and Fe concentrations in oyster tissues (*Crassostrea brasiliana* Lamarck, 1819) from Sepetiba Bay, Brazil. *Environmental Technology Letters* 7:453–460.

Maltby, L., J.O.H. Snart, and P. Calow. 1987. Acute toxicity tests on the freshwater isopod, *Asellus aquaticus* using $FeSO_4 \cdot 7H_2O$, with special reference to techniques and the possibility of intraspecific variation. *Environmental Pollution* 43:271–279.

Martin, T.R., and D.M. Holdich. 1986. The acute lethal toxicity of heavy metals to peracarid crustaceans (with particular reference to fresh-water asellids and gammarids). *Water Research* 20:1137–1147.

Mason, C.F., and S.M. Macdonald. 1988. Metal concentration in mosses and otter distribution in a rural Welsh river receiving mine drainage. *Chemosphere* 17:1159–1166.

Moffett, J.W., and R.G. Zika. 1987. Reaction kinetics of hydrogen peroxide with copper and iron in seawater. *Environmental Science and Technology* 21:804–810.

Mohapatra, S.P. 1988. Distribution of heavy metals in polluted creek sediment. *Environmental Monitoring and Assessment* 10:157–163.

Moore, J.N., W.H. Ficklin, and C. Johns. 1988. Partitioning of arsenic and metals in reducing sulfidic sediments. *Environmental Science and Technology* 22:432–437.

Mudroch, A., L. Sarazin, and T. Lomas. 1988. Summary of surface and background concentrations of selected elements in the Great Lakes sediments. *Journal of Great Lakes Research* 14:241–251.

Naquadat. 1985. *National water quality data bank.* Environment Canada, Ottawa.

Reader, J.P., N.C. Everall, M.D.J. Sayer, and R. Morris. 1989. The effects of eight trace metals in acid soft water on survival, mineral uptake and skeletal calcium deposition in yolk-sac fry of brown trout, *Salmo trutta* L. *Journal of Fish Biology* 35:187–198.

Sadig, M., and I. Alam. 1989. Metal concentrations in pearl oyster, *Pinctada radiata,* collected from Saudi Arabian coast of the Arabian Gulf. *Bulletin of Environmental Contamination Toxicology* 42:111–118.

Samanidou, V., and K. Fytianos. 1987. Partitioning of heavy metals into selective chemical fractions in sediments from rivers in northern Greece. *Science of the Total Environment* 67:279–285.

Samhan, O., M. Zarba, and V. Anderlini. 1987. Multivariate geochemical investigation of trace metal pollution in Kuwait marine sediments. *Marine Environmental Research* 21:31–48.

Saxena, D.K., R.C. Murthy, V.K. Jain, and S.V. Chandra. 1986. Influence of cadmium on the distribution of Cu, Zn and Fe in different regions of central and peripheral nervous system of rats. *Chemosphere* 15:373–377.

Soderlund, S., A. Forsberg, and M. Pedersen. 1988. Concentrations of cadmium and other metals in *Fucus vesiculosus* L. and *Fontinalis dalecarlica* Br. Eur. from the northern Baltic Sea and the southern Bothnian Sea. *Environmental Pollution* 51:197–212.

Sridhar, M.K.C. 1986. Trace element composition of *Pistia stratiotes* L. in a polluted lake in Nigeria. *Hydrobiologia* 131:273–276.

Stauber, J.L., and T.M. Florence. 1985. The influence of iron on copper toxicity to the marine diatom, *Nitzschia closterium* (Ehrenberg) W. Smith. *Aquatic Toxicology* 6:297–305.

Stull, J.K., and R.B. Baird. 1985. Trace metals in marine surface sediments of the Palos Verdes Shelf, 1974–1980. *Journal of the Water Pollution Control Federation* 57:833–840.

Subramanian, V., P.K. Jha, and R. van Grieken. 1988. Heavy metals in the Ganges Estuary. *Marine Pollution Bulletin* 19:290–293.

Sugawara, N., B.Q. Chen, C. Sugawara, and H. Miyake. 1988. Effect of cadmium on Fe^{+3}-transferrin formation in the rat intestinal mucosa. *Bulletin of Environmental Contamination and Toxicology* 41:50–55.

Sullivan, M.F., and P.S. Ruennler. 1987. Effect of excess Fe on Cd or Pb absorption by rats. *Journal of Toxicology and Environmental Health* 22:131–139.

Sung, J.F.C., A.E. Nevissi, and F.B. Dewalle. 1986. Concentration and removal efficiency of major and trace elements in municipal wastewater. *Journal of Environmental Science and Health* 21:435–448.

Taylor, G.J., and A.A. Crowder. 1983. Accumulation of atmospherically deposited metals in wetland soils of Sudbury, Ontario. *Water, Air, and Soil Pollution* 19:29–42.

Tessier, A., P.G.C. Campbell, J.C. Auclair, and M. Bisson. 1984. Relationships between the partitioning of trace metals in sediments and their accumulation in the tissues of the freshwater mollusc *Elliptio complanata* in a mining area. *Canadian Journal of Fisheries and Aquatic Sciences* 41:1463–1472.

US Environmental Protection Agency. 1930–1989. Bureau of Mines, US Department of the Interior, Washington, DC.

US Minerals Yearbooks. 1930–1989. Bureau of Mines, US Department of the Interior, Washington, DC.

Vas, P., and J.D.M. Gordon. 1988. Trace metal concentrations in the scyliorhinid shark *Galeus melastomus* from the Rockall Trough. *Marine Pollution Bulletin* 19:396–398.

V.-Balogh, K., D.S. Fernandez, and J. Salanki. 1988. Heavy metal concentrations of *Lymnaea stagnalis* L. in the environs of Lake Balaton (Hungary). *Water Research* 22:1205–1210.

Villarreal-Trevino, C.M., M.E. Obregon-Morales, J.F. Lozano-Morales, and A. Villegas-Navarro. 1986. Bioaccumulation of lead, copper, iron and zinc by fish in a transect of the Santa Catarina River in Cadereyta Jimenez, Nuevo Leon, Mexico. *Bulletin of Environmental Contamination and Toxicology* 37:395–401.

Vymazal, J. 1984. Short-term uptake of heavy metals by periphyton algae. *Hydrobiologia* 119:171–179.

Wai, C.M., D.E. Reece, B.D. Trexler, D.R. Ralston, and R.E. Williams. 1980. Production of acid water in a lead-zinc mine, Coeur d'Alene, Idaho. *Environmental Geology* 3:159–162.

Wang, W. 1986. Toxicity tests of aquatic pollutants by using common duckweed. *Environmental Pollution* 11:1–14.

Warnick, S.L., and H.L. Bell. 1969. The acute toxicity of some heavy metals to different species of aquatic insects. *Journal of the Water Pollution Control Federation* 4:280–284.

Watling, H.R., and R.J. Watling. 1983. Sandy beach molluscs as possible bioindicators of metal pollution. 1. Field survey. *Bulletin of Environmental Contamination and Toxicology* 31:331–338.

Westall, J., and W. Stumm. 1980. The hydrosphere. *In: The handbook of environmental chemistry,* ed. O. Hutzinger, 17–49. Springer-Verlag, New York.

14
Lead

Lead is the 36th most abundant element in the earth's crust, with an average concentration of 15 mg/kg. Although found in over 200 minerals, lead is concentrated (30 to 80 g/kg) in galena (PbS), gelesite ($PbSO_4$), and cerrusite ($PbCO_3$). The input of anthropogenically derived lead to the environment now outweighs all natural sources, and is likely to remain so for the foreseeable future.

Production, Sources, and Residues

Production

World production of lead was 1,700 × 10^3 metric tons in 1930, increasing to 2,400 × 10^3 tons in 1960 and 3,100 × 10^3 metric tons in 1980 (US Minerals Yearbooks, 1930–1989). Production in recent years has stayed around the 3,100 × 10^3 metric tons mark, reflecting concern about the health effects of lead, particularly in young children. The world's leading mine producers are the USSR, the USA, Canada, Peru, and Mexico; the leading consumers of refined lead are the USA, the USSR, Japan, the FRG and the UK (Table 14.1).

Lead continues to be used in large amounts in storage batteries, metal products, pigments, and chemicals. Of these, storage batteries typically account for 60% of total consumption in Western nations. Lead use in fuel has decreased dramatically in recent years, as discussed in the section on Health Effects.

Table 14.1. World's leading producers and consumers of lead.

Producing nation	Quantity (1000 metric tons/yr)	Consuming nation	Quantity (1000 metric tons/yr)
USSR	570 (approx.)	USA	1,067
USA	422	USSR	780
Canada	285	Japan	395
Peru	216	FRG	350
Mexico	182	UK	274
China	165 (approx.)	Italy	244
Yugoslavia	113	China	230 (approx.)
North Korea	110 (approx.)	France	207
Morocco	101	Spain	116
Spain	92	Yugoslavia	116

Source: Cordero (1988).

Sources

The total amount of lead discharged to freshwaters from anthropogenic sources amounts to 97–180 × 10^3 metric tons per year (Table 14.2). Primary sources include manufacturing processes (particularly metals), atmospheric deposition, and domestic wastewater. Approximately 96% of all lead emissions originate from anthropogenic sources (Nriagu, 1989), particularly combustion of leaded fuels, pyrometallurgical nonferrous metal production, and coal combustion (Tables 14.2, 14.3).

Hutton and Symon (1986) reported that the total input of lead to the

Table 14.2. Worldwide anthropogenic input of lead to freshwaters.

Source	Input (thousand metric tons/yr)
Atmospheric fallout	87–113
Manufacturing processes	
metals	2.5–22
chemicals	0.4–3.0
pulp and paper	0.01–0.9
petroleum products	0–0.1
Dumping of sewage sludge	2.9–16
Domestic waste water	
central	0.9–7.2
noncentral	0.6–4.8
Smelting and refining	
nonferrous metals	1.0–6.0
iron and steel	1.4–2.8
Steam electrical production	0.2–1.2

Source: Nriagu and Pacyna (1988).

Table 14.3. Worldwide emissions of lead to the atmosphere.

Source	Input (thousand metric tons/yr)
Leaded fuel combustion	248 (approx.)
Pyrometallurgical nonferrous metal production	
mining	1.7–3.4
lead production	11.7–31.2
copper-nickel production	11.1–22.1
zinc-cadmium production	5.5–11.5
secondary metal production	0.1–1.4
Steel and iron manufacturing	1.1–14.2
Cement production	<0.01–14.2
Coal combustion	
electrical utilities	0.8–4.7
industrial and domestic	1.0–9.9
Wood combustion	1.2–3.0
Refuse incineration	
municipal	1.4–2.8
sewage sludge	<0.3
Phosphate fertilizer production	<0.3
Other emissions	5 (approx.)
Total emissions	289–376

Source: Nriagu and Pacyna (1988).

coastal waters of the United Kingdom amounted to 2,400 metric tons per year. Of this, 878 metric tons came from rivers, 789 metric tons from dredged material, 300 metric tons from sewage and sewage sludge, and 245 metric tons from industrial waste dumping. Input into The Netherlands' part of the North Sea is dominated by atmospheric deposition and fluvial transport (Table 14.4).

Table 14.4. Annual input of total lead to The Netherlands' part of the North Sea in 1980 (actual) and 1990 (projected).

Source	Input 1980 (metric tons/yr)	Input 1990 (metric tons/yr)
Total input	1,800	1,300–1,400
Atmospheric deposition	630	630
Rivers	870	550–640
Coastal discharges	24	4
Dredging sludges	310	80
Industrial wastes	7	0
Incineration at sea	not known	not known
Offshore mining	2.5	3

Source: Beukema et al. (1986).

Lead is emitted in large amounts from municipalities, both from the incineration of waste products and the discharge of waste water. Gounon and Milhau (1986), for example, reported that the concentration of lead in fly ash from incinerators in the city of Paris ranged from 1,700 to 31,300 mg/kg and that residues in atmospheric emissions ranged from 50.0 to 56.1 mg/m^3. Similarly high levels (1,500–25,000 mg/kg) were found in ash from incinerators in Sweden (Carlsson, 1986), and emissions from incinerators in all of the United Kingdom amounted to 115 metric tons per year (Wadge and Hutton, 1987). Other studies (Hutton et al., 1988) have reported that lead in street dusts downwind of municipal incinerators ranged from 100 to 3,300 mg/kg compared to residues of 51–269 mg/kg in soils.

Lead concentrations in municipal sludge typically exceed 300 mg/kg (Table 14.5). This material may be deposited in landfills or deposited into surface/coastal waters. Residues in waste waters are also periodically elevated and may exceed 0.04 mg/L in the final effluent (Sung et al., 1986). Equally high levels, averaging 0.09 mg/L, have been found in urban runoff (Marsalek and Schroeter, 1988). Lema et al. (1988), having examined studies on landfill leachates from nine countries, found that lead residues ranged from 0.01 to 1.6 mg/L.

Because lead continues to be used as an antiknock agent in gasoline in many nations, residues are correspondingly elevated in urban air. Royset and Thomassen (1987) found that inorganic lead in Olso (Norway) was as high as 0.003 mg/m^3 whereas Moura et al. (1988) reported residues of up to 0.001 mg/m^3 in the air of Oporto (Portugal). Although equally high levels have been found in rainwater in England (Radojevic and Harrison, 1987), much lower residues (<0.001 mg/m^3) have been found over the Atlantic Ocean and Antarctica (Volkening et al., 1988; Barrie et al., 1987).

Lead generally occurs in high concentrations in urban snow, a reflection of the combustion of leaded gasoline. Residues as high as 475 mg/L were found in snow in Winnipeg (Canada) during the early 1980s (Lock-

Table 14.5. Concentration of lead in municipal sludge and wastewater.

Sample	Location	Concentration average (range)
Sludge[1]	Tripoli, Libya	638 (84–1,579) mg/kg
Sludge[2]	Rome, Italy	380 (NR[a]) mg/kg
Sludge[3]	Washington state, USA	437 (383–491) mg/kg
Sludge[4]	Redditch, England	3.8 (NR) mg/L
Wastewater[3]	Washington state, USA	<0.07 (<0.07) mg/L

[a]Not reported.
Sources: [1]Abuzkhar et al. (1987), [2]Campanella et al. (1987), [3]Nevissi et al. (1988), [4]Lake et al. (1989).

ery et al., 1983). These concentrations, recorded prior to the introduction of lead controls in gasoline, have produced significant adulteration of surface water supplies during the spring melt. More recent studies (e.g., Sakai et al., 1988) have reported lower residues (<0.1 mg/L) in urban snow, a reflection of tighter controls on automobile exhausts.

Lead is obviously elevated in tailings and waste water from lead- zinc and other mines. Wong (1986), having reviewed a number of studies, reported that residues in such wastes range from 5,700 to 40,500 mg/kg. Similarly, lead in waste water from lead–zinc mines in Idaho ranged from 0.55 to 2.1 mg/L (Wai et al., 1980). Such discharges often produce elevated concentrations in the sediments and water of nearby lakes and rivers (Arafat, 1985).

Residues

Water. Total Pb in surface water is highly variable, but is typically <0.05 mg/L (Table 14.6). Although maximum residues are often associated with poorly treated industrial or mining effluents, the input of lead into remote waters is due primarily to atmospheric deposition (Urban et al., 1987). Particulate Pb generally accounts for >75% of residues in flowing waters and >50% in standing waters (Scoullos and Hatzianestis, 1989; Zingde et al., 1988; Whitehead et al., 1988).

Sediments. Approximately 99% of the lead entering the oceans with the suspended load of rivers is deposited in the sediments of estuaries and continental shelves (Craig, 1980). Residues in the 15–50 mg/kg dry weight range are frequently reported for coastal/estuarine sediments (Araujo et al., 1988; Samhan et al., 1987; Subramanian et al., 1988), and may exceed 400 mg/kg near waste outfalls (Stull et al., 1986; Stull and Baird, 1985; Luoma and Phillips, 1988). The major fraction of dissolved lead

Table 14.6. Concentration (mg/L) of total lead in surface water.

Region	Number of samples	Concentration
Pacific coast rivers, Canada[1]	12	0.001–0.004
Prairie lakes and rivers, Canada[1]	1,848	0.001–0.077
Atlantic coast lakes and rivers, Canada[1]	229	0.001–0.041
Awba River, Nigeria[2]	NR[a]	0.001–0.006
Hindon River, India[3]	9	0.010–0.143
Kandy Canal, Sri Lanka[4]	51	0.02–0.85
Hamilton Harbor, Canada[5]	42	<0.003–0.031
Blanca Bay, Argentina[6]	13	0.008–0.048
Mississippi River, USA[7]	9	<0.001–0.009

[a]Not reported.

Sources: [1]Naquadat (1985), [2]Sridhar (1986), [3]Ajmal et al. (1987), [4]Dissanayake et al. (1987). [5]Poulton (1987), [6]Zubillaga and Pucci (1986), [7]DeLeon et al. (1986).

Table 14.7. Concentration (mg/kg dry weight) of total lead in sediments of depositional basins of the Great Lakes.

Lake	Surface	Background
Ontario	7.0–285	18–32
Erie	6.0–299	21–49
Huron	3.0–151	14–36
Michigan	10–130	8–10
Superior	75–138	21–68

Source: Mudroch et al. (1988).

reaches the open sea, and has a residence time of approximately 100–200 years (Veron et al., 1987). Until recently, it has been assumed that residues in the deep ocean are unchanged from prehistoric levels, despite the vast increase in lead inputs in the last century. However, Veron et al. (1987) showed that concentrations in two short cores from the northeast Atlantic were greater in the uppermost 1 cm (15–21 mg/kg) than in the uppermost 10 cm (2.8–6.0 mg/kg). This is of considerable significance in the study of lead cycles and needs to be confirmed through additional investigation.

Extremely high concentrations of lead are periodically reported for freshwater sediments receiving industrial or municipal wastes. In one example, residues in the sediments of two lakes in the Sudbury mining district (Canada) were as high as 250–350 mg/kg in the upper 10 cm of cores, falling to 50 mg/kg below 15 cm (Nriagu and Rao, 1987). In a similar coring study of the Calcasieu River (Louisiana), Mueller et al. (1989) found maximum residues (16 mg/kg) at 10 cm, deposited in the year 1971; the corresponding concentrations at 2 cm (1982) and 38 cm (1933) were 12 and 10 mg/kg, respectively. Mudroch et al. (1988), reviewing a number of studies on lead in the deposition basins of the Great Lakes, found that surface residues were up to 13 times greater than background levels (Table 14.7). Jones (1986) reported that lead in the River Rheidol (Wales) was extremely high (up to 2,000 mg/kg) in surficial sediments, due to discharge from a lead–zinc mine.

Chemistry

Lead is a member of the Group IV elements (C, Si, Ge, Sn, and Pb) but, unlike C and Si, does not bind with another identical atom, shows a marked decrease in covalency, and has stable +2 and +4 oxidation states. In freshwater, lead forms a number of complexes of low solubility with many of the major anions, including hydroxides, carbonates, sulfides, and (less commonly) sulfates. Lead also partitions favorably with humic and fulvic acids, forming moderately strong chelates. At pH 10, $Pb(OH)^+$ dominates all other species. Speciation shifts to favor chloride and hydroxide complexes in saltwater. Approximately 75% of lead in rivers is in suspension and 25% in solution, but in saltwater, the corresponding ratio is approximately 50:50

Lead undergoes methylation in the environment to form several organic derivatives. The process is mediated by bacteria in sediments to form $(CH_3)_3Pb^+$ and related compounds. $(CH_3)_3Pb^+$ disproportionates slowly as follows:

$$3(CH_3)_3PbX \longrightarrow 2(CH_3)_4Pb + PbX_2 + CH_3X \ldots$$

Chemical alkylation of lead has been reported under laboratory conditions, apparently in the absence of bacterial methylation.

Numerous organic complexes are found in the atmosphere and rainwater due to the use of lead in fuels and concomitant breakdown of the parent compounds. Allen et al. (1988), for example, recorded at least 12 alkyllead compounds in the atmosphere over the UK; the main species in that study were Me_4Pb, Me_3EtPb, Me_3Pb^+, and Et_4Pb. Van Cleuvenbergen et al. (1986) also found a predominance of those species in rainwater over Belgium and, significantly, in a lake and river near Antwerp (Table 14.8). The environmental significance of such compounds, derived from the atmosphere, has yet to be determined.

Sorption to sediments plays a key role in the fate of lead complexes. Huang et al. (1977) showed that, under laboratory conditions, most lead compounds exist in the solid phase above pH 7; this relationship has been reported on numerous occasions under actual field conditions (e.g., White and Driscoll, 1985). The rate of adsorption is initially rapid and follows the Freundlich isotherm, before reaching a plateau after 24–48 h. Although desorption is a slow process, McKee et al. (1989) suggested that sediment-bound lead in Lake Superior may ultimately appear in the pore water and be recycled to the overlying water. The presence of Cu^{2+}, Zn^{2+}, and other metals is likely to hinder the uptake of Pb^{2+} (Saikia et al., 1987).

Mudroch and Duncan (1986) showed that lead in the Niagara River

Table 14.8. Concentration (ng/L) of tri- and dialkyllead compounds in surface waters in Belgium.

Location	$PbMe_3^+$	$PbMe_2^{2+}$	$PbEt_3^+$	$PbEt_2^{2+}$
Lake, near Antwerp	3.8–5.0	<0.3–0.3	0.6–1.2	<0.4–0.5
River Scheldt	4.1	0.6	5.3	0.7

Source: Van Cleuvenbergen et al. (1986).

(Canada) preferentially bound to the smallest size fractions of the sediments, as follows:

	Particle size (μm)					
	<13	13–19	19–27	27–40	40–54	54–150
Concentration (mg/kg dry weight)	518	184	130	106	73	67

This has been reported in many other studies and means that monitoring programs should focus on the smallest fractions of sediments to obtain an accurate picture of potential lead contamination.

Lead typically complexes with sulfides and Fe–Mn hydrous oxides in sediments. In one study of a Greek river, over 50% of sediment-bound lead was associated with these two components (Table 14.9), as also reported for sediments in Lake Superior (McKee et al., 1989). Lead–carbonate complexes are common only in surficial sediments (Purchase and Fergusson, 1986) whereas lead sulfide is dominant in anaerobic sediments. Lin et al. (1988) showed that Pb^{2+} readily complexed with sulfide mineral surfaces (chalcocite, sphalerite) and precipitated onto pyrite. Lead typically desorbs from sediments and suspended solids in estuaries, owing to competition with chlorides, producing an appreciable increase in residues in the water column (Ferrari and Ferrario, 1989).

The majority of lead complexes in water are not subject to photolysis. Although volatilization is also an unimportant fate process in most cases, some compounds such as tetramethyllead are volatile and, presumably, are removed from the water column by that process.

Bioaccumulation

Plants

Total Pb is often found in high concentrations in aquatic plants, particularly those growing in freshwaters and receiving mine or other industrial waste. Residues in six macroscopic species in streams from an industrial-

Table 14-9. Partitioning of lead in the sediments of the Axios River and Estuary (Greece).

	Distribution (%)	
Fraction	River	Estuary
Cation-exchangeable	1.0	1.2
Carbonates	15.5	13.3
Fe–Mn hydrous oxides	25.5	23.8
Organic sulfides	29.8	33.1
Residual	28.2	28.6

Source: Samanidou and Fytianos (1987).

ized area of West Germany ranged from 100 to 5,300 mg/kg dry weight (Table 14.10), whereas concentrations of up to 16,000 mg/kg were found in aquatic bryophytes growing in streams receiving mine wastes (Burton and Peterson, 1979). Substantially lower residues have been found in nonpolluted or mildly polluted rivers and estuaries (Table 14.10). In the study by Mason and Macdonald (Table 14.10), lead in aquatic mosses was as low as 22 mg/kg in unpolluted parts of the Mawddach basin but increased to 462 mg/kg near mine discharge sites.

The rate of uptake of inorganic lead is generally rapid and increases with exposure concentration. Sorption is generally suppressed by H^+ and humic acids (Vymazal, 1984). Concentration factors (plant residue/water residue) for inorganic lead are often high and variable, ranging from 5,000 to >15,000. Wong et al. (1987) reported that the corresponding concentration factors for the freshwater green alga *Ankistrodesmus falcatus* exposed to trialkyllead and dialkyllead were only 100 and 2,000, respectively. The concentration factor for that species exposed to inorganic lead was as high as 20,000. Wong et al. (1987) also showed that *Ankistrodesmus falcatus* metabolized trimethyllead through a dealkylation sequence to form dimethyllead and Pb^{2+} compounds.

Invertebrates

Inorganic lead rarely concentrates in invertebrate tissues. In unpolluted and mildly polluted waters, residues are typically <5 mg/kg wet weight regardless of species (Hungspreugs et al., 1989; Pugsley et al., 1988; Lopez-Artiguez et al., 1989). However, because residues are directly related to exposure concentration, relatively high total Pb levels are periodically found under some conditions. For example, total Pb was as high as 49 mg/kg dry weight in the soft tissues of the mollusc *Mya arenia* from Gdansk Harbor (Szefer, 1986), and ranged from nondetectable to 39 mg/kg dry weight in the freshwater mollusc *Elliptio complanata* from a mining area in Canada (Tessier et al., 1984).

Table 14.10. Concentration (mg/kg dry weight) of total Pb in plants collected from marine and freshwaters.

Species	Average (range)	Location
Attached plants, 6 species[1]	973 (100–5,300)	Polluted streams, FRG
Moss, *Fontinalis squamosa*[2]	116 (22–462)	River Mawddach, Wales
Benthic diatoms[3]	NR[a] (15–95)	Contraband Bayou, Louisiana
Seaweeds, 4 species[4]	9.5 (0.9–79)	Gulf of Maine
Fucus vesiculosus[5]	3 (2.0–11.7)	Swedish coastal waters
Seagrass, 9 species[6]	3 (1.1–6.1)	Flores Sea, Indonesia

[a]Not reported.
Sources: [1]Dietz (1973), [2]Mason and Macdonald (1988), [3]Ramelow et al., (1987), [4]Sears et al. (1985), [5]Soderlund et al. (1988), [6]Nienhuis (1986).

Relatively little is known about the uptake of organic lead compounds by invertebrates. Although it is assumed that most alkyllead compounds originate from the combustion of gasoline, in vivo metabolism of parent compounds likely occurs in some species. Krishnan et al. (1988) found the following compounds in periwinkles *Littorina irrorata* from the east coast of the United States: $EtMe_2Pb^+$, Et_2MePb^+, Et_3Pb^+, Et_2Pb^{2+}, and Me_3Pb^+. The last compound apparently arose from the demethylation of Me_4Pb in gasoline and also, possibly, from an environmentally mediated methylation of ethyllead salts.

Fish

Total-Pb is not normally a significant contaminant of fish tissues, except in cases of extreme ambient pollution. In the case of Dutch fisheries products, for example, average residues ranged from 0.07 to 0.2 mg/kg wet weight as follows (Hagel, 1986): sole (*Solea solea*), 0.07 mg/kg; cod (*Gadus morhua*), 0.08 mg/kg; herring (*Clupea harengus*), 0.1 mg/kg; pikeperch (*Stizostedion lucioperca*), 0.08 mg/kg; eel (*Anguilla anguilla*), 0.2 mg/kg.

Slightly higher values (up to 0.4 mg/kg) have been found in recent years in cod from the Baltic Sea and Gulf of Finland, as also found for herring from the same waters (United Nations, 1987). Ashraf and Jaffar (1988) noted that residues in six species from the Arabian Sea near Pakistan ranged from only 0.02 to 0.13 mg/kg.

Although total Pb residues are generally low in freshwater fish (e.g., Saiki and May, 1988; Villarreal-Trevino et al., 1986), site-specific anthropogenic inputs periodically give rise to relatively high lead levels. Some species from the United Kingdom carry residues of up to 8.7 mg/kg, far beyond the maximum allowable concentration in fish tissues of 2.0 mg/kg (Table 14.11). Such high levels are due, to a large degree, to the discharge of poorly treated mining wastes, particularly in former years. Similarly,

Table 14.11. Concentration (mg/kg wet weight) of total lead in eel (*Anguilla anguilla*) and roach (*Rutilus rutilus*) from different regions of the UK.

	Eel		Roach	
Region	Average	Range	Average	Range
Southwest England	0.44	ND[a]–2.20	1.57	ND–8.70
Wales	0.92	ND –7.40	0.32	0.20–0.60
East Anglia	NR[b]	0.40–1.00	1.11	ND–8.70
Northeast England	NR	NR	0.89	ND–1.90
Northeast Scotland	2.06	0.02–4.80	NR	NR

[a]Not detected.
[b]Not reported.
Source: Mason (1987).

residues in two species from the Periyar River (India) averaged 2.0 mg/kg near waste outfalls from ore processing plants, and were lower by approximately 50% in downstream areas (Borkar et al., 1984).

Organolead compounds have been detected in fish from several freshwater systems and coastal marine waters, as well as in fish-eating birds such as the herring gull (*Larus argentatus*) which, in recent years, have been used as a biomonitor of toxic contaminants in the Great Lakes and elsewhere (Forsyth and Marshall, 1986). Although residues are generally low (<0.001 mg/kg wet weight) in these species, the continued use of leaded gasoline, particularly in third world countries, may result in additional contamination of potential foodstuffs. Hence special considerations should be given to the formulation of guidelines for total lead and organoleads in fishery products.

Toxic Effects to Aquatic Organisms

Plants

Inorganic lead is moderately toxic to aquatic plants. Under many test conditions, it is more toxic than chromium, manganese, barium, zinc, and iron, but is less toxic than cadmium, mercury, and copper. Although growth limitation typically occurs at lead concentrations of 0.1 to 8 mg/L, some species, such as the green alga *Chlorella saccharophila*, can tolerate concentrations of over 63 mg/L (reviewed by US Environmental Protection Agency, 1985). Since aquatic animals are generally more sensitive to lead than plants, any surface water quality value aimed at protecting animals will also protect plants.

Lead acts synergistically with combinations of copper and zinc, and with copper, zinc, and H^+ (Starodub et al., 1987). As with most metals, complexation with humic acid and other organic molecules, and inorganic ligands, reduces toxicity to most plant species studied to date. It is generally assumed that organoleads, particularly tetraethyllead, are more toxic to aquatic plants than either the methylated derivatives or inorganic compounds.

Invertebrates

Although there is considerable variability among species and test conditions, the acute toxicity of inorganic lead to marine and freshwater invertebrates is generally less than that of cadmium, copper, mercury, and zinc. Acute toxic effects have been recorded at concentrations in the 0.5–5.0 mg/L range in both marine and freshwater species (Oladimeji and Offem, 1989; Bodar et al., 1989). Some species, such as amphipods, are much more resistant, showing LC_{50}s of 28–124 mg/L (Table 14.12). Chronic toxic effects, such as reduced reproductive capacity, have been reported in freshwater and marine invertebrate species at concentrations

Table 14.12. Acute toxicity (LC_{50}) of Pb^{2+} to two crustacean species.

48-h LC_{50} (mg/L)		96-h LC_{50} (mg/L)	
Average	95% Confidence limits	Average	95% Confidence limits
Asellus aquaticus			
120	94–201	124	108–144
Crangonyx pseudogracilis			
44	40–49	28	25–31

Water hardness, 50 mg/L; pH 6.75; water temperature, 13°C.
Source: Martin and Holdich (1986).

as low as 0.019–0.025 mg/L (reviewed by US Environmental Protection Agency, 1985).

Fish

Although the LC_{50} for freshwater fish exposed to inorganic lead typically ranges from 0.5 to 10 mg/L, increasing water hardness may increase resistance to >400 mg/L, particularly if water hardness exceeds 350 mg $CaCO_3$/L. Chronic toxic effects following long-term exposure to inorganic lead have been reported at concentrations of <0.010 mg/L. These effects include (1) impairment of calcium, sodium, potassium, and skeletal calcification in brown trout *Salmo trutta* exposed to 60 nmol Pb/L (Reader et al., 1989); (2) retarded ovarian maturation in the catfish *Clarias batrachus* following exposure to lead nitrate (5 mg/L) for 9 months (Katti and Sathyanesan, 1987); (3) inhibition of Na, K-ATPase activity in rainbow trout *Oncorhynchus mykiss* exposed to dietary lead at 10 mg/kg daily (Crespo et al., 1986); and (4) extensive (90%) spinal deformities in rainbow trout exposed to lead nitrate at 0.055 mg/L for 570 days (Davies et al., 1976).

Although the LC_{50} of inorganic lead compounds is relatively low in saltwater, generally <5 mg/L (Taylor et al., 1985), organic complexes are far more toxic to most marine fish. Some studies have placed the LC_{50} of tetramethyllead and tetraethyllead at <0.25 mg/L. No definitive data are available on the chronic effects of inorganic and organic lead compounds to marine fish.

Many nations have promulgated guidelines and standards for the protection of freshwater and marine life. Examples of some of these are listed below:

Canada
- 0.001 mg/L (water hardness 0–60 mg/L)
- 0.002 mg/L (water hardness 61–120 mg/L)
- 0.004 mg/L (water hardness 121–180 mg/L)
- 0.007 mg/L (water hardness >180 mg/L)

USA

0.0013 mg/L (water hardness 50 mg/L; 4-day average)
0.0032 mg/L (water hardness 100 mg/L; 4-day average)
0.0077 mg/L (water hardness 200 mg/L; 4-day average)

European Community

0.004 mg/L (salmonids, water hardness 0–50 mg/L)
0.010 mg/L (salmonids, water hardness 51–150 mg/L)
0.020 mg/L (salmonids, water hardness >151 mg/L)
0.050 mg/L (coarse fish, water hardness 0–50 mg/L)
0.125 mg/L (coarse fish, water hardness 51–151 mg/L)
0.250 mg/L (coarse fish, water hardness >151 mg/L)
0.025 mg/L for the protection of saltwater fish and shellfish.

Health Effects

Intake

The average worldwide intake of lead through food is approximately 0.2 mg/day for a 70-kg reference adult, but, depending on culture, intake may range from 0.1 to 0.5 mg/day (World Health Organization, 1984). Drinking water may also be an important source of lead, contributing 0.1 mg/day, on the assumption that daily water intake is 2 L at an average concentration of 0.05 mg Pb/L. Although only about 10% of the lead in the gastrointestinal tract of adults is absorbed, uptake in young children is as high as 50% (Landrigan, 1988). Inhalation as a source is highly variable, averaging <0.001 mg/day for rural dwellers but up to 0.009 mg/day in urban areas (World Health Organization, 1984). These calculations assume that the daily respired volume of air is 15 to 23 m^3 and that 40% of the inhaled lead is retained in the lungs.

The total body burden of lead in a 70-kg reference male ranges from 100 to 400 mg. Of this, approximately 90% is tied up in the mineral matrix of the skeleton. Since the half-life of mineral-bound lead approaches 20 years, the body burden increases with age (reviewed by Carson et al., 1987; Schutz et al., 1987).

Acute Toxicity

Acute poisoning by lead is relatively rare and is generally restricted to occupational settings. The primary symptoms are fatigue, colic anemia, neuritis, seizures, and other neurologic disorders.

Chronic Toxicity

Chronic poisoning produces the following symptoms: loss of appetite, constipation, metallic taste, anemia, weakness, insomnia, muscle and joint pains, and colic. Other effects include hypertension, renal dysfunction, minor congenital malformations, sperm count suppression, and damage to the peripheral nervous system, principally affecting large myelinated motor fibers (Landrigan, 1988; Assennato et al., 1987; Greenberg et al., 1986). These symptoms are, once again, restricted largely to occupational exposure, and are not normally related to the ingestion of contaminated drinking water.

Children are particularly sensitive to low levels of lead, particularly from automobile exhaust, but also potentially from contaminated drinking water. Chronic neuropathy in children is evident at a blood lead (PbB) level of 70 to 100 μg/dL, while colic and other gastrointestinal symptoms are evident at 60 μg/dL. Similarly, anemia may appear at PbB of 70 μg/dL, and reduced hemoglobin synthesis at 40 μg/dL. All of these values are significantly lower than those reported for adults.

There is some evidence that PbB levels of <30 μg/dL induce cognitive and behavioral disorders in children. No single study has shown these effects (e.g., Ernhart et al., 1988), yet the collective neurobehavioral studies of CNS function indicate a correlation between neuropsychological deficits and low PbB levels (Federal Register, 1985).

Organolead compounds, by virtue of their lipophilic properties, readily penetrate the blood–brain barrier, making them more neurotoxic than inorganic compounds. The organic moiety is, in fact, only a vehicle for the effective transport of lead into the tissues.This is of considerable importance in establishing drinking water guidelines, given the fact that most of the lead in water is inorganic and less toxic than than the organic forms.

Carcinogenicity

Experimental evidence exists for the carcinogenicity (renal tumors) of inorganic lead, following oral ingestion at high doses in the rat. There is suggestive evidence for the carcinogenicity of some lead compounds in occupationally exposed workers, but because of confounding exposures from other carcinogens and the lack of lead exposure data, it is not possible to prove or disprove carcinogenicity (US Environmental Protection Agency, 1989). The International Agency for Research on Cancer classifies lead into Group 3: inadequate evidence for carcinogenicity in humans and sufficient evidence of carcinogenicity for some salts in animals.

Drinking Water

Residues

Lead is generally detected in finished drinking water, reflecting its presence in raw water and dissolution of solder and other material in the distribution system. In a survey of the American Water Works Association

(1985) of drinking water in 39 states and 3 territories, there were 55 cases of noncompliance with the Maximum Contaminant Limit of 0.05 mg/L. For comparison, fluoride and nitrates were in noncompliance in 907 and 369 episodes, respectively. Ajmal and Uddin (1986a) reported that residues in running water at the Aligarh Muslim University (India) ranged from 0.0005 to 0.0025 mg/L, compared to 0.0008 to 0.0055 mg/L for standing tap water. Hand pump water from the city of Aligarh had higher lead residues, up to 0.025 mg/L (Ajmal and Uddin, 1986b). Water treatment plants in Rio de Janeiro (Brazil) produced water with dissolved Pb of up to 0.003 mg/L and particulate Pb of up to 0.002 mg/L (Azcue et al., 1988).

A major source of lead in drinking water is dissolution from soldered joints and brass fittings. Birden et al. (1985) showed that pipes with liquid solder and tin–lead (50/50) solder produced water with residues of approximately 1 mg/L; however, when tin–antimony solder or copper pipe with no solder joints were used, concentrations were generally below 0.05 mg/L. Similarly, Schock and Neff (1988) showed that lead was as high as 0.10 mg/L in the first draw from standing tap water, decreasing to <0.005 mg/L in later draws (Figure 14.1). Lead is generally highest in new plumbing systems; in homes older than 10 years, residues are typically <0.010 mg/L.

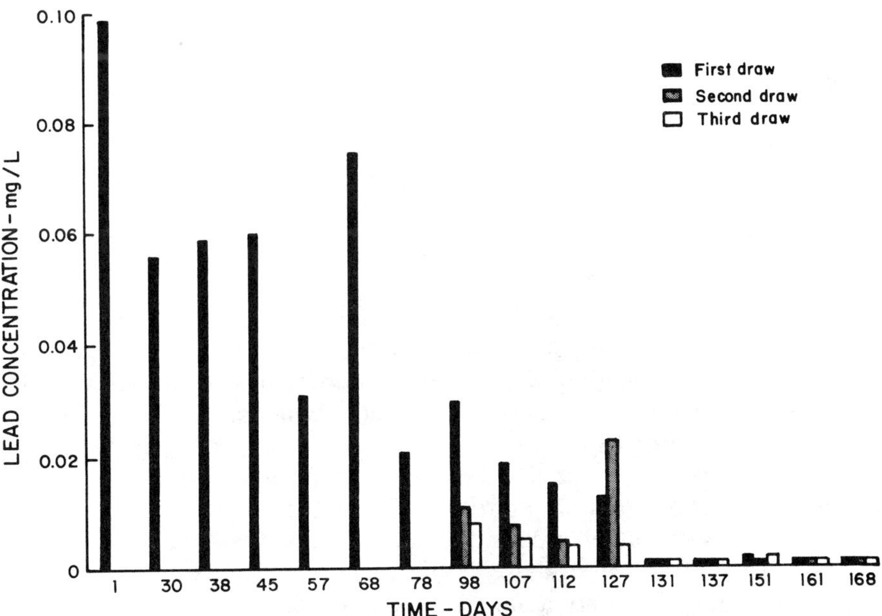

Figure 14.1. Concentration of lead in successive standing water samples taken from Illinois water treatment systems. (From Schock and Neff, 1988.) Reprinted from *Journal American Water Works Association,* Vol. 80, No. 11 (November 1988), by permission. Copyright © 1988, American Water Works Association.

Consumption Guidelines

Most nations, plus the World Health Organization, currently use a guideline/standard of 0.05 mg/L for the protection of drinking water. This value was developed using a provisional tolerable weekly intake of 3 mg Pb per person. Assuming a drinking water lead level of 0.05 mg/L and a weekly water intake of 14 L for a 70-kg reference man, total lead intake comes to 0.7 mg/week. Since only about 33% of total lead intake comes from water, the value of 0.05 mg/L apparently provides adequate protection for consumers. However, recent studies have indicated that the weekly tolerance of 3 mg is too high, and that the guideline/standard for drinking water should be reduced to at least 0.02 mg/L.

It must be pointed out that criteria setting has often neglected to consider differences in the transport of the lead compounds across the blood–brain barrier. These differences make a major impact on toxicity and thus on the actual guideline/standard value.

Treatment

Lead in the source water can be easily controlled by conventional coagulation or lime softening. Both alum and ferric sulfate are effective treatment agents, particularly if the pH of the water is high (9–11). It is likely that the use of lead-containing solders and fittings will be discontinued in the near future to eliminate contamination in new distribution systems. Lead dissolution in older systems can be minimized by pH adjustment to at least 7.5, enhancement of $CaCO_3$ deposition to seal the pipe, and control of chloride and nitrate (Schock, 1989).

Recommendations

An enormous amount of lead has been mobilized from anthropogenic sources. Although some nations have reduced emissions, particularly by controlling the combustion of leaded gasoline, worldwide mobilization of lead will remain high for many years to come. One key question then is, "What is the fate of all of this material?" Because lead is readily transported by rivers, the ocean's sediments are a primary sink, but the rate and extent of deposition are poorly known at best.

Another poorly studied area is the fate and toxicity of organolead compounds. Although many of these complexes are more toxic than their inorganic counterparts, only a handful of studies are available on fate and toxicity. The environmental fate of many organic derivatives is almost completely unknown.

The final area cf concern centers around the health effects of lead. Potential carcinogenic effects following oral administration of lead salts need to be proved or disproved. In addition, the current drinking water guideline/standard of 0.05 mg/L needs to be reevaluated in light of recent tolerable weekly intake data and differences in the toxicity of the various lead species.

The following recommendations reflect these considerations.

1. Accumulation of lead in deep ocean sediments.
2. Environmental demethylation and methylation of lead compounds.
3. Metabolism of organic lead compounds by plant and animal species.
4. Chronic toxicity of inorganic and organic compounds to marine fish.
5. Carcinogenicity of lead salts via oral administration.
6. Reevaluation of the current drinking water guideline of 0.05 mg/L used in many nations.

References

Abuzkhar, A.A., A.S. Gibali, Y.I. Elmehrik, and R. Ahmatullah. 1987. Chemical monitoring of sewage wastes for their use in crop production. *Environmental Monitoring and Assessment* 8:127–133.

Ajmal, M., R. Khan, and A.U. Khan. 1987. Heavy metals in water, sediments, fish and plants of river Hindon, U.P., India. *Hydrobiologia* 148:151–157.

Ajmal, M., and R. Uddin. 1986a. Quality of drinking water in the Aligarh Muslim University campus, Aligarh, U.P. (India) with respect to heavy metals. *Environmental Monitoring and Assessment* 6:195–205.

Ajmal, M., and R. Uddin. 1986b. Studies on heavy metals in the ground waters of the City of Aligarh, U.P. (India). *Environmental Monitoring and Assessment* 6:181–194.

Allen, A.G., M. Radojevic, and R.M. Harrison. 1988. Atmospheric speciation and wet deposition of alkyllead compounds. *Environmental Science and Technology* 22:517–522.

American Water Works Association. 1985. An AWWA survey of inorganic contaminants in water supplies. *Journal of the American Water Works Association* 77:67–72.

Arafat, N.M. 1985. The impact of mining and smelting on trace metal distribution in lake sediments around Rouyn-Noranda, Quebec. *Water Pollution Research Journal of Canada* 20:1–8.

Araujo, M.F.D., P.C. Bernard, and R.E. van Grieken. 1988. Heavy metal contamination in sediments from the Belgian coast and Scheldt Estuary. *Marine Pollution Bulletin* 19:269–273.

Ashraf, M., and M. Jaffar. 1988. Correlation between some selected trace metal concentrations in six species of fish from the Arabian Sea. *Bulletin of Environmental Contamination and Toxicology* 41:86–93.

Assennato, G., C. Paci, M.E. Baser, R. Molinini, R.G. Candela, B.M. Altamura, and R. Giorgino. 1987. Sperm count suppression without endocrine dysfunction in lead-exposed men. *Archives of Environmental Health* 42:124–127.

Azcue, J.M.P., W.C. Pfeiffer, M. Fiszman, and O. Malm. 1988. Heavy metal removal by different water treatment plants, in Rio de Janeiro State, Brazil. *Environmental Technology Letters* 9:429–436.

Barrie, L.A., S.E. Lindberg, W.H. Chan, H.B. Ross, R. Arimoto, and T.M. Church. 1987. On the concentration of trace metals in precipitation. *Atmospheric Environment* 21:1133–1135.

Beukema, A.A., G.P. Hekstra, and C. Venema. 1986. The Netherlands' environmental policy for the North Sea and Wadden Sea. *Environmental Monitoring and Assessment* 7:117–155.

Birden, H.H., E.J. Calabrese, and A. Stoddard. 1985. Lead dissolution from soldered joints. *Journal of the American Water Works Association* 77:66–70.

Bodar, C.W.M., A.V.D. Zee, P.A. Voogt, H. Wynne, and D.I. Zandee. 1989. Toxicity of heavy metals to early life stages of *Daphnia magna*. *Ecotoxicology and Environmental Safety* 17:333–338.

Borkar, M.D., A.C. Paul, and K.C. Pillai. 1984. Pb and ^{210}Pb in a tropical river. *Science of the Total Environment* 34:279–288.

Burton, M.A.S., and P.J. Peterson. 1979. Metal accummulation by aquatic bryophytes from polluted mine streams. *Environmental Pollution* 19:39–46.

Campanella, L., E. Cardarelli, T. Ferri, B.M. Petronio, and A. Pupella. 1987. Evaluation of heavy metals speciation in an urban sludge. I. batch method. *Science of the Total Environment* 61:217–228.

Carlsson, K. 1986. Heavy metals from ''energy from waste'' plants— comparison of gas cleaning systems. *Waste Management and Research* 4:15–20.

Carson, B.L., H.V. Ellis, and J.L. McCann. 1987. *Toxicology and biological monitoring of metals in humans*. Lewis Publishing, Chelsea, MI. 328 pp.

Cordero, R. 1988. *Metal Bulletin's prices and data 1988*. Metal Bulletin Books Ltd. Surrey, England. 375 pp.

Craig, P.J. (1980). Metal cycles and biological methylation. *In: The natural environment and the biogeochemical cycles,* ed. O. Hutzinger, 169–227. Springer-Verlag, New York.

Crepso, S., G. Nonnotte, D.A. Colin, C. Leray, L. Nonnotte, and A. Aubree. 1986. Morphological and functional alterations induced in trout intestine by dietary cadmium and lead. Journal of Fish Biology 28:69–80.

Davies, P.H., J.P. Goettl, J.R. Sinley, and N.F. Smith. 1976. Acute and chronic toxicity of lead to rainbow trout, *Salmo gairdneri,* in hard and soft water. *Water Research* 10:199–206.

DeLeon, I.R., C.J. Byrne, E.A. Peuler, S.R. Antoine, J. Schaeffer, and R.C. Murphy. 1986. Trace organic and heavy metal pollutants in the Mississippi River. *Chemosphere* 15:795–805.

Dietz, F. 1973. The enrichment of heavy metals in submerged plants. *In: Advances in water pollution research,* ed. S.H. Jenkins, 53–62. Pergamon Press, Oxford, UK.

Dissanayake, C.B., J.M. Niwas, and S.V.R. Weerasooriya. 1987. Heavy metal pollution of the mid-canal of Kandy: an environmental case study from Sri Lanka. *Environmental Research* 42:24–35.

Ernhart, C.B., M. Morrow-Tlucak, and A.W. Wolf. 1988. Low level lead exposure and intelligence in the preschool years. *Science of the Total Environment* 71:453–459.

Federal Register. 1985. National primary drinking water regulations. *Federal Register* 50:46936–47022.

Ferrari, G.M., and P. Ferrario. 1989. Behavior of Cd, Pb, and Cu in the marine deltaic area of the Po River (North Adriatic Sea). *Water, Air, and Soil Pollution* 43:323–343.

Forsyth, D.S., and W.D. Marshall. 1986. Ionic alkylleads in herring gulls from the Great Lakes region. *Environmental Science and Technology* 20:1033–1038.

Gounon, J., and A. Milhau. 1986. Analysis of inorganic pollutants emitted by the city of Paris garbage incineration plants. *Waste Management and Research* 4:95–104.

Greenberg, A., D.K. Parkinson, D.E. Fetterolf, J.B. Puschett, K.J. Ellis, L. Wielopolski, A.N. Vaswani, S.H. Cohn, and P.J. Landrigan. 1986. Effects of elevated lead and cadmium burdens on renal function and calcium metabolism. *Archives of Environmental Health* 41:69–76.

Hagel, P. 1986. Monitoring of pollutants in Dutch fishery products. *Environmental Monitoring and Assessment* 7:257–262.

Huang, C.P., H.A. Elliott, and R.M. Ashmead. 1977. Interfacial reactions and the fate of heavy metals in solid-water system. *Journal of the Water Pollution Control Federation* 49:745–756.

Hungspreugs, M., W. Utoomprurkporn, S. Dharmvanij, and P. Somopongchaiyakul. 1989. The present status of the aquatic environment of Thailand. *Marine Pollution Bulletin* 20:327–332.

Hutton, M., and C. Symon. 1986. The quantities of cadmium, lead, mercury and arsenic entering the U.K. environment from human activities. *Science of the Total Environment* 57:129–150.

Hutton, M., Wadge, A., and P.J. Milligan. 1988. Environmental levels of cadmium and lead in the vicinity of a major refuse incinerator. *Atmospheric Environment* 22:411–416.

Jones, K.C. 1986. The distribution and partitioning of silver and other heavy metals in sediments associated with an acid mine drainage stream. *Environmental Pollution* 12:249–263.

Katti, S.R., and A.G. Sathyanesan. 1987. Lead nitrate-induced nuclear inclusions in the oocytes of the catfish *Clarias batrachus* (L). *Environmental Research* 44:238–240.

Krishnan, K., W.D. Marshall, and W.I. Hatch. 1988. Ionic alkylleads in salt marsh periwinkles *(Littorina irrirata). Environmental Science and Technology* 22:806–811.

Lake, D.L., P.W.W. Kirk, and J.N. Lester. 1989. Heavy metal solids association in sewage sludges. *Water Research* 23:285–291.

Landrigan, P.J. 1988. Lead: assessing its health hazards. *Health and Environment Digest* 2:1–5.

Lema, J.M., R. Mendez, and R. Blazquez. 1988. Characteristics of landfill leachates and alternatives for their treatment: a review. *Water, Air, and Soil Pollution* 40:223–250.

Lin, Y., G.W. Bailey, and A.T. Lynch. 1988. *Metal interactions at sulfide mineral surfaces: III. Metal affinities in single and multiple ion adsorption reactions.* US Environmental Protection Agency, EPA/600/01, Athens, GA.

Lockery, A.R., T. Gavrailoff, and D. Hatcher. 1983. Lead levels in snow dumping sites along rivers in downtown Winnipeg, Manitoba, Canada. *Journal of Environmental Management* 17:185–190.

Lopez-Artiguez, M., M.L. Soria, and M. Repetto. 1989. Heavy metals in bivalve molluscs in the Huelva Estuary. *Bulletin of Environmental Contamination and Toxicology* 42:634–642.

Luoma, S.N., and D.J.H. Phillips. 1988. Distribution, variability, and impacts of trace elements in San Francisco Bay. *Marine Pollution Bulletin* 19:413–425.

Marsalek, J., and H. Schroeter. 1988. Annual loadings of toxic contaminants in urban runoff from the Canadian Great Lakes basin. *Water Pollution Research Journal of Canada* 23:360–378.

Martin, T.R., and D.M. Holdich. 1986. The acute lethal toxicity of heavy metals to peracarid crustaceans (with particular reference to fresh-water asellids and gammarids). *Water Research* 20:1137–1147.

Mason, C.F. 1987. A survey of mercury, lead and cadmium in muscle of British freshwater fish. *Chemosphere* 16:901–906.

Mason, C.F., and S. Macdonald. 1988. Metal contamination in mosses and otter distribution in a rural Welsh river receiving mine drainage. *Chemosphere* 17:1159–1166.

McKee, J.D., T.P. Wilson, D.T. Long, and R.M. Owen. 1989a. Geochemical partitioning of Pb, Zn, Cu, Fe, and Mn across the sediment-water interface in large lakes. *Journal of Great Lakes Research* 15:46–58.

McKee, J.D., T.P. Wilson, D.T. Long, and R.M. Owen. 1989b. Pore water profiles and early diagenesis of Mn, Cu, and Pb in sediments from large lakes. *Journal of Great Lakes Research* 15:68–83.

Moura, M.J.M.P., S.M.R. Sousa, M.T.S.D. Vasconselos, and A.A.S.C. Machado. 1988. Lead and other heavy metals in atmospheric aerosols of Oporto. *Chemosphere* 17:2093–2106.

Mudroch, A., and G.A. Duncan. 1986. Distribution of metals in different size-fractions of sediment from the Niagara River. *Journal Great Lakes Research* 12:117–126.

Mudroch, A., L. Sarazin, and T. Lomas. 1988. Summary of surface and background concentrations of selected elements in the Great Lakes sediments. *Journal of Great Lakes Research* 14:241–251.

Mueller, C.S., G.J. Ramelow, and J.N. Beck. 1989. Spatial and temporal variation of heavy metals in sediment cores from the Calcasieu River/Lake complex. *Water, Air, and Soil Pollution* 43:213–230.

Naquadat. 1985. *National water quality data bank*. Environment Canada, Ottawa.

Nevissi, A.E., F.B. DeWalle, J.F.C. Sung, K. Mayer, and R. Dalsey. 1988. Heavy metal variability of different municipal sludges as measured by atomic absorption and inductively coupled plasma emission spectroscopy. *Journal of Environmental Science and Health* 23:823–841.

Nienhuis, P.H. 1986. Background levels of heavy metals in nine tropical seagrass species in Indonesia. *Marine Pollution Bulletin* 17:508–511.

Nriagu, J.O. 1989. A global assessment of natural sources of atmospheric trace metals. *Nature* 338:47–49.

Nriagu, J.O., and J.M. Pacyna. 1988. Quantitative assessment of worldwide contamination of air, water and soils by trace metals. *Nature* 333:134–139.

Nriagu, J.O., and S.S. Rao. 1987. Response of lake sediments to changes in trace metal emission from the smelters at Sudbury, Ontario. *Environmental Pollution* 44:211–218.

Oladimeji, A.A., and B.O. Offem. 1989. Toxicity of lead to *Clarias lazera, Oreochromis niloticus, Chironomus tentans* and *Benacus* sp. *Water, Air, and Soil Pollution* 44:191–201.

Poulton, D.J. 1987. Trace contaminant status of Hamilton harbour. *Journal of Great Lakes Research* 13:193–201.

Pugsley, C.W., P.D.N. Herbert, and P.M. McQuarrie. 1988. Distribution of contaminants in clams and sediments from the Huron-Erie corridor. II. Lead and cadmium. *Journal of Great Lakes Research* 14:356–368.
Pugsley, C.W., P.D.N. Herbert, and P.M. McQuarrie. 1988. Distribution of contaminants in clams and sediments from the Huron-Erie corridor. II. Lead and cadmium. *Journal of Great Lakes Research* 14:356–368.
Purchase, N.G., and J.E. Fergusson. 1986. The distribution and geochemistry of lead in river sediments, Christchurch, New Zealand. *Environmental Pollution* 12:203–216.

Radojevic, M., and R.M. Harrison. 1987. Concentrations, speciation and decomposition of organolead compounds in rainwater. *Atmospheric Environment* 21:2403–2411.

Ramelow, G.J., R.S. Maples, R.L. Thompson, C.S. Mueller, C. Webre, and J.N. Beck. 1987. Periphyton as monitors for heavy metal pollution in the Calcasieu River Estuary. *Environmental Pollution* 43:247–261.

Reader, J.P., N.C. Everall, M.D.J. Sayer, and R. Morris. 1989. The effects of eight trace metals in acid soft water on survival, mineral uptake and skeletal calcium deposition in yolk-sac fry of brown trout, *Salmo trutta* L. *Journal of Fish Biology* 35:187–198.

Royset, O., and Y. Thomassen. 1987. Presence of alkyllead in rural and urban air: evaluation of some sources of alkyllead pollution of the atmosphere in Norway. *Atmospheric Environment* 21:655–658.

Saiki, M.K., and T.W. May. 1988. Trace element residues in bluegills and common carp from the lower San Joaquin River, California, and its tributaries. *Science of the Total Environment* 74:199–217.

Saikia, D.K., R.P. Mathur, and S.K. Srivastava. 1987. Adsorption of copper, zinc and lead by bed sediments of River Ganges in India. *Environmental Technology Letters* 8:149–152.

Sakai, H., T. Sasaki, and K. Saito. 1988. Heavy metal concentrations in urban snow as an indicator of air pollution. *Science of the Total Environment* 77:163–174.

Salim, R. 1986. Adsorption of lead on mud. *Journal of Environmental Science and Health* 21:551–560.

Samanidou, V., and K. Fytianos. 1987. Partitioning of heavy metals into selective chemical fractions in sediments from rivers in northern Greece. *Science of the Total Environment* 67:279–285.

Samhan, O., M. Zarba, and V. Anderlini. 1987. Multivariate geochemical investigation of trace metal pollution in Kuwait marine sediments. *Marine Environmental Research* 21:31–48.

Schock, M.R. 1989. Understanding corrosion control strategies for lead. *Journal of the American Water Works Association* 81:88–100.

Schock, M.R., and C.H. Neff. 1988. Trace metal contamination from brass fittings. *Journal of the American Water Works Association* 80:47–56.

Schutz, A., S. Skerfving, S. Mattson, J. Christoffersson, and L. Ahlgren. 1987. Lead in vertebral bone biopsies from active and retired lead workers. *Archives of Environmental Health* 42:340–346.

Scoullos, M.J., and J. Hatzianestis. 1989. Dissolved and particulate trace metals in a wetland of international importance: Lake Mikri Prespa, Greece. *Water, Air, and Soil Pollution* 44:307–320.

Sears, J.R., K.J. Pecci, and R.A. Cooper. 1985. Trace metal concentrations in offshore, deep-water seaweeds in the western North Atlantic Ocean. *Marine Pollution Bulletin* 16:325–328.

Soderlund, S., A. Forsberg, and M. Pedersen. 1988. Concentrations of cadmium and other metals in *Fucus vesiculosus* L. and *Fontinalis dalecarlica* Br. Eur. from the northern Baltic Sea and the southern Bothnian Sea. *Environmental Pollution* 51:197–212.

Sridhar, M.K.C. 1986. Trace element composition of *Pistia stratiotes* L. in a polluted lake in Nigeria. *Hydrobiologia* 131:273–276.

Starodub, M.E., P.T.S. Wong, C.I. Mayfield, and Y.K. Chau. 1987. Influence of complexation and pH on individual and combined heavy metal toxicity to a freshwater green alga. *Canadian Journal of Fisheries and Aquatic Sciences* 44:1173–1180.

Stull, J.K., and R.B. Baird. 1985. Trace metals in marine surface sediments of the Palos Vedes shelf, 1974–1980. *Journal of the Water Pollution Control Federation* 57:833–840.

Stull, J.K., R.B. Baird, and T.C. Heesen. 1986. Marine sediment core profiles of trace constituents offshore of a deep wastewater outfall. *Journal of the Water Pollution Control Federation* 58:985–991.

Subramanian, V., P.K. Jha, and R. van Grieken. 1988. Heavy metals in the Ganges estuary. *Marine Pollution Bulletin* 19:290–293.

Sung, J.F.C., A.E. Nevissi, and F.B. DeWalle. 1986. Concentration and removal efficiency of major and trace elements in municipal wastewater. *Journal of Environmental Science and Health* 21:435–448.

Szefer, P. 1986. Some metals in benthic invertebrates in Gdansk Bay. *Marine Pollution Bulletin* 17:503–507.

Taylor, D., B.G. Maddock, and G. Mance. 1985. The acute toxicity of nine "grey list" metals (arsenic, boron, chromium, copper, lead, nickel, tin, vanadium and zinc) to two marine fish species: dab *(Limanda limanda)* and grey mullet *(Chelon labrosus)*. *Aquatic Toxicology* 7:135–144.

Tessier, A., P.G.C. Campbell, J.C. Auclair, and M. Bisson. 1984. Relationships between the partitioning of trace metals in sediments and their accumulation in the tissues of the freshwater mollusc *Elliptio complanata* in a mining area. *Canadian Journal of Fisheries and Aquatic Sciences* 41:1463–1472.

United Nations. 1987. *Environment statistics in Europe and North America*. United Nations, New York. 187 pp.

Urban, N.R., S.J. Eisenreich, and E. Gorham. 1987. Aluminum, iron, zinc and lead in bog waters of northeastern North America. *Canadian Journal of Fisheries and Aquatic Sciences* 44:1165–1172.

US Environmental Protection Agency. 1985. *Ambient water quality criteria for lead—1984*. US Environmental Protection Agency, EPA 440/5–84–027. Washington, DC. 81 pp.

US Environmental Protection Agency. 1989. *Evaluation of the potential carcinogenicity of lead and lead compounds: in support of reportable quantity adjustments pursuant to CERCLA (Comprehensive Environmental Response, Compensation, and Liability Act) Section 102*. US Environmental Protection Agency, EPA/600/8–89/045A, Washington, DC. 142 pp.

US Minerals Yearbooks. 1930–1989. Bureau of Mines, US Department of the Interior, Washington, DC.

Van Cleuvenbergen, J.A., D. Chakraborti, and F.C. Adams. 1986. Occurrence of tri–and dialkyllead species in environmental water. *Environmental Science and Technology* 20:589–593.

Veron, A., C.E. Lambert, A. Isley, P. Linet, and F. Grousset. 1987. Evidence of recent lead pollution in deep north-east Atlantic sediments. *Nature* 326:278–281.

Villarreal-Trevino, C.M., M.E. Obregon-Morales, J.F. Lozano-Morales, and A. Villegas-Navarro. 1986. Bioaccumulation of lead, copper, iron, and zinc by fish in a transect of the Santa Catarina River in Cadereyta Jimenez, Nuevoo Leon, Mexico. *Bulletin of Environmental Contamination and Toxicology* 37:395–401.

Volkening, J., H. Baumann, and K.G. Heumann. 1988. Atmospheric distribution of particulate lead over the Atlantic Ocean from Europe to Antarctica. *Atmospheric Environment* 22:1169–1174.

Vymazal, J. 1984. Short-term uptake of heavy metals by periphyton algae. *Hydrobiologia* 119:171–179.

Wadge, A., and M. Hutton. 1987. The cadmium and lead content of suspended particulate matter emitted from a U.K. refuse incinerator. *Science of the Total Environment* 67:91–95.

Wai, C.M., D.E. Reece, B.D. Trexler, D.R. Ralston, and R.E. Williams. 1980. Production of acid water in a lead-zinc mine, Coeur d'Alene, Idaho. *Environmental Geology* 3:159–162.

White, J.R., and C.T. Driscoll. 1985. Lead cycling in an acidic Adirondack lake. *Environmental Science and Technology* 19:1182–1187.

Whitehead, N.E., L. Huynh-Ngoc, and S.R. Aston. 1988. Trace metals in two north Mediterranean rivers. *Water, Air, and Soil Pollution* 42:7–18.

Wong, M.H. 1986. Reclamation of wastes contaminated by copper, lead, and zinc. *Environmental Management* 10:707–713.

Wong, P.T.S., Y.K. Chau, J.L. Yaromich, and O. Kramar. 1987. Bioaccumulation and metabolism of tri–and dialkyllead compounds by a freshwater alga. *Canadian Journal of Fisheries and Aquatic Sciences* 44:1257–1260.

World Health Organization. 1984. *Guidelines for drinking-water quality*. World Health Organization, Geneva, Switzerland.

Zingde, M.D., M.A. Rokade, and A.V. Mandalia. 1988. Heavy metals in Mindhola River estuary, India. *Marine Pollution Bulletin* 10:538–540.

Zubillaga, H.V., and A.E. Pucci. 1986. Cu, Cd, Pb and Zn in tributaries to Blanca Bay, Argentina. *Marine Pollution Bulletin* 17:230–232.

15
Manganese

Manganese occurs in the Earth's crust at an average concentration of 950 mg/kg, principally in ores: pyrolusite (MnO_2), rhodocrosite ($MnCO_3$), manganite ($Mn_2O_3 \cdot H_2O$), hausmannite (Mn_3O_4), biotite mica ($K(Mg,Fe)_3(AlSi_3O_{10})(OH)_2$), and amphibole ($((Mg,Fe)_7Si_8O_{22}(OH)_2)$). Manganese is an essential trace element required by both plants and animals. In some waters it may limit, either directly or indirectly, the growth of algae; it is also an essential component of several enzyme systems in animals. Although manganese is of little direct toxicologic significance, it may control the concentration of other elements, including toxic heavy metals, in surface waters.

Production, Sources, and Residues

Production

World production of manganese (as ore concentrate) was 3.5×10^6 metric tons in 1930, increasing to 13.6×10^6 metric tons in 1960 and 26.7×10^6 metric tons in 1980 (US Minerals Yearbooks, 1930–1989). Production in recent years has continued to increase and is currently above 30×10^6 metric tons. The world's leading exporters of refined manganese are South Africa, Gabon, the USSR, Australia, and Brazil; the major importers are Japan, France, Norway, Poland, and Czechoslovakia (Table 15.1). The main use of manganese is in iron alloys, nonferrous alloys, and dry cells.

Table 15.1. World's leading exporters and importers of refined manganese.

Exporting nation	Quantity (10^3 metric tons/yr)	Importing nation	Quantity (10^3 metric tons/yr)
South Africa	898	Japan	742
Gabon	639	France	326
USSR	486	Norway	280
Australia	452	Poland	259
Brazil	328	Czechoslovakia	228

Source: Cordero (1988).

Total environmental flux of manganese is enormous, greater than that of all other metals except iron and aluminum. Each year, approximately 16×10^6 metric tons of manganese (mostly from natural sources) are transported by rivers to the world's oceans (Westall and Stumm, 1980). Anthropogenic discharges are relatively small, amounting to $109–414 \times 10^3$ metric tons/yr for freshwater. Of that total, municipal wastewater is the No. 1 source, followed by dumping of sewage sludge, smelting and refining, and metal manufacturing processes (Table 15.2).

Anthropogenic emissions to the atmosphere are relatively small, amounting to $11–66 \times 10^3$ metric tons/yr, whereas natural source emissions are much greater—$52–582 \times 10^3$/yr (Nriagu, 1989). The primary natural sources of atmospheric manganese are windborne soil particles

Table 15.2. Anthropogenic anthropogenic input of manganese to freshwaters.

Source	Input (thousand metric tons/yr)
Domestic wastewater	
central	18–81
noncentral	30–90
Dumping of sewage sludge	32–106
Smelting and refining	
iron and steel	14–36
nonferrous metals	2–15
Manufacturing processes	
metals	2.5–20
chemicals	2.0–15
pulp and paper	<0.1–1.5
Atmospheric fallout	3.2–20
Steam electrical production	5–18
Base metal mining and dressing	0.8–12

Source: Nriagu and Pacyna (1988).

Table 15.3. Worldwide natural emissions of manganese to the atmosphere.

Source	Input (thousand metric tons/yr)
Windborne soil particles	42–400
Volcanoes	4–80
Biogenic	
continental particles	4–50
continental volatiles	0.03–2.5
marine sources	0.08–3
Forest fires	1.2–45
Sea salt spray	0.02–1.7
Total emissions	52–582

Source: Nriagu (1989).

and volcanoes (Table 15.3), and the major anthropogenic sources are coal burning and incineration of municipal waste.

Sung et al. (1986), working on waste water from 25 plants in the state of Washington, found manganese residues of 14 mg/L wet weight in primary sludge and 0.38 mg/L in raw sewage (Table 15.4), whereas dried sludge from Rome (Italy) contained manganese at 480 mg/kg (Campanella et al., 1987). For comparison, Lema et al. (1988), having reviewed data from seven countries, noted that residues from landfill leachates ranged from 0.05 to 125 mg/L. Abuzkhar et al. (1987), studying the suitability of sewage sludge from Tripoli (Libya) as a soil amendment, found that manganese averaged 121 mg/kg dry weight, with a maximum of 312 mg/kg. Similarly, incineration of sewage sludge, the third largest anthropogenic source of manganese, contributed up to 10×10^3 metric tons to the atmosphere in 1983 (Nriagu and Pacyna, 1988).

Table 15.4. Concentration (mg/L wet weight) of manganese in municipal waste from treatment plants in the state of Washington.

Sample	Concentration	
	Average	Range
Raw sewage	0.38	0.04–4.41
Primary effluent	0.21	0.02–0.35
Secondary effluent	0.17	0.02–0.46
Discharge	0.16	0.01–0.34
Primary sludge	14.0	1.53–44.6
Secondary sludge	5.22	1.11–10.2

Source: Sung et al. (1986).

Manganese is often extremely high near base and precious metal mines. For example, a gold mine in California produced wastewater with residues of up to 4.4 mg/L (Fillpek et al., 1987) whereas manganese in sediments immediately downstream of a copper mine in England ranged up to 10,000 mg/kg dry weight (Davison et al., 1985). Comparably high concentrations have been found in wastes from some peat and coal mines, strip mines, and coal-fired generating plants. In fact, the No. 1 anthropogenic source of manganese to the atmosphere comes from secondary nonferrous metal production (max. 28.4×10^3 metric tons/yr), followed by combustion of coal (max. 18.9×10^3 metric tons/yr) (Nriagu and Pacyna, 1988).

Residues

Water. Total manganese in freshwater is extremely variable, ranging from 0.002 to >4 mg/L (Table 15.5). Typically, particulate manganese accounts for >90%, and often >95% of the total waterborne residue. Similarly variable residues are found in marine estuaries and harbors (Table 15.5), but concentrations are uniformly lower in offshore marine areas (Nolting, 1986).

Sediments. Manganese in the sediments of the North Sea near Belgium averaged 261 mg/kg dry weight and was enriched (438 mg/kg) in the clay/silt fraction (Araujo et al., 1988); the corresponding values for the Scheldt Estuary (The Netherlands) were also reported by Araujo et al. (1988) to be 117 and 434 mg/kg, respectively. Comparable residues, ranging from 216 to 528 mg/kg, were found in bulk samples from the Arabian Gulf (Samhan et al., 1987), whereas sediments from the Palos Verdes Shelf off the coast of Los Angeles carried manganese burdens of 273 to 403 mg/kg

Table 15.5. Concentration (mg/L) of total manganese in surface waters.

Location	Average (range)	Number of samples
Pacific coast rivers, Canada[1]	NR[a] (0.01–1.70)	340
Prairie rivers, Canada[1]	NR (0.01–4.8)	3,777
Atlantic coast rivers, Canada[1]	NR (0.002–3.80)	7,976
Awba stream, Nigeria[2]	NR (0.34–0.68)	NR
River Hindon, India[3]	0.11 (0.03–0.20)	9
Hamilton Harbor, Canada[4]	0.07 (0.02–0.88)	90
Lake Mikri Prespa, Greece[5]	0.08 (0.02–0.15)	11
Mindhola Estuary, India[6]	1.1 (0.9–1.36)	34
Bombay Harbor, India[7]	0.008 (0.007–0.009)	56

[a]Not reported.
Source: [1]Naquadat (1985), [2]Sridhar (1986), [3]Ajmal et al. (1987), [4]Poulton (1987), [5]Scoullos and Hatzianestis (1989), [6]Zingde et al. (1988), [7]Patel et al. (1985).

(Stull and Baird, 1985). Much greater residues, up to 1,467 mg/kg, were found in Thane Creek (India), adjacent to the city of Bombay (Mohapatra, 1988).

Chemistry

Although manganese may exist in oxidation states ranging from −3 to +7, the Mn^{2+} (manganous) and Mn^{4+} (manganic) states are most important in aqueous systems. Mn^{7+} (permanganate) is not persistent because of rapid reduction, which is mediated through oxidation with organic complexes. The nitrate, chloride, and sulfate salts of manganese are soluble in water whereas the corresponding hydroxides, sulfides, carbonates, oxides, and phosphates are only sparingly soluble. These last five species are typically found in the particulate fraction (>0.40 μm) of the water column (Luther et al., 1986). The oxidation–reduction cycle is important in controlling the fate of manganese in most surface waters. The cycle varies seasonally, particularly in lakes that develop an anoxic hypolimnion during the summer. Oxygen concentrations at the water–sediment interface often approach zero. This causes the reduction of Mn^{4+} to soluble Mn^{2+}, which is then transported upward in the water column. The oxygenated water results in reoxidation to insoluble Mn^{4+}, which settles to the bottom to repeat the cycle.

The oxidation of Mn^{2+} is autocatalytic and may be represented as follows:

$$Mn^{2+} + \tfrac{1}{2}\,O_2 \longrightarrow MnO_2(s)$$
$$Mn^{2+} + MnO_2(s) \longrightarrow Mn^{2+}{\cdot}MnO_2(s)$$
$$Mn^{2+}{\cdot}MnO_2(s) + \tfrac{1}{2}\,O_2 \longrightarrow 2MnO_2(s)$$

At circumneutral pH, oxidation leads to considerable sorption of Mn^{2+} from solution. The rate of oxidation of Mn^{2+} increases through the presence of manganese-oxidizing bacteria in bottom sediments (Richardson and Nealson, 1989) and mining wastes (Francis et al., 1989). Microbiologically mediated reduction is exemplified as

$$\tfrac{1}{4}CH_2O + \tfrac{1}{2}MnO_2(s) + H^+ \longrightarrow \tfrac{1}{4}CO_2 + \tfrac{1}{2}Mn^{2+} + \tfrac{3}{4}H_2O$$

The oxidation–reduction cycle also controls the fate of iron through the same processes (more or less) as those described for manganese. Mn^{4+} oxides are, however, reduced to dissolved Mn^{2+} at higher redox potentials than Fe^{3+} oxides can be reduced to dissolved Fe^{2+}. In addition, oxidation of Mn^{2+} to Mn^{4+} occurs much more slowly than the oxidation of Fe^{2+} to Fe^{3+}. As a result, soluble Mn in lakes is often supplied almost entirely from in situ reduction in the water column whereas soluble Fe is supplied by reduction in the sediments. The seasonal concentration of manganese and iron in surface waters is often different, a result of these two factors.

Mn^{2+} and Mn^{4+} follow essentially the same cycle in coastal and estuarine waters where oxygen may be restricted during one or more seasons. In offshore areas, Mn^{2+} reaches its greatest concentration where oxygen is at a minimum. Because pH and oxygen conditions are more or less constant in the deep sea, extensive ferromanganese deposits have developed that feature high concentrations of other elements such as molybdenum, cobalt, nickel, and copper.

Many trace metals sorb to Fe–Mn hydrous oxides in bottom sediments, making these deposits either the primary or secondary site of metal scavenging in lakes and rivers (e.g., Samanidou and Fytianos, 1987). In addition, manganese oxides in the water column have a high adsorption capacity for some metal ions, providing a local surface environment for metal oxidation and manganese oxide reduction. Eary and Rai (1987) found that the oxidation of aqueous Cr^{3+} under laboratory conditions was not appreciably affected by dissolved oxygen, but that Cr^{3+} reacted directly with MnO_2 to produce Cr^{6+}. Hence, the presence of MnO_2 is likely to increase the transport of Cr as Cr^{6+}. It is not known at this time if other metals react similarly to oxidation on the surface of MnO_2.

Bioaccumulation

Plants

Because manganese is found in relatively high concentrations in the environment, it is also plentiful in marine and freshwater plants. Ho (1987), for example, showed that residues in three species of macroalgae from Hong Kong ranged from 9 to 1,704 mg/kg dry weight, whereas *Fucus vesiculosus* from 17 sites along Sweden's east coast contained concentrations ranging from 79 to 290 mg/kg (Soderlund et al., 1988). That latter study also showed that manganese in the moss *Fontinalis dalecarlica* varied from 127 to 264 mg/kg. Vymazal (1984) reported that uptake by two periphytic freshwater species (*Cladophora glomerata* and *Oedogonium rivulare*) occurred continuously over a 4-h exposure period and that complexation with humic substances limited uptake.

Invertebrates

Total Mn in invertebrates, from both marine and freshwaters, is generally low, posing little or no threat to human consumers (Table 15.6). The occasional record of elevated levels (>1,000 mg/kg dry weight) is usually a reflection of anthropogenic inputs, particularly from mines (Nicolaidou

Table 15.6. Concentration (mg/kg dry weight) of total Mn in the soft tissues of marine and freshwater invertebrates.

Species	Average (range)	Location
Bivalve molluscs, 5 species[1]	48 (26–90)	Pacific Ocean, Fiji
Pearl oyster, *Pinctada radiata*[2]	2 (0.3–4.9)[a]	Arabian Gulf
Bivalve mollusc, *Donax serra*[3]	2 (0.6–3.7)[a]	South African coast
Gastropod, *Lymnaea stagnalis*[4]	NR[b] (170–310)	Lake Balaton, Hungary
Asiatic clam, *Corbicula fluminea*[5]	2 (1.8–3.0)	Shatt al-Arab River, Iraq
Bivalve mollusc, *Macoma balthica*[6]	11 (7–14)	Gdansk Bay, Poland
Gastropod mollusc, *Cerithium vulgatum*[7]	1,359 (1,307–1,396)	Evoikos Gulf, Greece

[a]Wet weight
[b]Not reported.
Sources: [1]Dougherty (1988), [2]Sadig and Alam (1989), [3]Watling and Watling (1983), [4]V.-Balogh et al. (1988), [5]Abaychi and Mustafa (1988), [6]Szefer (1986), [7]Nicolaidou and Nott (1989).

and Nott, 1989). Molluscs generally accumulate more manganese, particularly in the gills and mantle, than other invertebrates. For example, Tessier et al. (1984) reported that residues in those two tissues in the bivalve *Elliptio complanata* ranged up to 8,800 mg/kg dry weight, compared to a maximum concentration of 166 mg/kg for the foot.

Fish

Total Mn in the muscle tissue of fish is generally low, often <0.1 mg/kg wet weight. Legorburu et al. (1988), working with eel *Anguilla anguilla* from the Urola River (Spain), noted that muscle residues averaged approximately 0.1 mg/kg, and, in 54% of the fish, were <0.1 mg/kg; by comparison, the average manganese concentration in the liver of the same fish was 0.6 mg/kg, and in the gills, the corresponding concentration was 21 mg/kg. Relatively high residues were also found in the gills, vertebrae, and skin of the shark *Galeus melastomus* collected from the Rockall Trough near Scotland (Table 15.7). Capelli et al., (1987) showed that levels were lower in the white muscle than in the dark muscle of Atlantic bonito *Sarda sarda* from the Gulf of Genoa. This has been noted in other species and is related to the different muscle functions and presence of metal-binding proteins.

There do not appear to be any cases in which the human consumption of fish is limited by manganese residues in the muscle tissue.

Table 15.7. Average concentration (mg/kg wet weight) of total Mn in various tissues from the shark *Galeus melastomus* from the Rockall Trough.

Tissue	Concentration	Tissue	Concentration
Spleen	<0.02	Vertebrae	0.29
Skin	0.27	Muscle	<0.02
Liver	0.04	Heart	0.08
Gonads	<0.02	Kidney	0.07
Gills	0.05		

Source: Vas and Gordon (1988).

Toxic Effects to Aquatic Organisms

Plants

Mn^{2+} is only slightly to moderately toxic to most species of aquatic plants (e.g., Wang, 1986). The main significance of Mn^{2+} centers around its protective effect against more toxic heavy metals. Mn^{2+} acts by saturating extracellular metal-binding sites, thereby restricting sorption by other metals. Stauber and Florence (1985), working with the marine diatom *Nitzschia closterium*, noted that manganese was much more effective than iron in protecting against copper toxicity. This has been noted in a number of other species and reflects the greater affinity of copper for manganese in salt water than iron. Manganese was unable to reverse copper toxicity, nor did it inhibit the toxicity of lipid-soluble copper complexes, such as copper oxinate.

It should be noted that the protective mechanism of manganese against toxic agents is still poorly known and should be considered a research priority.

Invertebrates

Mn^{2+} is, at most, moderately toxic to the majority of invertebrate species, both marine and freshwater. In a study of two freshwater crustaceans, Martin and Holdich (1986) found that the LC_{50} was always above 300 mg/L in both species (Table 15.8). Although this is typical of toxicity data for many species, Martin and Holdich (1986) did show that Mn^{7+} was far more toxic, with an LC_{50} of <1 mg/L.

As is the case with aquatic plants, manganese offers a protective effect against many toxic metals, likely owing to competition for uptake or binding sites. Although it is assumed that manganese oxides compete as effectively as hydrous iron oxides for other metals, particularly in freshwater, the data base concerning competition is relatively incomplete at this time and should be expanded.

Table 15.8. Acute toxicity (LC_{50}) of manganese to two crustacean species.

	48-h LC_{50} (mg/L)		96-h LC_{50} (mg/L)	
Species	Average	95% Confidence limits	Average	95% Confidence limits
Asellus aquaticus				
Mn^{2+}	771	563–1,050	333	216–502
Crangonyx pseudogracilis				
Mn^{2+}	1,389	1,162–1,694	694	575–843
Mn^{7+}	0.99	0.86–1.12	0.50	0.41–0.57

Water hardness, 50 mg/L; pH, 6.75; water temperature, 13°C.
Source: Martin and Holdich (1986).

Fish

Mn^{2+} is not acutely toxic to fish, with reported LC_{50}s often exceeding 1,000 mg/L in freshwater species. Chronic effects have been reported at a wide range of concentrations and include (1) initial increase in liver glycogen followed by a prolonged decrease in the freshwater teleost *Colisa fasciatus* exposed to 2,584 mg Mn/L for 96 h (Nath and Kumar, 1987); (2) depletion of muscle glycogen but elevation of blood lactic acid in the freshwater teleost *Colisa fasciatus* exposed to 2,548 mg Mn/L for 96 h (Nath and Kumar, 1988); (3) impairment of net calcium uptake and calcium deposition in the skeleton in brown trout *Salmo trutta* exposed to 3.4 mg Mn/L for 30 days (Reader et al., 1988); (4) impairment of skeletal calcification in brown trout exposed to a mixture of Mn, Al, Fe, Cd and Pb (Reader et al., 1989); and (5) reproductive failure of the American flagfish *Jordanella floridae* exposed to a mixture of Mn, Al, Fe, Ni, Zn, Cu and Pb at pH 5.8 (Hutchinson and Sprague, 1986).

Because manganese is relatively nontoxic and often ameliorates the hazard posed by other metals, most nations have not promulgated guidelines for the protection of freshwater and marine life. A guideline of 0.1 mg/L was proposed by the US Environmental Protection Agency in 1973 for the protection of consumers of marine molluscs, but no similar criterion has been developed in recent years by other nations.

Health Effects

Intake

The typical Western diet yields approximately 3.7 mg Mn/day in a 70-kg reference man. Another 0.002 mg/day comes through inhalation in the nonoccupationally exposed subject. Although oral absorption of manganese in the diet is slow and incomplete, inhaled manganese is rapidly absorbed through the lungs. The total body burden is approximately 12 mg in a 70-kg reference man, of which 5 mg is in the skeleton.

Manganese is an essential trace element in animals, forming part of several important enzyme systems involved in protein and energy metabolism and in mucopolysaccharide formation. The recommended daily allowance for adults is at least 1.2 mg. Insufficient dietary manganese may result in abnormal carbohydrate metabolism and impaired insulin production in humans, and a host of ailments in experimental animals.

Acute Toxicity

Inhalation of large doses of manganese results in localized respiratory necrosis. Acute effects in humans, following oral ingestion of manganese compounds, have not been reported.

Chronic Toxicity

Chronic effects are well known in miners, millworkers, and other occupationally exposed workers, and involve central nervous system toxicity. The neurological symptoms of manganese poisoning are progressive, affect the neurotransmitter function of the central nervous system, and include three stages (Seth and Chandra, 1988):

Prodromal phase:	Apathy, anorexia, insomnia, hallucinations, impaired memory, compulsive actions
Intermediate phase:	Speech disturbance, abnormal gate, altered balance, adiodokinesis, fine tremors
Established phase:	Muscular rigidity in the extremities, irregular gate, fine tremors, excessive sweating

The prodromal symptoms have been reported, on occasion, for subjects consuming manganese-tainted water or fish products, but only in areas where there was insufficient environmental control of mining/industrial wastes.

Carcinogenicity

No data are available on the carcinogenicity of manganese and its compounds.

Drinking Water

Residues

Manganese is routinely detected in finished drinking water, primarily reflecting its presence in raw water. In a survey of the American Water Works Association (1985) of drinking water in 39 states and 3 territories, there were 3,992 cases of noncompliance with the Maximum Contaminant

Limit of 0.05 mg/L. For comparison, fluoride and nitrates were in non-compliance in 907 and 369 episodes, respectively. Ajmal and Uddin (1986a) reported that running tap water at the Aligarh Muslim University (India) contained manganese residues of 0.004–0.051 mg/L, compared to 0.005–0.067 mg/L for standing water. In a related study, Ajmal and Uddin (1986b) noted far higher levels (up to 0.425 mg/L) in water from hand pumps (fed by untreated groundwater) in the city of Aligarh. Mobilization of manganese in groundwater occurs worldwide and is due to reduction (both chemically and bacterially) of Mn^{4+} into the Mn^{2+}-soluble form (Jaudon et al., 1989). Khoe and Waite (1989), working in New South Wales (Australia), found that raw water entering one treatment plant contained total Mn residues of up to 0.35 mg/L, and that in more than 50% of the samples, levels were over 0.05 mg/L; such concentrations had the potential to significantly taint tap water, and required special treatment procedures including flocculation and filtration. Water treatment plants in Rio de Janeiro (Brazil) produced water with a dissolved Mn content of 0.018 mg/L, and a suspended Mn content of 0.015 mg/L (Azcue et al., 1988).

Consumption Guidelines

The primary concerns about manganese in drinking water are its objectionable taste and its capacity to stain plumbing and laundry. The taste of manganese can be readily detected at 0.15 mg/L in drinking water, and encrustation of the distribution system may begin at concentrations as low at 0.02 mg/L. The drinking water guideline/standard of 0.05 mg Mn/L (used by many nations) is based on these aesthetic considerations rather than health concerns. Even at 0.05 mg/L, the intake of manganese from drinking water amounts to 0.10 mg/day, far below the intake from food.

Treatment

Because elevated manganese concentrations are often associated with high iron content, water treatment processes generally aim at control of both agents. Treatment of manganese generally requires conversion of the soluble species into insoluble precipitates, followed by filtration. Manganese can be oxidized, albeit slowly, with chlorine, ozone, or potassium permanganate. An increase in pH to approximately 9 enhances the process of coagulation, using agents such as alum and ferric sulfate.

Recommendations

The main interest in manganese lies in its interaction with other agents, particularly heavy metals. Manganese itself is not acutely toxic to the majority of freshwater and marine species. Although it may taint drinking water, manganese has little or no toxicologic significance in most water supplies.

One of the priorities for research on manganese is its effect on the fate of more toxic metals. Manganese concentrations in surface waters fluctuate seasonally with the redox cycle, so the concomitant effects on the chemistry of toxic metals must also vary. The specific areas of interest are sequestration of metals by manganese and its related compounds, and the oxidation of these same metals on the surface of manganese compounds.

Another priority area is the antagonistic effect manganese has on the toxicity of many heavy metals. At present, relatively little is known about the actual mechanism by which manganese protects plant and animal species in water; there is also an insufficient data base on the effects of seasonal fluxes in manganese on the toxicity of other metals.

The following recommendations reflect these considerations.

1. Environmental fate of sequestered and remobilized metals in the manganese-redox cycle.
2. Oxidation of metals on the surface of MnO_2.
3. Protective mechanism of Mn^{2+} against toxic metals, using a number of plant and animal species (both freshwater and marine).
4. Seasonal flux in manganese and concomitant change in the toxicity of manganese-bound metals.

References

Abaychi, J.K., and Y.Z. Mustafa. 1988. The Asiatic clam, *Corbicula fluminea:* an indicator of trace metal pollution in the Shatt-Arab River, Iraq. *Environmental Pollution* 54:109–122.

Abuzkhar, A.A., A.S. Gibali, Y.I. Elmehrik, and R. Ahmatullah. 1987. Chemical monitoring of sewage wastes for their use in crop production. I. Solid sludge as a soil amendment. *Environmental Monitoring and Assessment* 8:127–133.

Ajmal, M., R. Khan, and A.U. Khan. 1987. Heavy metals in water, sediments, fish and plants of river Hindon, U.P., India. *Hydrobiologia* 148:151–157.

Ajmal, M., and R. Uddin. 1986a. Quality of drinking water in the Aligarh Muslim University campus, Aligarh, U.P. (India) with respect to heavy metals. *Environmental Monitoring and Assessment* 6:195–205.

Ajmal, M., and R. Uddin. 1986b. Studies on heavy metals in the ground waters of the City of Aligarh, U.P. (India). *Environmental Monitoring and Assessment* 6:181–194.

American Water Works Association. 1985. An AWWA survey of inorganic contaminants in water supplies. *Journal of the American Water Works Association* 77:67–72.

Araujo, M.F.D., P.C. Bernard, and R.E. Van Grieken. 1988. Heavy metal contamination in sediments from the Belgian coast and Scheldt Estuary. *Marine Pollution Bulletin* 19:269–273.

Azcue, J.M.P., W.C. Pfeiffer, M. Fiszman, and O. Malm. 1988. Heavy metal removal by different water treatment plants, in Rio de Janeiro State, Brazil. *Environmental Technology Letters* 9:429–436.

Campanella, L., E. Cardarelli, T. Ferri, B.M. Petronio, and A. Pupella. 1987. Evaluation of heavy metals speciation in an urban sludge. I. Batch method. *Science of the Total Environment* 61:217–228.

Capelli, R., V. Minganti, and M. Bernhard. 1987. Total mercury, organic mercury, copper, manganese, selenium, and zinc in *Sarda sarda* from the Gulf of Genoa. *Science of the Total Environment* 63:83–99.

Cordero, R. 1988. *Metal Bulletin's prices and data 1988*. Metal Bulletin Books, Surrey, England. 375 pp.

Davison, W., J. Hilton, J.P. Lishman, and W. Pennington. 1985. Contemporary lake transport processes determined from sedimentary records of copper mining activity. *Environmental Science and Technology* 19:356–360.

Dougherty, G. 1988. Heavy metal concentrations in bivalves from Fiji's coastal waters. *Marine Pollution Bulletin* 19:81–84.

Eary, L.E., and D. Rai. 1987. Kinetics of chromium (III) oxidation to chromium (VI) by reaction with manganese dioxide. *Environmental Science and Technology* 21:1187–1193.

Fillpèk, K.H., D.K. Nordstrom, and W.H. Ficklin. 1987. Interaction of acid mine drainage with waters and sediments of West Squaw Creek in the West Shasta Mining District, California. *Environmental Science and Technology* 21:388–396.

Francis, A.J., C.J. Dodge, A.W. Rose, and A.J. Ramirez. 1989. Aerobic and anaerobic microbial dissolution of toxic metals from coal wastes: mechanism of action. *Environmental Science and Technology* 23:435–441.

Ho, Y.B. 1987. Metals in 19 intertidal macroalgae in Hong Kong waters. *Marine Pollution Bulletin* 18:564–567.

Hutchinson, N.J., and J.B. Sprague. 1986. Toxicity of trace metal mixtures to American flagfish *(Jordanella floridae)* in soft, acidic water and implications for cultural acidification. *Canadian Journal of Fisheries and Aquatic Science* 43:647–655.

Jaudon, P., C. Massiani, J. Galea, and J. Rey. 1989. Groundwater pollution by manganese. Manganese speciation: application to the selection and discussion of an in situ groundwater treatment. *Science of the Total Environment* 84:169–183.

Khoe, G.H., and T.D. Waite. 1989. Manganese and iron related problems in Australian water supplies. *Environmental Technology Letters* 10:479–490.

Legorburu, I., L. Canton, E. Millan, and A. Casado. 1988. Trace metal levels in fish from Urola River (Spain). Anguillidae, Mugillidae and Salmonidae. *Environmental Technology Letters* 9:1373–1378.

Lema, J.M., R. Mendez, and R. Blazquez. 1988. Characteristics of landfill leachates and alternatives for their treatment: a review. *Water, Air, and Soil Pollution* 40:223–250.

Luther, G.W., Z. Wilk, R.A. Ryans, and A.L. Meyerson. 1986. On the speciation of metals in the water column of a polluted estuary. *Marine Pollution Bulletin* 17:535–542.

Martin, T.R., and D.M. Holdich. 1986. The acute lethal toxicity of heavy metals to peracarid crustaceans (with particular reference to fresh-water asellids and gammarids). *Water Research* 20:1137–1147.

Mohapatra, S.P. 1988. Distribution of heavy metals in polluted creek sediment. *Environmental Monitoring and Assessment* 10:157–163.

Naquadat. 1985. *National water quality data bank*. Environment Canada, Ottawa.

Nath, K., and N. Kumar. 1987. Toxicity of manganese and its impact on some aspects of carbohydrate metabolism of a freshwater teleost, *Colisa fasciatus*. *Science of the Total Environment* 67:257–262.

Nath, K., and N. Kumar. 1988. Impact of manganese intoxication on certain parameters of carbohydrate metabolism of a freshwater tropical perch, *Colisa fasciatus*. *Chemosphere* 17:617–624.

Nicolaidou, A., and J.A. Nott. 1989. Heavy metal pollution induced by a ferro-nickel smelting plant in Greece. *Science of the Total Environment* 84: 113–117.

Nolting, R.F. 1986. Copper, zinc, cadmium, nickel, iron and manganese in the Southern Bight of the North Sea. *Marine Pollution Bulletin* 17:113–117.

Nriagu, J.O. 1989. A global assessment of natural sources of atmospheric trace metals. *Nature* 338:47–49.

Nriagu, J.O., and J.M. Pacyna. 1988. Quantitative assessment of worldwide contamination of air, water and soils by trace metals. *Nature* 333:134–139.

Patel, B., V.S. Bangera, S. Patel, and M.C. Balani. 1985. Heavy metals in the Bombay Harbour area. *Marine Pollution Bulletin* 16:22–28.

Poulton, D.J. 1987. Trace contaminant status of Hamilton Harbour. *Journal of Great Lakes Research* 13:193–201.

Reader, J.P., T.R.K. Dalziel, and R. Morris. 1988. Growth, mineral uptake and skeletal calcium deposition in brown trout, *Salmo trutta* L., yolk-sac fry exposed to aluminum and manganese in soft acid water. *Journal of Fish Biology* 32:607–624.

Reader, J.P., N.C. Everall, M.D.J. Sayer, and R. Morris. 1989. The effects of eight trace metals in acid soft water on survival, mineral uptake and skeletal calcium deposition in yolk-sac fry of brown trout, *Salmo trutta* L. *Journal of Fish Biology* 35:187–198.

Richardson, L.L., and K.H. Nealson. 1989. Distributions of manganese, iron, and manganese-oxidizing bacteria in Lake Superior sediments of different organic carbon content. *Journal of Great Lakes Research* 15:123–132.

Sadig, M., and I. Alam. 1989. Metal concentrations in pearl oyster, *Pinctada radiata,* collected from Saudi Arabian coast of the Arabian Gulf. *Bulletin of Environmental Contamination and Toxicology* 42:111–118.

Samanidou, V., and K. Fytianos. 1987. Partitioning of heavy metals into selective chemical fractions in sediments from rivers in northern Greece. *Science of the Total Environment* 67:279–285.

Samhan, O., M. Zarba, and V. Anderlini. 1987. Multivariate geochemical investigation of trace metal pollution in Kuwait marine sediments. *Marine Environmental Research* 21:31–48.

Scoullos, M.J., and J. Hatzianestis. 1989. Dissolved and particulate trace metals in a wetland of international importance: Lake Mikri Prespa, Greece. *Water, Air, and Soil Pollution* 44:307–320.

Seth, P.K., and S.V. Chandra. 1988. Neurotoxic effects of manganese. *In: Metal Neurotoxicity,* eds. S.C. Bondy and K.N. Prasad, 19–33. CRC Press, Boca Raton, FL.

Soderlund, S., A. Forsberg, and M. Pedersen. 1988. Concentrations of cadmium and other metals in *Fucus vesiculosus* L. and *Fontinalis dalecarlica* Br. Eur.

from the northern Baltic Sea and the southern Bothnian Sea. *Environmental Pollution* 51:197–212.

Sridhar, M.K.C. 1986. Trace element composition of *Pistia stratiotes* L. in a polluted lake in Nigeria. *Hydrobiologia* 131:273–276.

Stauber, J.L., and T.M. Florence. 1985. Interactions of copper and manganese: a mechanism by which manganese alleviates copper toxicity to the marine diatom, *Nitzschia closterium* (Ehrenberg) W. Smith. *Aquatic Toxicology* 7:241–254.

Stull, J.K., and R.B. Baird. 1985. Trace metals in marine surface sediments of the Palos Verdes Shelf, 1974 to 1980. *Journal of the Water Pollution Control Federation* 57:833–840.

Sung, J.F.C., A.E. Nevissi, and F.B. Dewalle. 1986. Concentration and removal efficiency of major and trace elements in municipal wastewater. *Journal of Environmental Science and Health* 21:435–448.

Szefer, P. 1986. Some metals in benthic invertebrates in Gdansk Bay. *Marine Pollution Bulletin* 17:503–507.

Tessier, A., P.G.C. Campbell, J.C. Auclair, and M. Bisson. 1984. Relationships between the partitioning of trace metals in sediments and their accumulation in the tissues of the freshwater mollusc *Elliptio complanata* in a mining area. *Canadian Journal of Fisheries and Aquatic Sciences* 41:1463–1472.

US Minerals Yearbooks. 1930–1989. Bureau of Mines, US Department of the Interior, Washington, DC.

Vas, P., and J.D.M. Gordon. 1988. Trace metal concentrations in the scyliorhinid shark *Galeus melastomus* from the Rockall Trough. *Marine Pollution Bulletin* 19:396–398.

V.-Balogh, K., D.S. Fernandez, and J. Salanki. 1988. Heavy metal concentrations of *Lymnaea stagnalis* L. in the environs of Lake Balaton (Hungary). *Water Research* 10:1205–1210.

Vymazal, J. 1984. Short-term uptake of heavy metals by periphyton algae. *Hydrobiologia* 119:171–179.

Wang, W. 1986. Toxicity tests of aquatic pollutants by using common duckweed. *Environmental Pollution* 11:1–14.

Watling, H.R., and R.J. Watling. 1983. Sandy beach molluscs as possible bioindicators of metal pollution. 1. Field survey. *Bulletin of Environmental Contamination and Toxicology* 31:331–338.

Westall, J., and W. Stumm. 1980. The hydrosphere. In: The Handbook of Environmental Chemistry, ed. O. Hutzinger, 17–49. Springer-Verlag, New York.

Zingde, M.D., M.A. Rokade, and A.V. Mandalia. 1988. Heavy metals in Mindhola River estuary, India. *Marine Pollution Bulletin* 19:538–540.

16
Mercury

Mercury, with an average crustal abundance of 0.08 mg/kg, occurs in sedimentary, igneous, and metamorphic rocks. The most important ore is cinnabar (HgS), but there are at least another 30 common ores and gangue minerals that contain mercury in relatively high concentrations. Mercury, undergoing complex chemical reactions in the environment, forms a number of highly toxic organic derivatives. These compounds have been implicated in the adulteration of aquatic resources in numerous countries.

Production, Sources, and Residues

Production

World production of mercury was 3.8×10^3 metric tons in 1930, increasing to 8.3×10^3 metric tons in 1960 but decreasing to 7.1×10^3 metric tons in 1980 (US Minerals Yearbooks, 1930–1989). Production in recent years has continued to remain relatively low, approximately 6×10^3 metric tons/yr, a reflection of the environmental and health problems associated with the use of mercury. The world's leading producers of refined mercury are the USSR, Spain, the USA, China, and Algeria (Table 16.1).

Mercurials have found widespread application in biocides and pharmaceuticals. Because of their stereospecific properties, several oxides, chlorides, and sulfides have been used as catalysts, particularly in the manufacture of synthetic polymers. The manufacture of electrical apparatus and production of chlorine and caustic soda also continue to be major uses of mercury.

Table 16.1. World's leading producers of refined mercury.

Nation	Quantity (metric tons/yr)	Nation	Quantity (metric tons/yr)
USSR	2,200	Spain	1,539
USA	552	China	500
Algeria	400	Peru	372
Turkey	292	Czechoslovakia	148
Finland	125	Yugoslavia	75

Source: Cordero (1988).

Sources

The total amount of mercury discharged from anthropogenic sources to freshwaters amounts to 0.3–8.8 × 10^3 metric tons/yr (Table 16.2). The No. 1 source is discharge from coal-burning power plants, followed by atmospheric fallout from other sources, chemical manufacturing processes, and discharge of municipal wastes (Table 16.2). Hutton and Symon (1986) reported that the total amount of mercury discharged to coastal waters of the United Kingdom came to 38 metric tons annually, principally from river discharge (14 metric tons/yr), dumping of dredged materials (11.2 metric tons/yr), and industrial discharges (9.5 metric tons/yr). Beukema et al. (1986) reported that the total amount of mercury discharged to The Netherlands part of the North Sea was 23 metric tons in 1980, decreasing to a projected rate of 11–12 metric tons in 1990 (Table 16.3).

Atmospheric deposition is the largest nonpoint source of mercury in

Table 16.2. Worldwide anthropogenic input of mercury to surface waters.

Source	Input (thousand metric tons per year)
Coal-burning power plants	0–3.6
Atmospheric fallout	0.22–1.8
Manufacturing processes	
chemicals	0.02–1.5
metals	0–0.75
petroleum products	0–0.02
Domestic wastewater	
central	0–0.18
noncentral	0–0.42
Dumping of sewage sludge	0.01–0.31
Base metal mining and dressing	0–0.15
Smelting and refining	
nonferrous metals	0–0.04

Source: Nriagu and Pacyna (1988).

Table 16.3. Annual input of mercury to The Netherlands' part of the North Sea in 1980 (actual) and 1990 (projected).

Source	Input 1980 (metric tons/yr)	Input 1990 (metric tons/yr)
Total input	23	11–12
Atmospheric deposition	2.3	2.3
Rivers	13	6.2–7.4
Coastal discharges	1.0	0.3
Dredging sludges	6.6	2.1
Industrial wastes	0.1	0
Incineration at sea	0.004	0
Offshore mining	0.07	0.05

Source: Beukema et al. (1986).

both marine and freshwaters. Total anthropogenic emissions of mercury are approximately 3.6×10^3 metric tons/yr worldwide, compared to 2.5×10^3 metric tons/yr for natural sources (Nriagu, 1989). Mercury in rain and snowfall in industrialized/populated areas typically ranges from 10 to 3,400 ng/L (Glass et al., 1986; Ferrara et al., 1986).

Within the anthropogenic emission category, incineration of municipal refuse and the combustion of coal account for >75% of total emissions (Nriagu and Pacyna, 1988). Residues as high as 5.2 mg/m^3 can be found in scrubbed emissions, producing a mercury discharge rate of approximately 2 g Hg per metric ton of waste incinerated (examples in Table 16.4). Similarly, the mean mercury content of coal typically ranges from 0.2 to 0.6 mg/kg (reviewed by Mitra, 1986). About 90% of this mercury is released to the atmosphere on combustion, the remainder staying with the ash. Relatively high concentrations of mercury (approx. 0.5 mg/kg) have also been found in oil and other petroleum products (Patterson et al., 1987).

At one time, huge amounts of mercury were released from chlorine-

Table 16.4. Mercury residues in emissions from incinerators and power stations.

Source	Location	Residue
Garbage[1]	City of Paris	3.1–5.2 mg/m^3
Refuse-fired power station[2]	Bamberg, FRG	0.05–0.08 mg/m^3
Municipal waste incinerators[3]	Sweden	2 g/metric ton of waste
Municipal waste incinerator[4]	Stockholm	0.5–4.9 g/metric ton of waste

Sources: [1]Gounon and Milhau (1986), [2]Reimann (1986), [3]Lindqvist (1986), [4]Westergard (1986).

manufacturing facilities. Mercury in such plants is used according to the following equations:

$$NaCl\ (solution) \longrightarrow Cl_2\ (anode) + NaHg_x\ (cathode) \quad (1)$$

$$NaHg_x + H_2O \longrightarrow NaOH + xHg + (1/2)H_2 \quad (2)$$

A typical facility uses up to 5,000 kg Hg in a 30 m^2 Hg cell. Although the mercury is contained within a closed loop, there are usually losses, which need to be closely regulated to avoid contamination of surface waters. In extreme cases, mercury in the effluents of chlor-alkali plants has reached 2,000 mg/L (Shiber et al., 1978). Most nations have now promulgated control standards for chlorine-producing industries, greatly reducing the amount of mercury in the waste water (e.g., Drabkin and Rissmann, 1988).

Mercury residues are often relatively high in municipal waste waters and sludges.In a study of 25 sewage treatment plants in the state of Washington, Sung et al. (1986) reported that concentrations in primary sludge ranged up to 1.05 mg/L whereas residues in the liquid effluent reached a maximum of only 0.005 mg/L (Table 16.5). In another study, Marcovecchio et al. (1986) showed that there was a strong positive correlation between the concentration of mercury in the sediments (ranging up to 1.46 mg/kg dry weight) of a coastal canal in Argentina and the distance from the sewage outfall from the city of Bahia Blanca; a significant correlation was also observed between mercury residues and the silt/clay and protein content of the sediments. On the other hand, mercury concentrations in urban runoff are often extremely low ($<$50 ng/L), reflecting the restricted use of mercury in most general applications (Marsalek and Schroeter, 1988).

Mercury input to surface waters from mining, smelting, and other industries can be extremely high if appropriate regulatory controls are not

Table 16.5. Concentration (mg/L) of total mercury in wastes from 25 treatment plants in the state of Washington.

Sample	Average	Range
Raw sewage	0.007	ND[a]–0.088
Primary effluent	0.001	ND–0.010
Secondary effluent	0.001	ND–0.020
Discharge	0.001	ND–0.005
Primary sludge	0.191	0.060–1.050
Secondary sludge	0.044	0.019–0.075

[a]Not determined.
Source: Sung et al. (1986).

in place. This is partially due to the presence of mercury in ores and other raw products, plus the release of mercury used in the industrial processes. For example, Pfeiffer and De Lacerda (1988) reported that total losses of mercury to the Brazilian Amazon ecosystem from gold mining came to 1.32 kg Hg/kg Au. Of this, 45% was released to rivers and 55% to the atmosphere. Similarly, Semu et al. (1986), working near a battery factory in Dar es Salaam (Tanzania), reported that mercury residues in liquid effluent ranged from <0.2 to 5.2 mg/L while concentrations in soil near the factory were as high as 472 mg/kg.

Residues

Water. Concentration of dissolved Hg is typically low in unpolluted freshwaters, ranging from 10 to 100 ng/L, and in the open ocean from <10 to 30 ng/L. A significant portion of mercury is associated with suspended solids (see Chemistry section) and accounts for a major part of the downstream transport of mercury in rivers. Only a few studies are available on the concentration of methyl Hg in surface waters, reflecting the difficulties involved in separating and determining extremely low concentrations of such complexes. In one study, Schintu et al. (1989) found that methyl Hg accounted for 22–37% of total Hg (Table 16.6). Much higher proportions are found in aquatic biota (see Bioaccumulation section).

Sediment. Total Hg residues in sediments typically range from 0.1 to 0.5 mg/kg dry weight in unpolluted waterways, increasing to 5 mg/kg and higher in anthropogenically contaminated areas (Table 16.7). Although elevated residues are generally associated with the discharge of industrial and municipal wastewaters, Johnson et al. (1986) showed that atmospheric loadings to a series of lakes in Ontario (Canada) were several times greater than the background accumulation rates in sediments. In a related study, Johnson (1987) reported that anthropogenic loadings to 14 Ontario lakes were 1.8–2.6 times greater than background loadings; in that investigation, inputs came almost exclusively from precipitation and dry deposition.

Table 16.6. Concentration (ng/L) of mercury in surface waters near Ottawa, Canada.

Location	Inorganic Hg	Methyl Hg	Methyl Hg (%)
Ottawa River	6.0–6.7	1.7–2.3	22–26
Gatineau River	7.3	4.1	36
Black Lake	2.1	1.3	37

Source: Schintu et al. (1989).

Table 16.7. Concentration (mg/kg dry weight) of total mercury in freshwater and marine sediments.

Location	Average (range)	Polluting source
Estuary, Bombay Island[1]	2.48 (0.17–7.00)	chemical industries
Coastal waters, Bay of Bengal[2]	1.95 (0.75–55.2)	chlor-alkali plants
Arabian Gulf[3]	0.04 (0.02–0.12)	natural geological
Palos Verdes Shelf, Los Angeles[4]	2.0 (0.2–3.2)	municipal wastes
Lake Ontario, depositional basins[5]	NR[a] (0.14–3.95)	multiple industrial
Lake Erie, depositional basins[5]	NR (0.05–4.8)	multiple industrial
Lake Huron, depositional basins[5]	NR (0.01–0.81)	natural geologic, multiple industrial
Albegna River, Italy[6]	NR (63.5–688)	industrial, municipal

[a]Not reported.
Sources: [1]Mahajan and Srinivasan (1988), [2]Sasamal et al., (1987), [3]Samhan et al. (1987), [4]Stull and Baird (1985), [5]Mudroch et al. (1988), [6]Batti et al. (1975).

Chemistry

Mercury exists in three oxidation states in surface waters: Hg^o, Hg^+, and Hg^{2+}. In well-aerated waters (Eh > 0.5 V), Hg^{2+} should dominate, whereas under reducing conditions, Hg^o may develop. Ramamoorthy et al. (1983) also reported that Hg^o may be produced biologically and/or abiologically in microenvironments associated with suspended particulate matter. An excess of the sulfide ion stabilizes bivalent mercury as hydrosulfide or sulfide complexes, even at low redox potentials.

Mercury readily binds to a large number of inorganic and organic ligands. Among the inorganic anions, Hg^{2+} forms the strongest covalent bonds with the chloride ion. The mercuric ion hydrolyzes with $Hg(OH)_2$ to form the dominant species at pH > 4, assuming that the aqueous concentration of chloride is $< 10^{-5}$ *M*. When chloride occurs at higher concentrations (approx. 0.01 *M*), the region of predominance of $Hg(OH)_2$ shifts to pH > 6.

Mercury also forms stable complexes with a variety of organic ligands, particularly those containing S, such as cysteine. The next strongest are with amino acids and hydroxycarboxylic acids. Depending on environmental conditions, complexation with organic ligands may dominate complexation with inorganics such as chloride. In one study, Lovgren and Sjoberg (1989) examined the complexation of Hg^{2+} in acidic bog water with a number of organic and inorganic ligands. Organic complexes dominated the mercury speciation within the pH range 3 to 5 when organic material occurred in excess in the water. A number of complexes were formed, including a ternary species identified as $H^+–HgCl_2–H_2–H_2L–Cl^-$. Other studies have shown that competition with chloride enhances desorption, assuming that pH and the concentration of other ligands remain constant (Lodenius et al., 1987).

Inorganic Hg can be methylated in the environment to form highly soluble, toxic species. These compounds are readily absorbed and concentrated by aquatic plants and animals (see Bioaccumulation). The rate of methylation is typically greatest in the surficial sediments, and decreases with increasing sediment depth (Callister and Winfrey, 1986). Although methylation has also been recorded in the water column, the rate is relatively slow, typically near the level of detection. A number of factors influence the rate of methylation, the most important being temperature, the sulfide content of the sediments, and dissolved oxygen levels (Table 16.8). The effect of sulfide is represented as follows (Craig and Moreton, 1986):

$$2CH_3Hg^+ + S^{2-} \longrightarrow (CH_3Hg)_2S \longrightarrow (CH_3)_2Hg + HgS$$

Bacteria, molds, and fungi have all been implicated in the methylation of inorganic compounds. The concentration of methyl Hg typically represents <2% of total Hg in the water column and sediments.

The dominant fate process for mercury in both marine and freshwaters is adsorption to suspended solids and sediments. Ramamoorthy and Rust (1976) noted that the sorption maximum was correlated with surface area > organic content > cation exchange capacity > grain size, whereas the bonding constants followed the order organic content > grain size > cation exchange capacity > surface area. The association of mercury with sediments ranges from relatively weak Van der Waals forces to strong covalent bonding, coprecipitation with iron–manganese oxides, and incorporation within crystal lattices. In many waters, mercury binding is predominantly associated with sulfur sites.

Desorption is a slow process, facilitated to a small degree by low pH of the overlying water, Eh, and the quantity of chloride and other ligands

Table 16.8. Factors influencing the rate of methylation of inorganic mercury.

Factor	Response
Temperature	Increases rate to a maximum at 35°C
Sulfide content of sediment	Decreases rate at sulfide levels of 0.9–7.1 mg/g dry weight of sediment
Dissolved oxygen	Decreases rate with increasing oxygenation
Organic content of sediment	Increases rate with increasing nitrogen content of sediment
Chloride content of sediment and water	Decreases rate under estuarine conditions
pH	Increases rate with decreasing pH

Sources: Berman and Bartha (1986), Callister and Winfrey (1986), Nagase et al. (1984).

in the water and sediments (Bjornberg et al., 1988). Because inorganic Hg and methyl Hg are soft acids, they prefer sulfide complexation. These complexes remain largely intact upon transport to estuarine waters, even though chloride is in excess.

Bioaccumulation

Plants

Methyl Hg concentrates through all levels of the food chain whereas inorganic Hg residues increase only modestly at the higher trophic levels. Among aquatic plants, the greatest residues are typically recorded for marine species that are older than 1 year. For example, total Hg in *Ascophyllum nodosum* from Hardangerfjord (Norway) averaged 3.1 mg/kg dry weight and ranged up to 20 mg/kg, a result of waste water discharges from a metal smelter (Myklestad et al., 1978). Much lower residues were found in deep-water seaweeds in the western North Atlantic Ocean, where there was no point source of mercury (Table 16.9). A rural Welsh river receiving mine drainage contained the moss *Fontinalis squamosa* with total Hg residues ranging from 0.12 to 0.48 mg/kg dry weight (Mason and Macdonald, 1988).

Uptake of mercury from water by plants is usually extremely rapid and efficient. In an experimental study, Mo et al. (1989) showed that duckweed *Lemna minor* accumulated residues of up to 2,000 mg/kg in only 3 days. Although the rate of uptake was little affected by pH of 4 and 5, the presence of the copper ion and EDTA suppressed uptake. Humic acid (concentration 25 mg/L) also reduced uptake by duckweed, a reflection of complexation with inorganic Hg. Several other studies (e.g., Darnall et al., 1986) have shown that relatively high concentrations reported for microscopic species such as *Chlorella* were due to adsorption of mercury onto the cell wall rather than absorption.

Table 16.9. Concentration (mg/kg dry weight) of total mercury in four seaweeds from the western North Atlantic Ocean.

Species	Average (range)	Species	Average (range)
Ptilota serrata	0.014 (0.005–0.027)	*Polysiphonia urceolata*	0.007 (ND[a])
Phycodrys rubens	0.34 (0.006–0.64)	*Laminaria saccharina*	0.03 (ND)

[a]No data.
Source: Sears et al. (1985).

Invertebrates

Although total Hg in invertebrates is highly variable, residues are typically <1 mg/kg in both marine and freshwaters (Table 16.10). Methyl Hg concentrations are also variable and depend on a number of factors including the age and size of the organism. For example, Mohlenberg and Riisgard (1988) showed that the percentages of methyl Hg in 2-, 3-, and 4-year old cockles (*Cardium edule* and *Cardium glaucum*) were 30%, 60%, and 90% of total Hg, respectively. These specimens were collected from a chronically polluted area (western Limfjord) in Denmark. Another study on two lakes and three reservoirs in Finland showed that methyl Hg accounted for 89% of total Hg in zoobenthos and 84% in zooplankton (Surma-Aho et al., 1986).

Methyl Hg is sorbed at a much faster rate from both food and water than inorganic Hg. Accumulation rates are temperature-dependent within the thermal optimum of the species, and generally increase at acidic pH (Rodgers et al., 1987). The rate of sorption also depends on the concentration of ligands, both organic and inorganic, in the water.

Fish

Mercury has caused more problems to the consumers of fish than any other inorganic contaminant. In extreme cases, consumption of mercury-tainted fish has led to the onset of a serious neurological disease, termed

Table 16.10. Concentration (mg/kg wet weight) of total mercury in the soft tissues of invertebrates.

Species	Average (range)	Location
Freshwater crayfish, 5 species[1]	0.117 (0.015–0.488)[a]	15 lakes, Ontario, Canada
Gastropod, *Lymnaea stagnalis*[2]	1.05 (0.2–3.0)[a]	Lake Balaton, Hungary
Benthic invertebrates, multiple species[3]	0.089 (0.008–0.472)	2 lakes and 3 reservoirs, Finland
Clam, *Tapes decussatus*[4]	0.98 (0.78–1.18)	Huelva Estuary, Spain
Oysters, *Crassostrea angulata*[4]	0.56 (0.02–0.90)	Huelva Estuary, Spain
Prawns, 9 species[5]	NR[b] (<0.004–0.17)	Coastal waters, India
Grass shrimp, *Palaemonetes pugio*[6]	NR (0.1–0.4)	2 estuaries, New Jersey

[a]Dry weight.
[b]Not reported.
Sources: [1]Allard and Stokes (1989), [2]V.-Balogh et al. (1988), [3]Surma-Aho et al. (1986), [4]Lopez-Artiguez et al. (1989), [5]Sanzgiry et al. (1988), [6]Khan et al. (1989).

Minamata disease (see Health Effects). In other cases, entire fisheries have been either restricted or significantly curtailed because of mercury contamination.

Almost all (>95%) of the mercury in the edible tissues of fish is methylated, and since methyl Hg is rapidly concentrated, residues are often relatively high, even in areas where anthropogenic inputs are minimal. Most nations apply a methyl Hg standard for the consumption of fish; depending on country, that standard ranges from 0.1 to 1.0 mg/kg wet weight. The lower values are generally applied in areas where fish comprises a staple part of the diet.

Methyl Hg residues typically increase in fish tissues following impoundment of rivers. This phenomenon has been reported in many parts of the world, yielding tissue concentrations in fish of 0.5 to >3 mg/kg wet weight, depending on species and location (Bodaly et al., 1984; Bruce and Spencer, 1979; Bodaly and Hecky, 1979). Apparently, flooding of the freshly inundated soil enhances the activity of methylating bacteria; however, reducing conditions eventually develop in the reservoirs, decreasing the rate of methylation and the corresponding concentrations in fish tissues. It should be noted that mercury residues do not always increase following impoundment; if inorganic Hg is strongly complex with sulfides, or if redox conditions remain unsatisfactory, mercury levels in fish show little if any increase.

Toxic Effects to Aquatic Organisms

Plants

All mercurials are highly toxic to plants, both marine and freshwater. Growth inhibition of the green alga *Chlorella vulgaris* and the blue-green alga *Anabaena flos-aquae* has been observed following exposure to methylmercuric chloride in the 0.001 to 0.010 mg Hg per liter range (reviewed by US Environmental Protection Agency, 1984). Most species are less sensitive to inorganic Hg; in fact, Mora and Fabregas (1980) reported that the minimum toxic concentration of $HgSO_4$ was 0.15–0.20 mg Hg/L.

Several factors ameliorate the toxicity of mercurials to algae. These factors include sequestration by organic chelators and partitioning by suspended solids. Nickel, selenium, and probably other metals reduce toxicity, likely through reducing the active uptake of mercury, whereas cadmium and zinc apparently act synergistically with mercury in their toxicity to algae. An excess of H^+ ions may also enhance toxic effects in some species (Campbell and Stokes, 1985).

Invertebrates

Although the acute toxicity of inorganic Hg to marine and freshwater invertebrates is highly variable, most reported LC_{50}s are < 0.05 mg/L. Microscopic species, such as protozoans and rotifers, are often most susceptible, reflecting their high surface volume:size ratio, which allows a relatively large amount of mercury to enter the cells. On the other hand, some species, such as the crustacean *Asellus aquaticus*, are less sensitive, with LC_{50}s ranging from 0.1 to 0.5 mg/L (Martin and Holdich, 1986). Organic Hg is highly toxic to most species, with LC_{50}s generally falling below 0.01 mg/L.

A large number of chronic effects have been reported for invertebrates exposed to both inorganic and organic Hg. These include (1) degenerative changes in the hepatopancreas and gill of the intertidal crab *Scylla serrata* exposed to phenyl mercuric acetate at 0.32 mg Hg/L (Krishnaja et al., 1987); (2) increase in tissue lactic acid and decrease in glycogen levels in the estuarine clam *Villorita cyprinoides* exposed to mercuric chloride at 0.6 mg Hg/L for 48 h (Satyhyanathan *et al.*, 1988); and (3) loss of weight in *Mytilus edulis* exposed to two organic mercurials at 0.003 mg Hg/g wet weight for 32 days (Pelletier, 1988).

Selenium is often implicated in the protection of aquatic invertebrates by selectively binding with mercury within tissues. This phenomenon has been noted for both inorganic Hg and organic Hg complexes (Pelletier, 1988; Micallef and Tyler, 1987). Other agents, such as organic ligands and Ca^{2+}, also ameliorate toxicity.

Fish

Organic Hg is considerably more toxic to fish, both marine and freshwater, than inorganic Hg. One of the earliest reports (Wobeser, 1975) indicated that the 24-h median tolerance limit of fingerlings to rainbow trout *Oncorhynchus mykiss* to methyl mercuric chloride was 0.125 mg/L whereas the corresponding value for mercuric chloride was 0.90 mg/L. Numerous other studies (reviewed by Mance, 1987) have placed the acute LC_{50} of inorganic Hg and organic Hg to both marine and freshwater fish in the 0.1–1.0 mg/L and 0.01–0.1 mg/L range, respectively.

A number of chronic effects have been reported in fish following long-term exposure to mercurials: (1) reduction in sperm motility in mummichog *Fundulus heteroclitus* exposed to mercuric chloride at 0.05 mg Hg/L for 2 min (Khan and Weis, 1987); (2) inhibition of Ca^{2+}-ATPase in catfish *Ictalurus punctatus* exposed to mercuric chloride 10 μM Hg/L (Reddy et al., 1988); (3) reduced rate of fertilization in rainbow trout *Oncorhynchus mykiss* exposed to mercuric chloride at >1 mg Hg/L (Billard and Roubaud, 1985); and (4) reduction in electric organ discharge in the weakly electric fish *Gnathonemus petersi* exposed to mercuric chloride at 0.5 mg Hg/L (Geller, 1984).

Many nations have promulgated guidelines/standards for the protection of aquatic organisms.

Canada
0.1 μg/L (total Hg)

USA
  0.025 μg/L (total Hg, 4-day average, not exceeded more than once every 3 years)
  2.1 μg/L (total Hg, 1-day average, not exceeded more than once every 3 years)
European Community
  0.2 μg/L (dissolved Hg, freshwater)
  0.3 μg/L (dissolved Hg, salt water)

These regulations are aimed at limiting both the accumulation of mercury in biological tissues and potential acute/chronic affects related to the discharge of mercury to receiving waters.

Health Effects

Intake

The typical Western diet yields approximately 0.015 mg Hg/day in a 70-kg reference man. An additional 0.001 mg/day comes from inhalation in the nonoccupationally exposed population. Once absorbed, mercury is generally distributed about the body, binding to sulfhydryl groups in many proteins. The soft tissue body burden is approximately 13 mg; kidneys are the major site of deposit following exposure to inorganic salts whereas kidney and brain are primary depots after mercury vapor exposure. Methyl Hg is found mainly in the erythrocytes and brain.

Elimination of methyl Hg follows first-order whole-body kinetics, with elimination occurring primary through the feces as as methyl Hg–glutathione complex. The biological half-life of methyl Hg is long and species-dependent. In humans, the $T_{1/2}$ is approximately 70 days, leading to accumulation of methyl Hg in tissues.

Acute Toxicity

Acute mercury poisoning is usually characterized by pharyngitis, abdominal pain, nausea, vomiting, bloody diarrhea, and shock. Nephritis, anuria, and hepatitis occur, followed by death from gastrointestinal and/or kidney lesions.

Chronic Toxicity

Humans exposed to excessive levels of methyl Hg may develop central nervous system symptoms collectively known as Minamata disease. The histopathologic changes seen in the cerebral and cerebellar cortices of the brain are particularly pronounced in the developing fetus and less so in adults (Komulainen, 1988). In the most severe cases, the brain becomes atrophic and the cortex is spongious owing to necrosis of the neurons.

Any group of people who eat large amounts of fish, such as native Indians, are potentially susceptible to methyl Hg intoxication. Analysis of mercury levels in blood and hair is widely used for populations at risk (Canadian Environmental Control Newsletter, 1986; Gonzalez et al., 1985). Tamashiro et al. (1987) used such techniques in evaluating the health of residents of five coastal towns in Japan; it was found that, although mercury levels in hair were three to six times greater than those of controls, there was no concomitant impact on life span. In a similar study, Valciukas et al. (1986) demonstrated that Mohawk Indians in New York State carried relatively low levels of mercury in their hair and blood, despite the consumption of large amounts of fish.

Carcinogenicity

Although methyl Hg may cause chromosome aberrations, there is inadequate evidence for carcinogenicity in animals and humans.

Drinking Water

Residues

Total Hg residues in drinking water are generally low, typically well below 0.005 mg/L. In a survey of systems from three Canadian provinces, concentrations averaged 0.0002–0.0003 mg/L (Canadian Water Quality Guidelines, 1987). Similarly, essentially all of the water wells tested in the USA contained residues below 0.002 mg/L (US Environmental Protection Agency, 1989).

Consumption Guidelines

The current drinking water guideline for total Hg in several nations is 0.001 mg/L. The World Health Organization also recommends that value, whereas a guideline of 0.002 mg/L is in place in the United States. Both guidelines were developed using the data of Andres (1984), Bernaudin et al. (1981), and Druet et al. (1978), which showed that the Lowest-Observed-Adverse-Effect Level in rats was 3 mg/kg. Using a factor of 0.74 to adjust the exposure from the mercuric ion to mercuric chloride in a mg/kg/day dose, the Reference Dose (RD) was calculated as follows:

$$RD = (3 \text{ mg/kg})(0.74)/(7 \text{ days})(1{,}000) = 0.0003 \text{ mg/kg/day}$$

The Drinking-Water-Equivalent Level (DWEL) was then calculated as

$$DWEL = (0.0003 \text{ mg/kg/day})(70 \text{ kg})/(2 \text{ L/day}) = 0.011 \text{ mg/L}$$

where 70 kg refers to a standard reference man and 2 L is the average water consumption of that same man.

The final guideline was calculated as

$$\text{Guideline} = (0.01\ \text{mg/L})(10\%) = 0.001\ \text{mg/L}$$

where 10% is the estimated dose of mercury from drinking water. The discrepancy between the recommendations of 0.001 and 0.002 mg/L among nations is due to differences in the estimated dose of mercury from drinking water (10% vs. 20%).

Treatment

Coagulation and filtration are moderately effective in removing inorganic Hg from drinking water. Ferric sulfate coagulation has achieved 97% removal at pH 8 and 66% removal at pH 7 from water containing 0.05 mg Hg/L (US Environmental Protection Agency, 1989). The corresponding percentages using alum are 38 and 47, respectively. Although coagulation/filtration is less effective in the removal of organic Hg, several studies have shown that powdered activated carbon is a useful pretreatment step which enhances overall removal efficiency.

Lime softening is ineffective in the removal of organic Hg but is moderately effective for the treatment of inorganic Hg. Other technologies, such as reverse osmosis and ion exchange, hold promise for effective treatment and are currently used on small-scale systems.

Recommendations

If left uncontrolled, mercury probably poses a greater environmental and health risk than any other inorganic contaminant. It concentrates rapidly in biological tissues and is highly toxic and persistent. Most nations now recognize the threat posed by mercury and have enacted tight regulations regarding its use and discharge.

Because the threat of mercury has been known for many years, all major chemical and toxicological problems have been well researched. This means that there are, in fact, no pressing research priorities for mercury. However, it is important to continue intensive monitoring of residues in water, sediments, and food products, simply to ensure that environmental concentrations remain low. The monitoring priorities are as follows:

1. Concentration of methyl Hg in surface waters.
2. Concentration of methyl Hg and total Hg in sediments.
3. Concentration of methyl Hg in fish tissues and other food products.

References

Allard, M., and P.M. Stokes. 1989. Mercury in crayfish species from thirteen Ontario lakes in relation to water chemistry and smallmouth bass (*Micropterus dolomieui*) mercury. *Canadian Journal of Fisheries and Aquatic Sciences* 46:1040–1046.

Andres, P. 1984. IgA-IgG disease in the intestine of brown Norway rats ingesting mercuric chloride. *Clinical Immunology and Immunopathology* 30:488–494.

Batti, R., R. Magnaval, and E. Lanzola. 1975. Methylmercury in river sediments. *Chemosphere* 1:13–14.

Berman, M., and R. Bartha. 1986. Control of the methylation process in a mercury-polluted aquatic sediment. *Environmental Pollution (Ser. B)* 11:41–53.

Bernaudin, J.F., E. Druet, P. Druet, and R. Masse. 1981. Inhalation or ingestion of organic or inorganic mercurials produces auto-immune disease in rats. *Clinical Immunology and Immunopathology* 20:129–135.

Beukema, A.A., G.P. Hekstra, and C. Venema. 1986. The Netherlands' environment policy for the North Sea and Wadden Sea. *Environmental Monitoring and Assessment* 7:117–155.

Billard, R., and P. Roubaud. 1985. The effect of metals and cyanide on fertilization in rainbow trout (*Salmo gairdneri*). *Water Research* 19:209–214.

Bjornberg, A., L. Hakanson, and K. Lundbergh. 1988. A theory on the mechanisms regulating the bioavailability of mercury in natural waters. *Environmental Pollution* 49:53–61.

Bodaly, R.A., and R.E. Hecky. 1979. *Post-impoundment increases in fish mercury levels in the Southern Indian Lake reservoir, Manitoba*. Fisheries and Marine Service, Manuscript Report No. 1531, Department of Fisheries and Oceans, Winnipeg: Manitoba. 15 pp.

Bodaly, R.A., R.E. Hecky, and R.J.P. Fudge. 1984. Increases in fish mercury levels in lakes flooded by the Churchill River diversion, northern Manitoba. *Canadian Journal of Fisheries and Aquatic Sciences* 41:682–691.

Bruce, W.J., and K.D. Spencer. 1979. Mercury levels in Labrador fish, 1977–78. *Canadian Industry Report of Fisheries and Aquatic Sciences* 111:1–12.

Callister, S.M., and M.R. Winfrey. 1986. Microbial methylation of mercury in upper Wisconsin river sediments. *Water, Air, and Soil Pollution* 29:453–465.

Campbell, P.G.C., and P.M. Stokes. 1985. Acidification and toxicity of metals to aquatic biota. *Canadian Journal of Fisheries and Aquatic Sciences* 42:2034–2049.

Canadian Environmental Control Newsletter. 1986. Natives in northern Quebec exposed to high mercury levels. *Canadian Environmental Control Newsletter* 317:2625.

Canadian Water Quality Guidelines. 1987. Canadian Council of Resource and Environment Ministers. Environment Canada, Ottawa.

Cordero, R. 1988. *Metal Bulletin's prices and data 1988*. Metal Bulletin Books, Surrey, England, 375 pp.

Craig, P.J., and P.A. Moreton. 1986. Total mercury, methyl mercury and sulphide levels in British estuarine sediments. III. *Water Research* 1111–1118.

Darnall, D.W., B. Greene, M.T. Henzl, J.M. Hosea, R.A. McPherson, J. Sneddon, and M.D. Alexander. 1986. Selective recovery of gold and other metal ions from an algal biomass. *Environmental Science and Technology* 20:206–208.

Drabkin, M., and E. Rissmann. 1988. *Waste minimization audit report: case studies of minimization of mercury-bearing wastes at a mercury cell chloralkali plant*. US Environmental Protection Agency, EPA/600/2–88/011, Cincinnati, OH.

Druet, P., E. Druet, F. Potdevin, and C. Sapin. 1978. Immune type of glomerulonephritis induced by $HgCl_2$ in the brown Norway rat. *Annals of Immunology* 129C:777–792.

Ferrara, R., B. Maserti, A. Petrosino, and R. Bargagli. 1986. Mercury levels in rain and air and the subsequent washout mechanism in a central Italian region. *Atmospheric Environment* 20:125–128.

Geller, W. 1984. A toxicity warning monitor using the weakly electric fish, *Gnathonemus petersi. Water Research* 18:1285–1290.

Glass, G.E., E.N. Leonard, W.H. Chan, and D.B. Orr. 1986. Airborne mercury in precipitation in the Lake Superior region. *Journal of Great Lakes Research* 12:37–51.

Gonzalez, M.J., M.C. Rico, L.M. Hernandez, and G. Baluja. 1985. Mercury in human hair: a study of residents in Madrid, Spain. *Archives of Environmental Health* 40:225–228.

Gounon, J., and A. Milhau. 1986. Analysis of inorganic pollutants emitted by City of Paris garbage incineration plants. *Waste Management and Research* 4:95–104.

Hutton, M., and C. Symon. 1986. The quantities of cadmium, lead, mercury and arsenic entering the U.K. environment from human activities. *Science of the Total Environment* 57:129–150.

Johnson, M.G. 1987. Trace element loadings to sediments of fourteen Ontario lakes and correlations with concentrations in fish. *Canadian Journal of Fisheries and Aquatic Sciences* 4:3–13.

Johnson, M.G., L.R. Culp, and S.E. George. 1986. Temporal and spatial trends in metal loadings to sediments of the Turkey Lakes, Ontario. *Canadian Journal of Fisheries and Aquatic Sciences* 43:754–762.

Khan, A.T., and J.T. Weis. 1987. Toxic effects of mercuric chloride on sperm and egg viability of two populations of mummichog, *Fundulus heteroclitus. Environmental Pollution* 48:263–273.

Khan, A.T., J.S. Weis, and L. D'Andrea. 1989. Bioaccumulation of four heavy metals in two populations of grass shrimp, *Palaemonetes pugio. Bulletin of Environmental Contamination and Toxicology* 42:339–343.

Komulainen, H. 1988. Neurotoxicity of methylmercury: cellular and subcellular aspects. *In: Metal neurotoxicity,* eds. S.C. Bondy and K.N. Prasad, 167–182. CRC Press, Boca Raton, FL.

Krishnaja, A.P., M.S. Rege, and A.G. Joshi. 1987. Toxic effects of certain heavy metals (Hg, Cd, Pb, As, and Se) on the intertidal crab *Scylla serrata. Marine Environmental Research* 21:109–119.

Lindqvist, O. 1986. Fluxes of mercury in the Swedish environment: contributions from waste incineration. *Waste Management and Research* 4:35–44.

Lodenius, M., A. Seppanen, and S. Autio. 1987. Leaching of mercury from peat soil. *Chemosphere* 16:1215–1220.

Lopez-Artiguez, M., M.L. Soria, and M. Repetto. 1989. Heavy metals in bivalve molluscs in the Huelva Estuary. *Bulletin of Environmental Contamination and Toxicology* 42:634–642.

Lovgren, L., and S. Sjoberg. 1989. Equilibrium approaches to natural water systems, 7. Complexation reactions of copper (II), cadmium (II) and mercury (II) with dissolved organic matter in a concentrated bog-water. *Water Research* 23:327–332.

Mahajan, B.A., and M. Srinivasan. 1988. Mercury pollution in the estuarine region around Bombay Island. *Environmental Technology Letters* 9:331–336.

Mance, G. 1987. *Pollution threat of heavy metals in aquatic environments*. Elsevier, London. 372 pp.

Marcovecchio, J.E., R.J. Lara, and E. Gomez. 1986. Total mercury in marine sediments near a sewage outfall. Relation with organic matter. *Environmental Technology Letters* 7:501–507.

Marsalek, J., and H. Schroeter. 1988. Annual loadings of toxic contaminants in urban runoff from the Canadian Great Lakes. *Water Pollution Research Journal of Canada* 23:360–378.

Martin, T.R., and D.M. Holdich. 1986. The acute lethal toxicity of heavy metals to peracarid crustaceans (with particular reference to fresh-water asellids and gammarids). *Water Research* 20:1137–1147.

Mason, C.F., and S.M. Macdonald. 1988. Metal contamination in mosses and otter distribution in a rural Welsh river receiving mine drainage. *Chemosphere* 17:1159–1166.

Micallef, S., and P.A. Tyler. 1987. Preliminary observations of the interactions of mercury and selenium in *Mytilus edulis*. *Marine Pollution Bulletin* 18:180–185.

Mitra, S. 1986. *Mercury in the ecosystem*. Trans Tech Publications, Lancaster, PA. 327 pp.

Mo, S.C., D.S. Choi, and J.W. Robinson. 1989. Uptake of mercury from aqueous solution by duckweed: the effects of pH, copper and humic acid. *Journal of Environmental Science and Health* 24:135–146.

Mohlenberg, F., and H.U. Riisgard. 1988. Partitioning of inorganic and organic mercury in cockles *Cardium edule* (L.) and *C. glaucum* (Bruguiere) from a chronically polluted area: influence of size and age. *Environmental Pollution* 55:137–148.

Mora, B., and F. Fabregas. 1980. The effect of inorganic and organic mercury on the growth kinetics of *Nitzschia acicularis* W. Sm. and *Tetraselmis suecica* Butch. *Canadian Journal of Microbiology* 26:930–937.

Mudroch, A., L. Sarazin, and T. Lomas. 1988. Summary of surface and background concentrations of selected elements in the Great Lakes sediments. *Journal of Great Lakes Research* 14:241–251.

Myklestad, S., I. Eidie, and S. Melsom. 1978. Exchange of heavy metals in *Ascophyllum nodosum* (L.) Le Jol in situ by means of transplanting experiments. *Environmental Pollution* 16:277–284.

Nagase, H., Y. Ose, T. Sato, and T. Ishikawa. 1984. Mercury methylation by compounds in humic material. *Science of the Total Environment* 32:147–156.

Nriagu, J.O. 1989. A global assessment of natural sources of atmospheric trace metals. *Nature* 338:47–49.

Nriagu, J.O., and J.M. Pacyna. 1988. Quantitative assessment of worldwide contamination of air, water and soils by trace metals. *Nature* 333:134–139.

Patterson, J.H., L.S. Dale, and J.F. Chapman. 1987. Trace element partitioning during the retorting of Julia Creek oil shale. *Environmental Science and Technology* 21:490–494.

Pelletier, E. 1988. Acute toxicity of some methylmercury complexes to *Mytilus edulis* and lack of selenium protection. *Marine Pollution Bulletin* 19: 213–219.

Pfeiffer, W.C., and L.D. de Lacerda. 1988. Mercury inputs into the Amazon region, Brazil. *Environmental Technology Letters* 9:325–330.

Ramamoorthy, S., and B.R. Rust. 1976. Mercury sorption and desorption characteristics of some Ottawa River sediments. *Canadian Journal of Earth Sciences* 13:530–536.

Ramamoorthy, S., T.C. Cheng, and D.J. Kushner. 1983. Mercury speciation in water. *Canadian Journal of Fisheries and Aquatic Sciences* 40:85–89.

Reddy, R.S., R.R. Jinna, J.E. Uzodinma, and D. Desaiah. 1988. In vitro effect of mercury and cadmium on brain Ca^{2+}-ATPase of the catfish *Ictalurus punctatus*. *Bulletin of Environmental Contamination and Toxicology* 41:324–328.

Reimann, D.O. 1986. Mercury output from garbage incineration. *Waste Management and Research* 4:45–56.

Rodgers, D.W., T.A. Watson, J.S. Langan, and T.J. Wheaton. 1987. Effects of pH and feeding regime on methylmercury accumulation within aquatic microcosms. *Environmental Pollution* 45:261–274.

Samhan, O., M. Zarba, and V. Anderlini. 1987. Multivariate geochemical investigation of trace metal pollution in Kuwait marine sediments. *Marine Environmental Research* 21:31–48.

Sanzgiry, S., A. Mesquita, and T.W. Kureishy. 1988. Total mercury in water, sediments, and animals along the Indian coast. *Marine Pollution Bulletin* 19:339–343.

Sasamal, S.K., B.K. Sahu, and R.C. Panigrahy. 1987. Mercury distribution in the estuarine and nearshore sediments of the western Bay of Bengal. *Marine Pollution Bulletin* 18:135–136.

Satyhyanathan, B., S.M. Nair, J. Chacko, and P.N.K. Nambisan. 1988. Sublethal effects of copper and mercury on some biochemical constituents of the estuarine clam *Villorita cyprinoides* var. *cochinensis* (Hanley). *Bulletin of Environmental Contamination and Toxicology* 40:510–516.

Schintu, M., T. Kauri, and A. Kudo. 1989. Inorganic and methyl mercury in inland waters. *Water Research* 23:699–704.

Sears, J.R., K.J. Pecci, and R.A. Cooper. 1985. Trace metal concentrations in offshore, deep-water seaweeds in the western North Atlantic Ocean. *Marine Pollution Bulletin* 16:325–328.

Semu, E., B.R. Singh, and A.R. Selmer-Olsen. 1986. Mercury pollution of effluent, air, and soil near a battery factory in Tanzania. *Water, Air, and Soil Pollution* 27:141–146.

Shiber, J., E. Washnburn, and A. Salib. 1978. Lead and mercury concentrations in the coastal waters of of North and South Lebanon. *Marine Pollution Bulletin* 9:109–111.

Stull, J.K., and R.B. Baird. 1985. Trace metals in marine surface sediments of the Palos Verdes Shelf, 1974 to 1980. *Journal of the Water Pollution Control Federation* 57:833–840.

Sung, J.F.C., A.E. Nevissi, and F.B. DeWalle. 1986. Concentration and removal efficiency of major and trace elements in municipal wastewaters. *Journal of Environmental Science and Health* 21:435–448.

Surma-Aho, K., J. Paasivirta, S. Rekolainen, and M. Verta. 1986. Organic and inorganic mercury in the food chain of some lakes and reservoirs in Finland. *Chemosphere* 15:353–372.

Tamashiro, H., K. Fukutomi, and E.S. Lee. 1987. Methylmercury exposure and mortality in Japan: a life table analysis. *Archives of Environmental Health* 42:100–107.

US Environmental Protection Agency. 1984. *Ambient water quality criteria for mercury—1984*. US Environmental Protection Agency, EPA 440/5–84–026, Washington, DC. 136 pp.

US Environmental Protection Agency. 1989. Office of Drinking Water Health Advisories. *Reviews of Environmental Contamination and Toxicology* 107:93–115.

US Minerals Yearbooks. 1930–1989. Bureau of Mines, US Department of the Interior, Washington, DC.

Valciukas, J.A., S.M. Levin, W.J. Nicholson, and I.J. Selikoff. 1986. Neurobehavioral assessment of Mohawk indians for subclinical indications of methyl mercury neurotoxicity. *Archives of Environmental Health* 41:269–272.

V.-Balogh, K., D.S. Fernandez, and J. Salanki. 1988. Heavy metal concentrations of *Lymnaea stagnalis* L. in the environs of Lake Balaton (Hungary). *Water Research* 22:1205–1210.

Westergard, B. 1986. Mercury from Hogdalen incineration plant in Stockholm, 1972–1985. *Waste Management and Research* 4:21.

Wobeser, G. 1975. Acute toxicity of methyl mercury chloride and mercuric chloride for rainbow trout *(Salmo gairdneri)* fry and fingerlings. *Journal of the Fisheries Research Board of Canada* 32:2005–2013.

17
Nickel

Nickel ranks as the 23rd most abundant element in the earth's crust, with an average concentration of 75 mg/kg. Although relatively low concentrations (1 mg/kg) are typically found in sedimentary rocks such as sandstone, much higher levels can be found in basalt (160 mg/kg) and peridotites and durites (2,000 mg/kg). The chief ores of nickel are pentlandite [$(Fe,Ni)_9S_8$], garnierite [$(Ni,Mg)_6(OH)_6(Si_4O_{11})H_2O$], and limonite [$(Ni,Fe)O(OH){\cdot}nH_2O$]. Nickel is also often associated with arsenide ores which, upon mining and smelting, release arsenic into the environment.

Production, Sources, and Residues

Production

World production of nickel was 22 $\times$ 10^3 metric tons in 1930, increasing to 326 $\times$ 10^3 metric tons in 1960 and 759 $\times$ 10^3 metric tons in 1980 (US Minerals Yearbooks, 1930–1989). Production in recent years has continued to increase and is now near 800 $\times$ 10^3 metric tons/yr. Although nickel is mined in many countries, the world's largest producers are the USSR, Canada, Australia, New Caledonia and Indonesia, whereas the major consumers are the USSR, the USA, Japan, the FRG and Italy (Table 17.1).

Nickel finds numerous applications in many industries because of its corrosion resistance, high strength and durability, pleasing appearance, good thermal and electrical conductivity, and alloying ability. The pro-

Table 17.1. World's leading producers and consumers of nickel.

Producing nation	Quantity (10^3 metric tons/yr)	Consuming nation	Quantity (10^3 metric tons/yr)
USSR	175	USSR	140
Canada	152	USA	129
Australia	77	Japan	99
New Caledonia	74	FRG	72
Indonesia	48	Italy	27
Cuba	38	France	27
Philippines	28	UK	25
Dominican Republic	25	China	20
South Africa	20	Sweden	15
Botswana	20		

Source: Cordero (1988).

duction of alloys accounts for approximately 75% of total nickel consumption. There are approximately 3,000 nickel–bearing alloys, falling into five major groups: stainless and heat-resisting steels, alloy steels (excluding stainless), super alloys, nickel–copper alloys, and permanent nickel alloys. Lesser uses of nickel are in cast irons, electroplating, and the production of chemicals.

Sources

The total amount of nickel discharged from anthropogenic sources to freshwaters amounts to 33–194 × 10^3 metric tons/yr (Table 17.2). The major source of discharge is municipal wastewater followed by smelting and refining of nonferrous metals (Table 17.2). Beukema et al. (1986) reported that the total amount of nickel discharged to The Netherlands' part of the North Sea amounted to 1,100 metric tons in 1980 (Table 17.3). Although projected rates for 1990 and beyond are not available, it is likely that there will only be a modest decrease in discharge for that time period.

Nickel residues are often relatively high in municipal waste waters and sludges. In a study of 25 sewage treatment plants in the state of Washington, Sung et al. (1986) reported that concentrations in primary sludge ranged up to 25.5 mg/L whereas residues in the liquid effluent reached a maximum of 0.21 mg/L (Table 17.4). In another study, Abuzkhar et al. (1987) assessed the suitability of solid waste samples collected from the Tripoli (Libya) sewage treatment plant as soil amendments; nickel residues in the samples averaged 83.8 mg/kg dry weight, with a range of 0.008–261.1 mg/kg. In another example, Schafer (1989) reported that the total amount of nickel discharged from the four largest southern Califor-

Table 17.2. Worldwide anthropogenic input of nickel to surface waters.

Source	Input (thousand metric tons/yr)
Domestic wastewater	
central	9–54
noncentral	12–48
Smelting and refining	
nonferrous metals	2–24
Dumping of sewage sludge	1.3–20
Coal-burning power plants	3–18
Atmospheric fallout	4.6–16
Manufacturing processes	
metals	0.2–7.5
chemicals	1–6
pulp and paper	0–0.1
petroleum products	0–0.1
Base metal mining and dressing	0.01–0.5

Source: Nriagu and Pacyna (1988).

nia municipal outfalls in coastal waters was approximately 50 metric tons in 1988; the corresponding values for 1980 and 1970 were approximately 160 and 250 metric tons, respectively. A large part of that decline was due to the increased rate of deposit of solid wastes in landfills.

Mining and smelting of nickel-bearing ores contribute substantially to the anthropogenic cycling of nickel in surface waters. Nicolaidou and

Table 17.3. Annual input of nickel to The Netherlands part of the North Sea in 1980 (actual) and 1990 (projected).

Source	Input 1980 (metric tons/yr)	Input 1990 (metric tons/yr)
Total input	1,100	ND[a]
Atmospheric deposition	110	110
Rivers	940	ND
Coastal discharges	5	ND
Dredging sludges	26	ND
Industrial wastes	2.5	ND
Incineration at sea	0.4	ND
Offshore mining	ND	ND

[a]Not determined.
Source: Beukema et al.

Table 17.4. Concentration (mg/L) of total nickel in wastes from 25 sewage treatment plants in the state of Washington.

Sample	Average	Range
Raw sewage	0.14	ND[a]–0.69
Primary effluent	0.15	ND –1.70
Secondary effluent	0.06	ND –0.20
Discharge	0.06	ND –0.21
Primary sludge	7.46	0.16–25.5
Secondary sludge	3.42	0.34–10.11

[a]Not detected.
Source: Sung et al. (1986).

Nott (1989), for example, examined metal levels near a ferronickel smelting plant in Greece.Nickel in the slag averaged 1,200 mg/kg, whereas sediments in the receiving waters (Evoikos Gulf) contained 600 mg/kg. For comparison, average residues in the sediments of the Aegean Sea were 28 mg/kg. Similarly, Nriagu and Rao (1987) showed that nickel was greatly enriched in the sediments of two lakes owing to emissions from Sudbury's (Canada) smelters; residues in the surficial sediments ranged from 5,000 to 6,500 mg/kg, compared to <50 mg/kg below 15 cm.

Use of fossil fuels, particularly coal, is widely implicated in the release of nickel to the atmosphere and surface waters (Wilson et al., 1986). Worldwide, 3–18 $\times$ 10^3 metric tons of nickel is discharged into freshwaters annually from the burning of coal (Table 17.2). Francis et al. (1989) showed that the residue from coal-cleaning contained 46–78 mg Ni/kg dry weight and that 11–58% of that concentration was mobilized by acidification at pH 1.6–2.2. The same study demonstrated that, under aerobic conditions, dissolution of metals from coal wastes was due mainly to bacterial oxidation of pyrite; under anaerobic conditions, bacterial reduction of iron and manganese oxides was the major mechanism for mobilization.

Residues

Water. Dissolved Ni levels in unpolluted freshwaters typically range from 0.001 to 0.003 mg/L (Scoullos and Hatzianestis, 1989), but may range up to 0.020 mg/L in areas where nickel-bearing ores reach the surface. The input of anthropogenic material may increase surface concentrations by 0.01–0.05 mg/L.Although dissolved Ni residues in uncontaminated coastal waters generally range from 0.001 to 0.005 mg/L (Seng et al., 1987), residues of up to 0.015 mg/L can be found in contaminated areas. Nolting (1986) noted that dissolved Ni in the Dutch Wadden Sea averaged approximately 0.003 mg/L, decreasing to <0.001 mg/L in the open waters of the North Sea.

Total Ni (dissolved + particulate) levels are often relatively high, particularly in estuaries and flowing waters. Zingde et al. (1988) reported that total Ni in the Mindhola Estuary (India) averaged 0.094 mg/L with a range of 0.078 to 0.131 mg/L. Similarly, average total Ni in the River Hindon (India) was 0.016 mg/L (range 0.010–0.021 mg/L) (Ajmal et al., 1987). For comparison, Nolting (1986) noted that dissolved Ni in the Dutch Wadden Sea averaged 0.003 mg/L whereas particulate Ni was lower at 0.0006 mg/L.

Sediments. Total Ni residues in sediments typically range up to 100 mg/kg dry weight but may fall below 1 mg/kg in some unpolluted coastal waters (Table 17.5). Maximum residues are typically associated with nickel-bearing geologic formations and the discharge of industrial and municipal wastes. Johnson et al. (1986) showed that atmospheric loadings to a series of lakes in Ontario (Canada) were not a significant source of nickel in the sediments of those same lakes. Enrichment factors (concentration in sediments/concentration in earth's crust) typically range from 1 to 12 (Sinex and Wright, 1988).

Table 17.5. Concentration (mg/kg dry weight) of total nickel in freshwater and marine sediments.

Location	Average (range)	Polluting source
Coastal waters, Georgia, South Carolina[1]	NR[a] (<1–42)	Multiple sources
Coastal waters, Florida[1]	NR (1–28)	Multiple sources
North Sea, Belgium[2]	27 (NR)	Multiple sources
Scheldt Estuary, The Netherlands[2]	60 (NR)	Multiple sources
Ganges Estuary, India[3]	34 (8–57)	Chemical industries
Palos Verdes Shelf, Los Angeles[4]	57 (30–79)	Municipal wastes
Arabian Gulf, Kuwait[5]	84 (36–102)	Multiple industrial
Lake Ontario, depositional basins[6]	NR (29–99)	Multiple industrial
Lake Erie, depositional basins[6]	NR (16–150)	Multiple industrial
Lake Huron, depositional basins[6]	NR (5–97)	Natural geologic, multiple industrial

[a]Not reported.

Sources: [1]Windom et al. (1989), [2]Araujo et al. (1988), [3]Subramanian et al. (1988), [4]Stull and Baird (1985), [5]Samhan et al. (1987), [6]Mudroch et al. (1988).

Chemistry

Nickel occurs principally as Ni^{2+} is surface waters, but oxidation states ranging from Ni^{-} to Ni^{4+} have been reported from time to time. Nickel is classified as a borderline element between hard and soft acid acceptors in chemical reactions with donor atoms. This behavior is reflected in nickel's abundance in the earth's crust as oxides, carbonates, and silicates with iron, magnesium, sulfides, and arsenides.

Under reducing conditions in surface water, nickel forms insoluble sulfides, provided that sulfur is present in excess. Under aerobic conditions and pH <9, nickel complexes with hydroxides, carbonates, sulfates, and naturally occurring organic ligands. This has been observed in both freshwaters and saline estuaries (Luther et al., 1986). Above pH 9 (rarely found in surface waters), the hydroxide and carbonate complexes precipitate.

Nickel reversibly binds with aluminum and manganese compounds. Windom et al. (1989), working on estuarine and coastal marine sediments of the southeastern United States, noted a significant correlation ($r^2 = 0.53$) between the concentrations of nickel and aluminum. In that study, nickel levels ranged from <1 to 42 mg/kg dry weight.Binding to clay and other fine particles is also a major, reversible fate process (e.g., Di Toro et al., 1985), but photolysis, volatilization, and biotransformation do not appear to be significant processes.

Bioaccumulation

Plants

Total Ni residues in aquatic plants, both marine and freshwater, are generally low (< 150 mg/kg dry weight) (Table 17.6). Substantially higher levels (150–700 mg/kg) have been reported for algae inhabiting water adja-

Table 17.6. Concentration (mg/kg dry weight) of total Ni in aquatic plants.

Species	Average (range)	Location
Fucus vesiculosus[1]	13 (4–46)	Swedish east coast
Fontinalis sp.[1]	5 (5–6)	Swedish east coast
Eicchornia crassipes[2]	11 (8–18)	River Hindon, India
Submerged plants, 7 species[3]	NR[a] (ND[b]–24)	Saginaw Bay, Lake Huron
Floating or emergent plants, 15 species[3]	NR (ND–144)	Saginaw Bay, Lake Huron
Seven freshwater species, Michigan[3]	NR (ND–3)	Small lakes, Michigan

[a]Not reported.
[b]Not detected.
Sources: [1]Soderlund et al. (1988), [2]Ajmal et al. (1987), [3]Estabrook et al. (1985).

cent to a UK smelter (Trollope and Evans, 1976), whereas burdens of up to 690 mg/kg wet weight were found in *Potamogeton* sp. from the Sudbury mining district (Canada) (Hutchinson et al., 1975).

Concentration factors relating nickel residues in plants to nickel residues in water typically range from 50 to 5,000 (Watras et al., 1985). Rate of uptake in laboratory cultures increases with exposure concentration and the application of phosphate to culture water. Vymazal (1984) noted that uptake of nickel by *Cladophora glomerata* and *Oedogonium rivulare* occurred continuously over a 4-h period, whereas other metals such copper, lead, and cadmium were rapidly sorbed during the first 1 h of exposure.

Invertebrates

Total Ni residues in the soft tissues of invertebrates are generally low, typically <5 mg/kg wet weight (Table 17.7). Higher concentrations are periodically recorded, particularly in the vicinity of metal mines and smelters. For example, Hutchinson et al. (1975) reported that maximum levels in molluscs and crustaceans collected from Sudbury (Canada) were 29–39 mg/kg wet weight. Overall, however, nickel cannot be considered a significant, widespread contaminant, except at site-specific points.

Nickel usually accumulates in tissues with high metabolic activity (Table 17.8). These include the hepatopancreas, kidney, and digestive gland; relatively high residues can also be found in the exoskeleton. In most cases, muscle residues are relatively low, which means that monitoring and impact assessment programs should not concentrate on such tissues. Several major, international programs currently use whole-body homoge-

Table 17.7. Concentration (mg/kg wet weight) of total Ni in the soft tissues of invertebrates.

Species	Average (range)	Location
Pearl oyster, *Pinctada radiata*[1]	0.2 (0.0–0.6)	Arabian Gulf, Saudi Arabia
Blood clam, *Anadara granosa*[2]	5.5 (3.9–10.8)[a]	Bombay harbor, India
Bivalve mollusc, *Donax serra*[3]	0.3 (0.1–0.5)	South African coast
Gastropod mollusc, *Bullia rhodostoma*[3]	0.1 (0.0–0.3)	South African coast
Asiatic clam, *Corbicula fluminea*[4]	0.2 (0.0–4.7)[a]	Shatt al-Arab River, Iraq

[a]Dry weight.
Sources: [1]Sadig and Alam (1989), [2]Patel et al. (1985), [3]Watling and Watling (1983), [4]Abaychi and Mustafa (1988).

Table 17.8. Concentration (mg/kg dry weight) of total Ni in tissues of crayfish (*Orconectes virilis* and *Cambarus bartoni*) collected from lakes near Sudbury (Canada).

Tissue	Average (range)	Tissue	Average (range)
Hepatopancreas	102 (0–402)	Exoskeleton	305 (0.4–1,243)
Muscle	45 (0–232)	Viscera	158 (2–493)
Digestive gut	479 (0–1,636)		

Source: Bagatto and Alikhan (1987).

nates of molluscs as biomonitors of contamination by nickel and other heavy metals (Galloway et al., 1983; Phillips, 1989). Because these homogenates include high metabolic tissues, representative tissue samples are included in the program.

Concentration factors (CF) relating nickel levels in tissues to those in water are generally <1,000. Watras et al. (1985), working with a model freshwater ecosystem, showed that the CF for the cladoceran *Daphnia magna* was only slightly above ambient; there was no biomagnification from *Scenedesmus obliquus* (the algal food source) to *Daphnia magna*. Uptake occurred principally through the water, rather than food, and ingested particulate Ni was eliminated in the feces.

Fish

Total Ni residues in the muscle tissue of fish, both marine and freshwater, are normally low, <0.5 mg/kg wet weight (e.g., Hagel, 1986; Saiki and May, 1988). Higher values in the 1–2 mg/kg wet weight range were reported for the shark *Galeus melastomus* collected from the coast of Scotland (Vas and Gordon, 1988), whereas Hutchinson et al. (1975) reported that residues in six freshwater species from the Sudbury region (Canada) ranged from 9.5 to 13.8 mg/kg. Those latter collections were made in an area of nickel-bearing ore deposits that are both mined and smelted.

Total Ni is deposited selectively in certain tissues, but not to the same extent as some metals such as cadmium. For example, nickel residues in the kidney of the shark *Galeus melastomus* collected from the coast of Scotland averaged only 1.61 mg/kg, essentially similar to that observed for muscle (Table 17.9).

Tjalve et al. (1988) exposed brown trout *Salmo trutta* to water containing 0.1 or 10 μg/L of $^{63}Ni^{2+}$. After 3 weeks of exposure, the whole body concentration of nickel was about eight times greater than that of water. There was rapid accumulation of nickel in the gills at both exposure concentrations (Table 17.10). In the second part of their experiment, a group of fish was exposed to nickel for 3 weeks and then allowed to swim in nickel-free water for 3 weeks. At the end of the experiment, nickel had

Table 17.9. Average concentration (mg/kg wet weight) of total Ni in the tissues of the shark *Galeus melastomus* collected from the coast of Scotland.

Tissue	Concentration	Tissue	Concentration
Spleen	1.33	Vertebrae	0.23
Skin	2.01	Muscle	1.41
Liver	1.88	Heart	2.83
Gonad	1.69	Kidney	1.61
Gill	1.71		

Source: Vas and Gordon (1988).

moved from the gills and was selectively deposited in the kidneys (Table 17.10).

There do not appear to be any cases in which human utilization of fisheries resources was limited by high nickel residues in tissues.

Toxic Effects to Aquatic Organisms

Plants

Ni^{2+} is moderately to highly toxic to most species of aquatic plants. Wang (1986), for example, found that the EC_{50} (Effective Concentration) for duckweed *Lemna minor* was 0.45 mg/L. In that study, only Cd was more toxic than Ni; Cu, Se, Fe, Pb, Zn, and Cr were less toxic. Patrick et al. (1975) noted that Ni^{2+} induced changes in the species composition of benthic algae at concentrations as low as 0.002 mg Ni/L.

Table 17.10. Percent distribution of $^{63}Ni^{2+}$ in two groups of brown trout *Salmo trutta* exposed to 0.1–10 μg/L of $^{63}Ni^{2+}$ for 3 weeks.[a]

Tissue	3-Week exposure only	3-Week exposure/ 3 weeks in Ni-free water
Gill	9.4–10.7	5.2–5.6
Kidney	2.9–4.0	11.7–11.8
Liver	1.2–2.1	2.3–2.6
Remaining viscera	8.8–12.6	8.8–9.3
Brain	0.5–0.6	1.1–1.5
Eye	1.9–2.0	2.4–3.2
Remaining body	68.0–75.3	66.1–68.4

[a]One group of fish was exposed to Ni prior to analysis whereas the second group of fish was allowed to swim in postexposure Ni-free water prior to analysis.
Source: Tjalve et al. (1988).

Combinations of Ni/Hg and Ni/Cu act synergistically to many algal species whereas antagonism has been noted for combinations of Ni/Cd, Ni/Zn, and Ni/Cd/Hg. H^+ has a variable effect on toxicity (Campbell and Stokes, 1985). The presence of ligands, both organic and inorganic, likely ameliorates toxicity, but the data base on this subject is incomplete.

Invertebrates

Ni^{2+} is one of the least toxic inorganic agents to invertebrates. The LC_{50}s typically range from 0.5 to 20 mg/L or more (Ewell et al., 1986). Certain crustacean species have an LC_{50} of >100 mg/L (Table 17.11). Khangarot et al. (1987), working with *Daphnia magna* in hard water (240 mg/L), showed that the 48-h LC_{50} of Ni^{2+} was 19.5 mg/L; this made Ni^{2+} the second least toxic agent investigated: Sn < Ni < Pb < Cr < Co < Cd < Zn < Cu < Ag < Hg. Saltwater organisms often develop an inability to osmoregulate when exposed to Ni^{2+} (and other heavy metals) (McLusky and Hagerman, 1987).

Fish

Ni^{2+} is relatively nontoxic to fish, both marine and freshwater. The 48- to 96-h LC_{50}s are almost always above 1 mg/L and may exceed 100 mg/L for some species and environmental conditions. Freshly fertilized eggs and fish fry may be slightly more sensitive than adults, but the corresponding LC_{50}s are still high (> 1 mg/L) compared to other metals.

Exposure of fish to low levels of Ni^{2+} over extended periods results in a number of toxicological effects. These include reduced skeletal calcification, increased hematocrit and hemoglobin, and reduced diffusion capacity of gills, which ultimately leads to asphyxiation.

Acute and chronic toxicity of Ni^{2+} is strongly related to water hardness

Table 17.11. Acute toxicity (LC_{50} of nickel to two crustacean species.

	48-h LC_{50} (mg/L)		96-h LC_{50} (mg/L)	
Nickel species	Average	95% confidence limits	Average	95% confidence limits
Asellus aquaticus				
Ni^{2+}	435	296–719	119	62–173
Crangonyx pseudogracilis				
Ni^{2+}	252	211–299	66	49–86

Source: Martin and Holdich (1986). Total hardness 50 mg/L, pH 6.75, temperature 13°C.

and excess H^+. Additive effects on toxicity have been reported for mixtures of Ni/Cu and Ni/Cu/Zn.

Many nations have promulgated guidelines/standards for the protection of aquatic organisms:

Canada
- 25 μg/L (water hardness 0 to 60 mg/L as $CaCO_3$)
- 65 μg/L (water hardness 61 to 120 mg/L as $CaCO_3$)
- 110 μg/L (water hardness 121 to 180 mg/L as $CaCO_3$)
- 150 μg/L (water hardness >180 mg/L as $CaCO_3$)

USA
- 56 μg/L (water hardness 50 mg/L as $CaCO_3$, 24-h average)
- 96 μg/L (water hardness 100 mg/L as $CaCO_3$, 24-h average)
- 160 μg/L (water hardness 200 mg/L as $CaCO_3$, 24-h average)

European Community
- 50 μg/L (water hardness 0 to 50 mg/L as $CaCO_3$)
- 150 μg/L (water hardness 100 to 200 mg/L as $CaCO_3$)
- 200 μg/L (water hardness >200 mg/L as $CaCO_3$)
- 30 μg/L (protection of saltwater fish and shellfish)

Health Effects

Intake

The typical Western diet yields 0.1–0.9 mg Ni/day (average 0.16–0.50 mg/day) in a 70-kg reference man. Intake from air is negligible (<0.001 mg/kg) in the nonoccupationally exposed population. Although only 1–10% of the nickel taken in from food is absorbed, uptake of inhaled organic compounds (such as nickel carbonyl) is rapid and efficient. The total body burden of nickel in the 70-kg reference man is approximately 10 mg, of which 5–6 mg is tied up in the soft tissues. In humans, highest concentrations are found in the liver, kidney, and brain.

Acute Toxicity

The primary acute effect of nickel exposure is dermatoses, including contact dermatitis, atopic dermatitis, and allergic sensitization (Coogan et al., 1989). No clinical or epidemiological studies are available on the acute effects of oral ingestion of nickel.

Chronic Toxicity

The primary issue associated with the chronic toxicity of nickel is its potential carcinogencity, as discussed in the following section.

Carcinogenicity

Nickel in some forms is likely to be carcinogenic in humans, and, in fact, statistically significant elevations in the incidence of respiratory cancers have been found in nickel refinery workers (Coogan et al., 1989). However, no evidence of carcinogenicity exists for ingestion of nickel in either food or water.

Drinking Water

Residues

Although total Ni levels in drinking water are generally low (<0.05 mg/L), some supplies contain residues of up to 0.50 mg/L. These high levels are due primarily to the dissolution of nickel in some plumbing materials. In a study of well water in the city of Aligarh (India), Ajmal and Uddin (1986) reported that total Ni residues averaged approximately 0.015 mg/L with a range of <0.005 to 0.025 mg/L.Similarly, residues in drinking water from Rio de Janeiro (Brazil) were also low, averaging only 0.002 mg/L (Azcue et al., 1988). Equally low levels (0.001 mg/L) were found in tap water from Hartford (Connecticut) compared to 0.20 mg/L for Sudbury (Canada), a major nickel-producing area (Carson et al., 1987).

Consumption Guidelines

Several nations, plus the World Health Organization, have not recommended any drinking water guideline for nickel; it is believed that nickel poses an insignificant toxicologic threat to drinking water. The US Environmental Protection Agency (1989) currently recommends a Lifetime Health Advisory of 0.20 mg/L. That value was developed using a 2-year rat feeding study (Ambrose et al., 1976). Rats were administered 0, 100, 1,000, or 2,500 mg/kg dietary nickel sulfate, giving an approximate daily dose of 0, 5, 50, or 125 mg Ni/kg. The No-Observed-Adverse-Effect Level (NOAEL) from that study was 5 mg/kg/day. The Lifetime Health Advisory was calculated as follows.

Step 1: Determination of the Reference Dose (RD):

$$RD = (5\ \text{mg/kg/day})/(100)(3) = 0.016\ \text{mg/kg/day}$$

where 100 is the uncertainty factor and 3 is an additional uncertainty factor used because of inadequacies in the study by Ambrose et al. (1976).

Step 2: Determination of the Drinking Water Equivalent Level (DWEL):

$$DWEL = (0.016\ \text{mg/kg/day})(70\text{-kg})/(2\ \text{L/day}) = 0.58\ \text{mg/L}$$

Step 3: Determination of the Lifetime Health Advisory:

$$\text{Lifetime Health Advisory} = 0.58 \times 30\% = 0.17 \text{ mg/L}$$

where 30% is the estimated dose from drinking water.

Treatment

A number of techniques, including lime softening, ion exchange, and reverse osmosis, are available for removing nickel from drinking water (US Environmental Protection Agency, 1989). Of these techniques, lime softening appears to be most efficient, with removals of 91–99.9% at moderate to high nickel concentrations (>5 mg/L). Cation exchange and reverse osmosis are also highly effective, but limited in use to small scale operations. Conventional coagulation, although widely used on some inorganic contaminants, is only moderately effective (removal efficiency 25–45%) in treating excess nickel.

Recommendations

Nickel is not a widespread contaminant of surface waters. Although nickel may reach relatively high concentrations under some geologic conditions, residues are generally low in both marine and freshwaters. There are only a few recorded cases in which anthropogenically derived nickel has threatened fish and other aquatic resources. Most nations have established guidelines or standards for the protection of such species. Also, nickel does not bioaccumulate through the food chain, and there do not appear to be any instances in which human utilization of fish has been limited by nickel contamination.

In short, then, nickel does appear at this time to pose a threat to the users of surface water. Although no major research deficiencies have been identified, Ni^{2+} should be routinely monitored in surface waters, particularly in those areas where geologic conditions may lead to significant mobilization of nickel.

References

Abaychi, J.K., and Y.Z. Mustafa. 1988. The Asiatic clam, *Corbicula fluminea*: an indicator of trace metal pollution in the Shatt al-Arab River, Iraq. *Environmental Pollution* 54:109–122.

Abuzkhar, A.A., A.S. Gibali, Y.I. Elmehrik, and R. Ahmatullah. 1987. Chemical monitoring of sewage wastes for their use in crop production. *Environmental Monitoring and Assessment* 8:127–133.

Ajmal, M., and R. Uddin. 1986. Studies on heavy metals in the ground waters of the City of Aligarh, U.P. (India). *Environmental Monitoring and Assessment* 6:181–194.

Ajmal, M., R. Khan, and A.U. Khan. 1987. Heavy metals in water, sediments, fish and plants of river Hindon, U.P., India. *Hydrobiologia* 148:151–157.

Ambrose, A.M., P.S. Larson, J.R. Borzelleca, and G.R. Hennigar. 1976. Long-term toxicologic assessment of nickel in rats and dogs. *Journal of Food Science and Technology* 13:181–187.

Araujo, M.F.D., P.C. Bernard, and R.E. Van Grieken. 1988. Heavy metal contamination in sediments from the Belgian coast and Scheldt Estuary. *Marine Pollution Bulletin* 19:269–273.

Azcue, J.M.P., W.C. Pfeiffer, M. Fiszman, and O. Malm. 1988. Heavy metal removal by different water treatment plants, in Rio de Janeiro State, Brazil. *Environmental Technology Letters* 9:429–436.

Bagatto, G., and M.A. Alikhan. 1987. Copper, cadmium, and nickel accumulation in crayfish populations near copper-nickel smelters at Sudbury, Ontario, Canada. *Bulletin of Environmental Contamination and Toxicology* 38:540–545.

Beukema, A.A., G.P. Hekstra, and C. Venema. 1986. The Netherlands' environmental policy for the North Sea and Wadden Sea. *Environmental Monitoring and Assessment* 7:117–155.

Campbell, P.G.C., and P.M. Stokes. 1985. Acidification and toxicity of metals to aquatic biota. *Canadian Journal of Fisheries and Aquatic Sciences* 42:2034–2049.

Carson, B.L., H.V. Ellis, and J.L. McCann. 1987. *Toxicology and biological monitoring in humans*. Lewis Publishing, Chelsea, MI. 328 pp.

Coogan, T.P., D.M. Latta, E.T. Snow, and M. Costa. 1989. Toxicity and carcinogenicity of nickel compounds. *CRC Critical Reviews in Toxicology* 19: 341–384.

Cordero, R. 1988. *Metal Bulletin's prices and data 1988*. Metal Bulletin Books Ltd., Surrey, England. 375 pp.

Di Toro, D., J.D. Mahony, P.R. Kirchgraber, A.L. O'Byrne, L.R. Pasquale, and D.C. Piccirill. 1985. Effects of nonreversibility, particle concentration, and ionic strength on heavy metal sorption. *Environmental Science and Technology* 20:55–61.

Estabrook, G.F., D.W. Burk, D.R. Inman, P.B. Kaufman, J.R. Wells, J.D. Jones, and N. Ghosheh. 1985. Comparison of heavy metals in aquatic plants on Charity Island, Saginaw Bay, Lake Huron, U.S.A., with plants along the shoreline of Saginaw Bay. *American Journal of Botany* 72:209–216.

Ewell, W.S., J.W. Gorsuch, R.O. Kringle, K.A. Robillard, and R.C. Spiegel. 1986. Simultaneous evaluation of the acute effects of chemicals on seven aquatic species. *Environmental Toxicology and Chemistry* 5:831–840.

Francis, A.J., C.J. Dodge, A.W. Pose, and A.J. Ramirez. 1989. Aerobic and anaerobic microbial dissolution of toxic metals from coal wastes: mechanism of action. *Environmental Science and Technology* 23:435–441.

Galloway, W.B., J.L. Lake, D.K. Phelps, P.F. Rogerson, V.T. Bowen, J.W. Farrington, E.D. Goldberg, J.L. Laseter, G.C. Lawler, J.H. Martin, and R.W. Risebrough. 1983. The mussel watch: intercomparison of trace level constituent determinations. *Environmental Toxicology and Chemistry* 2:395–410.

Hagel, P. 1986. Monitoring of pollutants in Dutch fishery products. *Environmental Monitoring and Assessment* 77:257–262.

Hutchinson, T.C., A. Fedorenko, J. Fitchko, A. Van Loon, and J. Lichwa. 1975. Movement and compartmentation of nickel and copper in an aquatic ecosystem. *In: Trace substances in environmental health. IX. A symposium,* ed. D.D. Hemphill (Ed.), 89–105. University of Missouri Press, Columbia.

Johnson, M.G., L.R. Culp, and S.E. George. 1986. Temporal and spatial trends in metal loadings to sediments of the Turkey Lakes, Ontario. *Canadian Journal of Fisheries and Aquatic Sciences* 43:754–762.

Khangarot, B.S., P.K. Ray, and H. Chandra. 1987. *Daphnia magna* as a model to assess heavy metal toxicity: comparative assessment with mouse system. *Acta Hydrochimica Hydrobiologica* 15:427–432.

Luther, G.W., Z. Wilk, R.A. Ryans, and A.L. Meyerson. 1986. On the speciation of metals in the water column of a polluted estuary. *Marine Pollution Bulletin* 17:535–542.

Martin, T.R., and D.M. Holdich. 1986. The acute lethal toxicity of heavy metals to peracarid crustaceans (with particular reference to fresh-water asellids and gammarids). *Water Research* 20:1137–1147.

McLusky, D.S., and L. Hagerman. 1987. The toxicity of chromium, nickel and zinc: effects of salinity and temperature, and the osmoregulatory consequences in the mysid *Praunus flexuosus. Aquatic Toxicology* 10:225–238.

Mudroch, A., L. Sarazin, and T. Lomas. 1988. Summary of surface and background concentrations of selected elements in the Great Lakes sediments. *Journal of Great Lakes Research* 14:241–251.

Nicolaidou, A., and J.A. Nott. 1989. Heavy metal pollution induced by a ferronickel smelting plant in Greece. *Science of the Total Environment* 84:113–117.

Nolting, R.F. 1986. Copper, zinc, cadmium, nickel, iron and manganese in the southern Bight of the North Sea. *Marine Pollution Bulletin* 17:113–117.

Nriagu, J.O., and J.M. Pacyna. 1988. Quantitative assessment of worldwide contamination of air, water and soils by trace metals. *Nature* 333:134–139.

Nriagu, J.O., and S.S. Rao. 1987. Response of lake sediments to changes in trace metal emission from the smelters at Sudbury, Ontario. *Environmental Pollution* 44:211–218.

Patel, B., V.S. Bangera, S. Patel, and M.C. Balani. 1985. Heavy metals in the Bombay Harbour area. *Marine Pollution Bulletin* 16:22–25.

Patrick, R.T., T. Bott, and R. Larson. 1975. *The role of trace elements in management of nuisance growths.* US Environmental Protection Agency, EPA 660/2–75–008, Corvallis, OR. 250 pp.

Phillips, D.J.H. 1989. Trace metals and organochlorines in the coastal waters of Hong Kong. *Marine Pollution Bulletin* 20:319–327.

Sadig, M., and I. Alam. 1989. Metal concentrations in pearl oyster, *Pinctada radiata*, collected from Saudi Arabian coast of the Arabian Gulf. *Bulletin of Environmental Contamination and Toxicology* 42:111–118.

Saiki, M.K., and T.W. May. 1988. Trace element residues in bluegills and common carp from the lower San Joaquin River, California, and its tributaries. *Science of the Total Environment* 74:199–217.

Samhan, O., M. Zarba, and V. Anderlini. 1987. Multivariate geochemical investigation of trace metal pollution in Kuwait marine sediments. *Marine Environmental Research* 21:31–48.

Schafer, H. 1989. Improving southern California's coastal waters. *Journal of the Water Pollution Control Federation* 61:1395–1401.

Scoullos, M.J., and J. Hatzianestis. 1989. Dissolved and particulate trace metals in a wetland of international importance: Lake Mikri Prespa, Greece. *Water, Air, and Soil Pollution* 44:307–320.

Seng, C.E., P.E. Lim, and T.T. Ang. 1987. Heavy metal concentrations in coastal seawater and sediments off Prai industrial Estate, Penang, Malaysia. *Marine Pollution Bulletin* 18:611–612.

Sinex, S.A., and D.A. Wright. 1988. Distribution of trace metals in sediments and biota of Chesapeake Bay. *Marine Pollution Bulletin* 19:425–431.

Soderlund, S., A. Forsberg, and M. Pedersen. 1988. Concentrations of cadmium and other metals in *Fucus vesiculosus* L. and *Fontinalis dalecarlica* Br. Eur. from the northern Baltic Sea and the southern Bothnian Sea. *Environmental Pollution* 51:197–212.

Stull, J.K., and R.B. Baird. 1985. Trace metals in marine surface sediments of the Palos Verdes Shelf, 1974 to 1980. *Journal of the Water Pollution Control Federation* 57:833–840.

Subramanian, V., P.K. Jha, and R. van Grieken. 1988. Heavy metals in the Ganges estuary. *Marine Pollution Bulletin* 19:290–293.

Sung, J.F.C., A.E. Nevissi, and F.B. DeWalle. 1986. Concentration and removal efficiency of major and trace elements in municipal wastewater. *Journal of Environmental Science and Health* 21:435–448.

Tjalve, H., J. Gottofrey, and K. Borg. 1988. Bioaccumulation, distribution and retentions of $^{63}Ni^{2+}$ in the brown trout (*Salmo trutta*). *Water Research* 9:1129–1136.

Trollope, D.R., and B. Evans. 1976. Concentrations of copper, iron, lead, nickel and zinc in freshwater algal blooms. *Environmental Pollution* 11:109–116.

US Environmental Protection Agency. 1989. Office of drinking water health advisories. *Reviews of Environmental Contamination and Toxicology* 107: 1–184.

US Minerals Yearbooks. 1930–1989. Bureau of Mines, US Department of the Interior, Washington, DC.

Vas, P., and J.D.M. Gordon. 1988. Trace metal concentrations in the scyliorhinid shark *Galeus melastomus* from the Rockall Trough. *Marine Pollution Bulletin* 19:396–397.

Vymazal, J. 1984. Short-term uptake of heavy metals by periphyton algae. *Hydrobiologia* 119:171–179.

Wang, W. 1986. Toxicity tests of aquatic pollutants by using common duckweed. *Environmental Pollution* 11:1–14.

Watling, H.R., and R.J. Watling. 1983. Sandy beach molluscs as possible bioindicators of metal pollution. 1. Field survey. *Bulletin of Environmental Contamination and Toxicology* 31:331–338.

Watras, C.J., J. MacFarlane, and F.M.M. Morel. 1985. Nickel accumulation in *Scenedesmus* and *Daphnia*: food-chain transport and geochemical implications. *Canadian Journal of Fisheries and Aquatic Sciences* 42:724–730.

Wilson, B.L., R.R. Schwarzer, and N. Etonyeaku. 1986. The evaluation of heavy metals (chromium, nickel, and cobalt) in the aqueous sediment surrounding a coal burning generating plant. *Journal of Environmental Science and Health* 21:791–808.

Windom, H.L., S.J. Schropp, F.D. Calder, J.D. Ryan, R.G. Smith, L.C. Burney, F.G. Lewis, and C.H. Rawlinson. 1989. Natural trace metal concentrations in estuarine and coastal marine sediments of the southeastern United States. *Environmental Science and Technology* 23:314–320.

Zingde, M.D., M.A. Rokade, and A.V. Mandalia. 1988. Heavy metals in Mindhola River Estuary, India. *Marine Pollution Bulletin* 19:538–540.

18
Nitrogen

Nitrogen occurs in the biosphere in oxygenation states ranging from 3 − (ammonia) to 5 + (nitrate). The most important inorganic forms of nitrogen are ammonia (NH_3), nitrate (NO_3^-), nitrite (NO_2^-), and molecular nitrogen (N_2). All of these forms are interrelated in the environment by the nitrogen cycle, a complicated series of transformations (described in Chemistry section). Naturally occurring organic nitrogen compounds occur in surface waters and contain amino and amide nitrogen, and some heterocyclic compounds such as purines and pyrimidines.

Production, Sources, and Residues

Production and Sources

Nitrogen-bearing compounds in surface waters originate from numerous sources, both natural and anthropogenic. The total input to aquatic systems is enormous, amounting to 73–248 $\times$ 10^6 metric tons per year (Table 18.1). The No. 1 source of nitrogen in the environment is still N_2 fixation, contributing 30–130 $\times$ 10^6 metric tons/year to freshwater and marine systems, followed by NH_4^+/NH_3 deposition, principally from agriculture.

Although most regulatory agencies see the need to reduce the total flux of nitrogen to surface waters, continued heavy use in the agricultural sector, plus naturally induced N_2 fixation, means that discharge to many waterways will not be curtailed for the foreseeable future. For example, total N discharge to The Netherlands' part of the North Sea amounted to

Table 18.1. Worldwide flux (10^6 metric tons/yr) of nitrogen in aquatic and marine systems.

Flux	Input
Input	
Biological N_2 fixation	30–130
NH_4^+ deposition	
wet	8–25
dry	11–25
NO_3^- /NO_x deposition	
wet	5–16
dry	6–17
River discharge	13–35
Output	
Denitrification	
NO_2 emission	0–179
NO_x emission	0–33
NH_3 emission	18–45
Sedimentation	30–40

Source: Robertson (1986).

650 × 10^3 metric tons in 1980, decreasing to only 510–620 × 10^3 metric tons in 1990 (Table 18.2).

Ammonia, nitrate, and nitrite all find commercial application, and are discharged to surface waters by user industries. The principal use of ammonia is in agriculture for the production of urea, nitric acid, ammonium sulfate, ammonium nitrate, and fertilizer mixes. Minor amounts of ammonia also go to use in mining and refining, pulp and paper production, and the manufacture of amines and nitriles.

Table 18.2. Annual input (10^3 metric tons) of nitrogen to The Netherlands' part of the North Sea in 1980 (actual) and 1990 (projected).

Source	Input 1980	Input 1990
Total input	650	510–620
Atmospheric deposition	57	57
Rivers	580	450–560
Coastal discharges	12	6
Dredging sludges	2.6	2
Offshore mining	0.1	0.1

Source: Beukema et al. (1986).

Nitrates are used mainly in the production of chemical fertilizers and oxidizing agents for use in the chemical industry. Ammonium nitrate, sodium nitrate, and calcium nitrate also find application in the production of explosives. Nitrite salts are mainly used in corrosion resistance.

Agriculture. Agriculture, together with municipal waste water, is the most important source of anthropogenically derived ammonia and nitrate to surface waters. In the case of agriculture, the major source is the increasing worldwide use of nitrogen-based fertilizers, followed by runoff from pastures and, to a lesser degree, feedlots. Robertson (1986) reported that nitrogen release due to fertilizer use in Wisconsin amounted to 127 $\times$ 10^3 metric tons/yr, equivalent to 22% of the total nitrogen flux for that area; for comparison, nitrogen release in West Africa was 74 $\times$ 10^3 metric tons/yr, only 0.4% of total flux. In another study, Kudeyarov and Bashkin (1980) showed that the input of nitrogen to the Skniga river basin (USSR) more than doubled between 1967 and 1977 (Table 18.3). During that period, the contribution of fertilizers to the nitrogen budget increased from 46.7% to 77.3%. Similarly, the rate of application of fertilizers to soils in the FRG increased from 50 kg N/ha in 1950 to 120 kg/ha in 1982 (Quentin, 1988).

Some of the nitrogen added to soils eventually finds its way to surface water and groundwater. Under aerobic conditions, typical of most surface waters, ammonia is rapidly oxidized to nitrate, thereby producing relatively high nitrate residues. In the Thames River, for example, nitrate concentrations averaged 7.5 mg N/L in 1978, up from 2.5 mg N/L in 1929 (Onstad and Blake, 1980). Nitrates are also elevated in some ground waters, often in excess of 25 mg/L, as exemplified in data from France (Table 18.4).

Table 18.3. Nitrogen inputs (metric tons) to the Skniga River basin (USSR) in 1967 and 1977.

Source	1967		1977	
	Quantity	% of total	Quantity	% of total
Fertilizers	573	46.7	2,121	77.3
Symbiotic N fixation	381	31.1	352	12.8
Nonsymbiotic N fixation	140	11.4	140	5.1
Precipitation	133	10.8	133	4.8
Total input	1,227	100	2,746	100

Source: Kudeyarov and Bashkin (1980).

Table 18.4. Nitrates in groundwater in France.

	Concentration (mg/L)			
	0–25	26–50	51–100	>100
Population affected				
Number (millions)	42.8	9.2	1.3	0.03
Percent	80.4	17.4	2.1	0.1
Water delivery systems affected				
Number	16,729	2,661	555	29
Percent	83.8	13.3	2.8	0.2

Source: Delavalle (1983).

Runoff from pastures is a secondary source of nitrogen-bearing compounds for surface waters. The main flux of nitrogen comes during rainfall, particularly after the first and second runoff events. In one study, McLeod and Hegg (1984) showed that 55–80% of the total flux occurred under these conditions. The resulting concentration of NO_3–N in surface runoff exceeded the permissible public water standard of 10 mg/L.

Municipal Wastewater. Nitrogen-bearing compounds, particularly ammonia and nitrate, are discharged in enormous amounts with municipal waste water. In one study, Nevissi et al. (1988) reported that the total amount of nitrogen discharged from two plants in the state of Washington ranged from 1,194 to 1,557 kg/day. In the absence of advanced removal systems, NH_3–N in final effluent typically ranges from 1 to >10 mg/L. Such levels are highly toxic to many aquatic species, and also lead to enrichment of nitrogen in the receiving waters and sediments.

Residues

Stull et al. (1986), working near municipal outfalls off the coast at Los Angeles, noted that organic N in the upper 50 cm of sediment ranged up to 9,000 mg/kg dry weight. Below that depth, residues declined to approximately 50 mg/kg. Kennedy and Bell (1986) determined nitrogen residues in the waters of the White River, which receives discharge from waste treatment plants from the city of Indianapolis. NH_3–N was as high as 10.1 mg/L near the discharge point but declined to 6.8 mg/L about 30 km downstream of that site. NO_3–N, on the other hand, increased from 0.3 to 1.8 mg/L, whereas NO_2–N increased from 0.4 to 0.7 mg/L. Other studies (Crumpton and Isenhart, 1987) have shown that NO_3–N from waste treatment plants varies diurnally in receiving waters, being greatest near midday.

Chemistry

The nitrogen cycle consists of six major processes as illustrated below (Canadian Water Quality Guidelines, 1987):

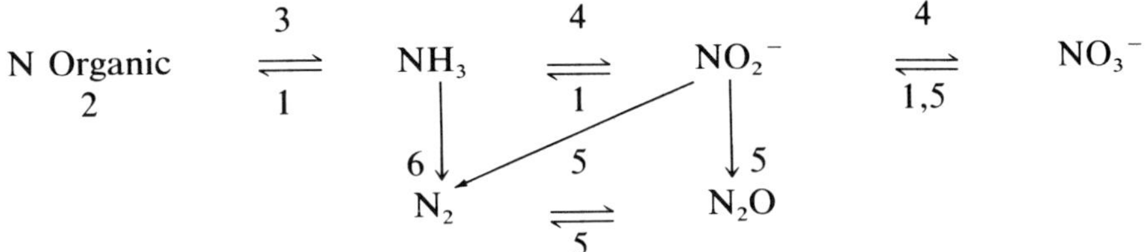

1. Assimilation of inorganic N (mainly NH_3–N and NO_3–N) by microorganisms and plants to form organic N such as proteins and amino acids.
2. Heterotrophic conversions, involving the transfer of organic N among organisms.
3. Ammonification, the breakdown of organic N to NH_3–N by bacteria and fungi.
4. Nitrification, the microbially mediated oxidation of NH_3–N to NO_2–N and NO_3–N.
5. Denitrification, the microbially mediated production of NO_2–N and N_2. This occurs mainly under anaerobic conditions and, to a lesser degree, under low oxygen conditions.
6. Biological nitrogen fixation, the conversion of N_2 to NH_3–N.

The nitrogen cycle in agricultural soils follows much the same pattern (Figure 18.1).

Ammonia has an oxidation state of 3 – and is the reduced form of inorganic N in surface water. The term total ammonia refers to the sum of ionized and un-ionized forms. The un-ionized form of ammonia exists in equilibrium with the ammonium ion (ionized form) per the following equation:

$$NH_3 + H_2O \rightleftharpoons NH_3{\cdot}nH_2O$$
$$(\text{aq.}) \rightleftharpoons NH^+{}_4 + OH^- + (n-1)H_2O$$

The concentration of un-ionized ammonia depends primarily on pH and, to a lesser extent, temperature (Table 18.5). Although the concentration of dissolved solids may also influence un-ionized ammonia levels, such affects are inconsequential in most surface waters compared to those of pH and temperature. From Table 18.5 it is apparent that the percent of un-ionized ammonia greatly increases beyond pH 7 and that in waters where pH exceeds 8, a major portion of ammonia exists as the un-ionized form.

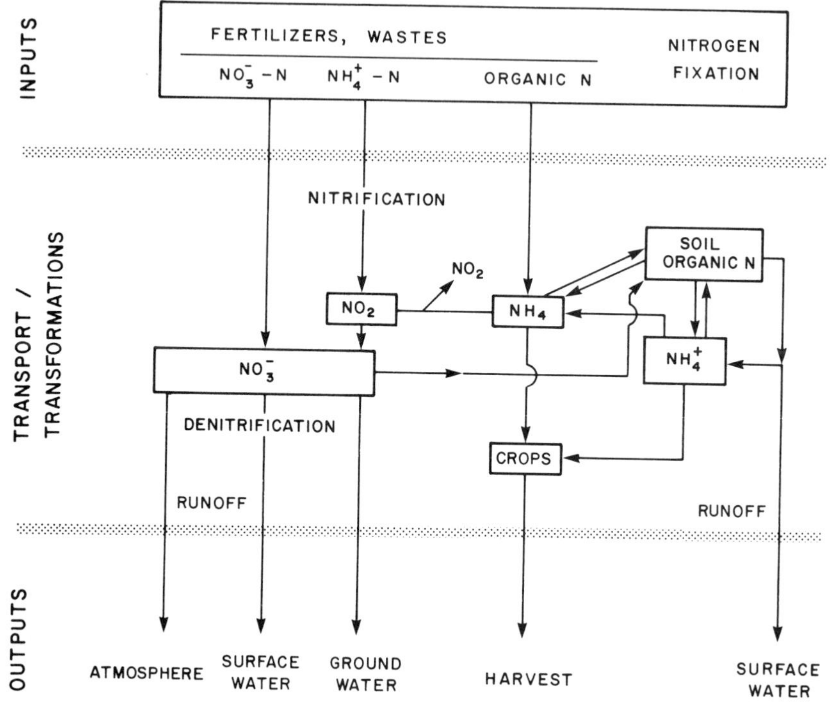

Figure 18.1. Example of a nitrogen cycle in agricultural soil.

Nitrite (NO_2^-) is rapidly converted to nitrate (NO_3^-) in most surface waters. Nitrite salts are soluble and can be detected in significant quantities in water when oxygen supply is limited. Nitrate salts, on the other hand, are soluble and exist under aerobic conditions. When dissolved oxygen concentration is low, nitrate may be reduced to nitrite via denitrification. Lindau et al. (1988) showed that, when nitrate was experimentally

Table 18.5. Percent un-ionized ammonia in aqueous ammonia solutions.

	Temperature (°C)						
pH	0	5	10	15	20	25	30
6.0	0.00827	0.0125	0.0186	0.0274	0.0397	0.0569	0.0805
6.5	0.0261	0.0314	0.0589	0.0865	0.125	0.180	0.254
7.0	0.0826	0.125	0.186	0.273	0.396	0.566	0.799
7.5	0.261	0.394	0.586	0.859	1.24	1.77	2.48
8.0	0.820	1.23	1.83	2.67	3.82	5.38	7.46
8.5	2.55	3.80	5.56	7.97	11.2	15.3	20.3
9.0	7.64	11.1	15.7	21.5	28.4	36.3	44.6

Source: Piper et al. (1982).

added to a Louisiana gulf coast swamp, denitrification produced large amounts of N_2O and N_2 that escaped to the atmosphere; ammonium was nitrified, ultimately resulting in the formation of NO_3^-, which was then denitrified to N_2O and N_2.

Bioaccumulation

NH_3–N, NO_3–N, and NO_2–N do not bioaccumulate in plant and animal tissues.

Toxic Effects to Aquatic Organisms

One of the key difficulties in obtaining reproducible data on ammonia toxicity is the effect of pH on the concentration of the un-ionized form. A flux of only 0.2 pH units produces a change of two to five times in un-ionized ammonia levels in basic waters. Because it is not practical in experimental systems to control the flux of pH to less than 0.2 units without using enormous amounts of buffer, highly variable data have been generated for many species.

Despite these difficulties, un-ionized ammonia is regarded as highly toxic to most species of aquatic invertebrates and fish, with 96-h LC_{50}s typically ranging from 0.1 to 0.5 mg/L. Some species, exhibiting LC_{50}s of 1 to 3 mg/L, are apparently more resistant to unionized ammonia (Table 18.6).

Table 18.6. Acute toxicity (96-h LC_{50}, mg/L) of un-ionized ammonia to several invertebrate species.

Species	LC_{50}
Oligochaete	
Limnodrilus hoffmeisteri	1.9
Gastropod	
Lymnaea stagnalis	1.0
Crustaceans	
Gammarus pulex	2.1
Asellus aquaticus	2.3
Mayfly (Ephemeroptera)	
Baetis rhodani (nymph)	1.7
Caddisfly (Trichoptera)	
Hydropsyche angustipennis (larva)	3.0
Chironomidae	
Chironomus riparus (larva)	1.7

Source: Williams et al. (1986).

Table 18.7. Recommended guidelines (mg/L) for the protection of aquatic life exposed to total ammonia.

	Ammonia concentration at following temperature (°C)						
pH	0	5	10	15	20	25	30
6.50	2.5	2.4	2.2	2.2	1.5	1.0	0.7
6.75	2.5	2.4	2.2	2.2	1.5	1.0	0.7
7.00	2.5	2.4	2.2	2.2	1.5	1.0	0.7
7.25	2.5	2.4	2.2	2.2	1.5	1.0	0.7
7.50	2.5	2.4	2.2	2.2	1.5	1.0	0.7
7.75	2.3	2.2	2.1	2.0	1.4	1.0	0.7
8.00	1.5	1.4	1.4	1.3	0.9	0.7	0.5
8.25	0.9	0.8	0.8	0.8	0.5	0.4	0.3
8.50	0.5	0.5	0.5	0.4	0.3	0.2	0.2
8.75	0.3	0.3	0.3	0.3	0.2	0.2	0.1
9.00	0.2	0.2	0.2	0.2	0.1	0.1	0.1

Source: US Environmental Protection Agency (1985).

Nitrates are essentially nontoxic is most aquatic species, both marine and freshwater. In the majority of cases, the LC_{50} for nitrate is >50 mg/L, and may exceed 1,000 mg/L under some conditions.

Nitrite, on the other hand, is much more toxic, and may cause mortality at concentrations of approximately 0.05 mg/L in sensitive species. Exposure to nitrite leads to the formation of methemoglobin, a derivative of hemoglobin that is incapable of binding with oxygen. This effect, together with liver hypoxia, is thought to be the cause of nitrite toxicity. Exposure to high concentrations of nitrite also causes gill damage, thereby reducing the passage of oxygen through the gill membrane (Ansari, 1987).

Because pH and temperature have such a marked effect on the concentration of un-ionized ammonia, guidelines aimed at protecting freshwater life are highly detailed (Table 18.7). The criteria used for saltwater life in the United States are 0.035 mg/L (3–yr average) and 0.233 mg/L (1-h average). No guideline is in place for nitrates, except that residues should be low enough to avoid the development of excessive plant growth. The guideline for nitrite is 0.06 mg/L in several nations.

Health Effects

Intake

The typical Western diet yields approximately 200–250 mg nitrate per day in a 70-kg reference man, with 85–90% coming from vegetables and 0–9% from cured meats. Inhalation is an insignificant route of exposure, contributing only 0.05–0.10 mg/day, even in cities where vehicular traffic is heavy. Approximately 20 mg of nitrate are ingested daily from drinking water, assuming a nitrate concentration of 10 mg/L and a daily water consumption of 2 L. Nitrites are ingested at a daily rate of 11–12 mg in adults and 3–4 mg in infants.

Acute and Chronic Toxicity

The toxicity of nitrate in humans is due to the body's reduction of nitrate to nitrite. Reduction is particularly produced (essentially 100% conversion) in the gastrointestinal tract of infants during the first 3 months of life and, to a lesser degree, in the saliva of people of all ages. At low doses, methemoglobin is produced, resulting in a condition called methemoglobinemia, in which hemoglobin is oxidized, resulting in asphyxia. From 1945 to 1970, approximately 2000 cases of infant methemoglobinemia were reported in the United States, with a mortality rate of 8% (Fan et al., 1987). Since methemoglobinemia is rapidly reversible, there are no long-term or cumulative effects.

At high nitrate doses, vasodilatory and cardiovascular effects may be observed. No adverse reproductive or teratogenic effects are normally reported following ingestion of nitrates (Fan et al., 1987). However, Arbuckle et al. (1988) noted that there was a small increase in risk of delivering an infant with a central nervous system disorder for mothers consuming water nitrates of up to 26 mg/L; that study involved groundwater supplies in New Brunswick (Canada).

Carcinogenicity

Nitrates, nitrites, and ammonia are not considered carcinogenic in humans or experimental animals.

Drinking Water

Residues

Nitrates are almost always detected in finished drinking water. In a survey by the American Water Works Association (1985) of systems in 39 states and three territories, nitrates were in noncompliance with the maximum contaminant limit (10 mg/L) in 369 cases. For comparison, fluoride and lead were in noncompliance in 907 and 55 cases, respectively. In those cases of noncompliance, nitrates averaged 15 mg/L with a range of 10–90 mg/L. Other studies have noted much higher levels in groundwater.

Consumption Guidelines

Most nations, plus the World Health Organization, have a guideline of 10 mg/L for nitrates and 1 mg/L for nitrites in drinking water. Both guidelines are aimed at preventing methemoglobinemia, primarily in infants, and are based on a study by Craun et al. (1981). In that work, levels of up to 0.111 mg nitrate/L failed to produce methemoglobin in children 1–8 years old. A provisional AADI (Adjusted Acceptable Daily Intake) was then calculated for a 4-kg infant using the work of Walton (1951). In that work, there were no cases of methemoglobinemia when drinking water contained 0.010 mg nitrate/L. Because there was no need to use an uncertainty factor, owing to the large number of subjects in Walton's study, the NOAEL (No-Observed-Adverse-Effect Level) was determined to be 10 mg/L.

The same studies were used for the derivation of the guideline for nitrite. However, an uncertainty factor of 10 has been applied to the NOAEL because of the direct effect of nitrites on infants, resulting in a concentration of 1 mg/L.

Treatment

Since nitrates are relatively common in drinking water, many water treatment facilities make special provision for their removal.Conventional water treatment methods are ineffective in removal, so the most commonly used process is anion exchange. One such facility in the San Joaquin Valley resulted in a nitrate reduction from 16 to 2.6 mg/L (Lauch and Guter, 1986). Sulfate also decreased (from 120 to 40 mg/L) whereas chloride increased (from 88 to 166 mg/L). The total cost of the system amounted to 24.2 cents/1000 gal.

Several other methods are potentially available for nitrate removal, including biological denitrification. However, the practice of prechlorination greatly reduces the effectiveness of such techniques (Bouwer and Crowe, 1988). Nitrites are, in most instances, rapidly oxidized by chlorine, so no special treatment is necessary for their removal.

Recommendations

The environmental threat posed by ammonia, nitrate, and nitrite to users of surface water has been known for a long time. Essentially all major research topics in the areas of environmental chemistry, aquatic toxicology, health effects, and drinking water have been adequately evaluated. Hence, research in any of these areas time cannot be considered a priority at the present time.

There are, however, two key aspects of monitoring that continue to remain important. One is the discharge of ammonia-bearing wastes to surface waters. At present, many regulatory agencies permit such discharges regardless of pH or temperature of the receiving water. Although it is widely assumed that the ammonia wastes are rapidly dissipated, high pH under summer conditions can result in the formation of an extended toxic, mixing zone. It is therefore important to monitor the formation of such zones with the aim of reducing discharges during unfavorable conditions.

The other important area is monitoring of nitrates and nitrites in drinking water.This will become increasingly important as the use of nitrogen-based fertilizers continues to grow. Although groundwater supplies are at particular risk, high nitrate levels are also found in surface waters.

References

American Water Works Association. 1985. An AWWA survey of inorganic contaminants in water supplies. *Journal of the American Water Works Association* 77:67–72.

Ansari, I.A. 1987. Acute toxicity of sodium nitrite on *Channa punctatus* (Bloch) and *Mystus (Mystus) vittatus* (Bloch). *Acta Hydrochimica Hydrobiolgica* 15:175–178.

Arbuckle, T.E., G.J. Sherman, P.N. Corey, D. Walters, and B. Lo. 1988. Water nitrates and CNS birth defects: a population-based case-control study. *Archives of Environmental Health* 43:162–167.

Beukema, A.A., G.P. Hekstra, and C. Venema. 1986. The Netherlands' environmental policy for the North Sea and Wadden Sea. *Environmental Monitoring and Assessment* 7:117–155.

Bouwer, E.J., and P.B. Crowe. 1988. Biological processes in drinking water treatment. *Journal of the American Water Works Association* 80:82–93.

Canadian Water Quality Guidelines. 1987. Canadian Council of Resource and Environment Ministers. Environment Canada, Ottawa.

Craun, G.F., D.G. Greathouse, and D.H. Zunderson. 1981. Methemoglobin levels in young children consuming high nitrate well water in the United States. *International Journal of Epidemiology* 10:309–317.

Crumpton, W.G., and T.M. Isenhart. 1987. Nitrogen mass balance in streams receiving secondary effluent: the role of algal assimilation. *Journal of the Water Pollution Control Federation* 59:821–824.

Delavalle, M. 1983. Prevention del'enrichissement des eaux souterraines en nitrates d'origine agricole. In: *Ground water in water resources planning,* IAHS Publication Number 142, 1005–1014. UNESCO, Paris.

Fan, A.M., C.C. Willhite, and S.A. Book. 1987. Evaluation of the nitrate drinking water standard with reference to infant methemoglobinemia and potential reproductive toxicity. *Regulatory Toxicology and Pharmacology* 7:135–148.

Kennedy, M.S., and J.M. Bell. 1986. The effects of advanced wastewater treatment on river water quality. *Journal of the Water Pollution Control Federation* 58:1138–1144.

Kudeyarov, V.N., and V.N. Bashkin. 1980. Nitrogen balance in small river basins under agricultural and forestry use. *Water, Air, and Soil Pollution* 14:23–27.

Lauch, R.P., and G.A. Guter. 1986. Ion exchange for the removal of nitrate from well water. *Journal of the American Water Works Association* 78:83–88.

Lindau, C.W., R.D. De Laune, and G.L. Jones. 1988. Fate of added nitrate and ammonium-nitrogen entering a Louisiana gulf coast swamp forest. *Journal of the Water Pollution Control Federation* 60:386–389.

McLeod, R.V., and R.O. Hegg. 1984. Pasture runoff water quality from application of inorganic and organic nitrogen sources. *Journal of Environmental Quality* 13:122–126.

Nevissi, A.E., F.B. DeWalle, J.F.C. Sung, K. Mayer, and R. Dalsey. 1988. Heavy metal variability of different municipal sludges as measured by atomic absorption and inductively coupled plasma emission spectroscopy. *Journal of Environmental Science and Health* 23:823–841.

Onstad, C.A., and J. Blake. 1980. Thames basin nitrate and agricultural relations. *In: Watershed management 1980,* 961–973. American Society of Civil Engineers, New York.

Piper, R.G., I.B. McElwain, L.E. Orme, J.P. McCraren, L.G. Fowler, and J.R. Leonard. 1982. *Fish hatchery management.* U.S. Department of the Interior, Washington, DC. 517 pp.

Quentin, K.E. 1988. Die Nitratsituation in der Bundersrepublik Deutschland. *Acta Hydrochimica Hydrobiologica* 16:385–396.

Robertson, G.P. 1986. Nitrogen: regional contributions to the global cycle. *Environment* 28:16–29.

Stull, J.K., R.B. Baird, and T.C. Heesen. 1986. Marine sediment core profiles of trace constituents offshore of a deep wastewater outfall. *Journal of the Water Pollution Control Federation* 58:985–991.

US Environmental Protection Agency. 1985. *Ambient water quality criteria for ammonia—1984.* US Environmental Protection Agency, EPA–440/5–84–026. Washington, DC.

Walton, G. 1951. Survey of literature relating to infant methemoglobinemia due to nitrate contaminated water. *American Journal of Public Health* 41:986–996.

Williams, K.A., D.W.J. Green, and D. Pascoe. 1986. Studies on the acute toxicity of pollutants to freshwater macroinvertebrates. *Archives fur Hydrobiolgie* 106:61–70.

19
Selenium

Selenium, with an average crustal abundance of 0.1 mg/kg, occurs in most types of rock, but is enriched in shale (approx. 0.6 mg/kg). Geochemically, selenium resembles sulfur, and the two are often found in association with one another. Although unweathered rocks may contain an abundance of metallic selenide, weathering enhances oxidation of that complex. Selenium is present in acidic soils as the selenite ion bound to iron oxide whereas in alkaline soils, it is oxidized to the selenate ion.

Production, Sources, and Residues

Production

World production of selenium is relatively small, currently at 1.5×10^3 metric tons/yr (excluding China and the Soviet Union). That rate makes selenium the 77th most important industrial mineral worldwide in terms of production weight (Noetstaller, 1988). Selenium's main uses are in photoelectric cells (particularly those involved in photocopying), glass and ceramic production, rectifiers, and metal alloy and rubber production. There are also some minor applications in insecticides and topical therapeutic agents.

Sources

Selenium is released in relatively large amounts as a by-product of various industrial and other anthropogenic sources. Total discharge from such sources to freshwaters amounts 10–72 × 10^3 metric tons/yr (Table 19.1). The No. 1 source is discharge from coal-burning power plants followed by discharge from the smelting and refining of nonferrous metals, and discharge of domestic wastewater (Table 19.1).

At present, natural sources of selenium to the atmosphere account for slightly more than 50% of all emissions to the atmosphere (Nriagu, 1989; Nriagu and Pacyna, 1988). The primary natural sources (in thousand metric tons/yr) are:

marine biogenic particles, 0.4–9.0
biogenic continental volatiles, 0.15–5.0
volcanoes, 0.1–1.8
sea salt spray, 0–1.1
wild forest fires, 0–0.5
windborne soil particles, 0.01–0.35
biogenic continental particles, 0–0.25

The primary anthropogenic sources (in thousand metric tons/yr) are:

coal combustion (industrial and domestic), 0.79–1.98
copper-nickel production, 0.43–1.28
coal combustion (electrical utilities), 0.11–0.78
oil combustion (industrial and electrical), 0.11–0.54
lead production, 0.20–0.39
oil combustion (electrical utilities), 0.04–0.29

Table 19.1. Worldwide anthropogenic input of selenium to freshwaters.

Source	Input (thousand metric tons/yr)
Coal-burning power plants	6–30
Smelting and refining nonferrous metals	3–20
Domestic wastewater	0–4.5
central	
noncentral	0–3.0
Manufacturing processes	0–5.0
metals	
chemicals	0.02–2.5
pulp and paper	0.01–0.9
petroleum products	0–0.09
Dumping of sewage sludge	0.3–3.8
Atmospheric fallout	0.5–1.1
Base metal mining and dressing	0.2–1.0

Source: Nriagu and Pacyna (1988).

The central role of combustion in the preceding dissemination patterns is due to the high concentration of selenium in coal and other fossil fuels (reviewed by Lemly, 1985). Selenium in coal typically ranges from 0.2 to 10 mg/kg. Upon combustion, 50–70% of that residue is bound to fly ash, with the remainder going to the atmosphere in the vapor phase. If SO_2 is present in excess, selenium is mobilized as follows:

$$SeO_2 + 2SO_2 \rightleftharpoons Se^o + 2SO_3$$

As might be anticipated, the concentration of selenium in air is closely related to the proximity of power plants (Dutkiewicz and Husain, 1988).

Since selenium residues in oil are approximately 1 order of magnitude lower than those found in coal, the mobilization of selenium from oil combustion is also relatively low. In addition, oil combustion is more efficient than that of coal, so atmospheric emissions are correspondingly lower. Patterson et al. (1987) noted that selenium partitioned favorably (79–95 mg/kg) to oil after retorting compared to the retort waters (0.02–25 mg/kg).

Mining and smelting of nonferrous metals continue to be the second most important source of selenium to freshwaters (Table 19.1). Most of the input comes from selenium's association with sulfur-bearing ores of copper, nickel, and lead. Under typical operating conditions in Western nations, selenium is discharged in the waste water and, to a lesser degree, the atmosphere (assuming that precipitators are working at 99% efficiency).

Selenium residues are often relatively high in municipal waste waters and sludges. In a study of 25 sewage treatment plants in the state of Washington, Sung et al. (1986) reported that concentrations in secondary sludge ranged up to 1.4 mg/L whereas residues in the liquid effluent reached a maximum of only 0.002 mg/L (Table 19.2). In a related study, Nevissi et al. (1988) showed that the total amount of selenium discharged from two sewage plants in the state of Washington was 0.04–0.07 kg/day.

Table 19.2. Concentration (mg/L) of total selenium in wastes from 25 treatment plants in the state of Washington.

Sample	Average	Range
Raw sewage	0.001	ND[a]–0.005
Primary effluent	0.001	ND–0.004
Secondary effluent	0.001	ND–0.002
Discharge	0.001	ND–0.002
Primary sludge	0.050	ND–0.092
Secondary sludge	0.047	0.013–0.140

[a]Not detected.
Source: Sung et al. (1986).

Selenium is also present in urban runoff, albeit in relatively low concentrations. For example, Marsalek and Schroeter (1988) noted that the total loadings to the Canadian Great Lakes comes to 1.1 metric tons/yr; the individual loadings (metric tons/yr) are as follows: Erie, 0.15; Huron, 0.12; Ontario, 0.60; St. Clair, 0.09; Superior, 0.15.

A final source of selenium, with enormous local consequences, is the movement of selenium-bearing groundwaters to the surface, as a result of agricultural activities. One of the best illustrations of this problem is the San Joaquin Valley (California), where intensive irrigation has resulted in the movement of naturally occurring selenium to Kesterson Reservoir, a major wildlife refuge (Presser and Ohlendorf, 1987). In that case, the concentration of dissolved Se in inflows ranged up to 1.4 mg/L while, in the reservoir, residues were as high as 0.35 mg/L. For comparison, selenium in control areas was <0.002 mg/L.

Residues

Water. In his review of several studies, Cutter (1989) noted that dissolved Se residues in freshwaters typically range from <0.1 μg/L to 5 μg/L when there are no major anthropogenic inputs. In another review, Siu and Berman (1989) reported that dissolved Se in seawater is normally <0.1 μg/L near the surface, increasing to 0.11–0.2 μg/L in deeper water (50–3,000 m). The lower concentrations near the surface are due, in part, to utilization of selenium by phytoplankton, and sorption to biogenic/nonbiogenic particles, followed by downward drift of these particles.

Sediments. Total Se in sediments is highly variable, depending primarily on underlying geologic conditions. Residues in the 0.2 to 30 mg/kg dry weight range have been been reported for many freshwaters (Cutter, 1989), whereas residues in marine sediments are typically below 1 mg/kg. Stull and Baird (1985) reported that sediments in coastal waters near Los Angeles contained total Se of up to 4.1 mg/kg, compared to background levels of 0.1–0.4 mg/kg. Those elevated levels were due to the input of municipal wastes (132,500 m^3/day) through a single point source.

Chemistry

Selenium exhibits a complicated chemistry because there are four oxidation states, and inorganic and organic compounds exist in surface waters. Dissolved inorganic selenium can exist as Se^{2-}, mainly as biselenide (HSe^-); Se^o, as colloidal elemental selenium; Se^{4+}, as the selenite oxyanion ($HSeO_3^-$ and SeO_3^{2-}); and Se^+, and as the selenate oxyanion (SeO_4^{2-}). Organic Se species are analogous to those of sulfur and include seleno amino acids and their derivatives, methyl selenones, and methylselenonium ions.

Selenite and selenate are the dominant species in most freshwaters, where they typically account for more than 90% of total Se in the water column (reviewed by Cutter, 1989). Unpolluted, aerobic rivers and lakes typically contain a predominance of selenate whereas selenite increases in concentration in freshwaters receiving anthropogenic inputs of selenium-bearing wastes. Selenite and selenate are also dominant in seawater (reviewed by Siu and Berman, 1989); although the relative proportion of these two species varies with depth, selenite typically reaches greatest importance in deep (>100 m) water.

A large percentage of selenium occurs in organic complexes, a result of sorption by biogenic particles and methylation. Cooke and Bruland (1987), working on surface waters from the San Joaquin and Imperial Valleys of California, detected the presence of dimethylselenonium ions ($DMSe^+$-R), volatile dimethyl selenides (DMSe), and volatile dimethyl diselenides (DMDSe). All three of these species occurred in relatively low concentrations, accounting for no more than 10% of total-Se in the samples. Dimethyl selenide was possibly produced as per the following reaction:

$$SeO_3^{2-} \text{ and } SeO_4^{2-} \rightarrow CH_3Se(CH_2)_2\underset{\displaystyle NH_3}{\underset{|}{C}}HCOOH \text{ (selenomethionine)}$$

$$\rightarrow 2CH_3Se^+{-}(CH_2)_2\underset{\displaystyle NH_2}{\underset{|}{C}}HCOOH \text{ (Se methyl selenomethionine)} \rightarrow$$

$$\overset{HOH}{\rightarrow} CH_3SeCH_3 \text{ (dimethyl selenide)} + H_2\underset{\displaystyle OH}{\underset{|}{C}}CH_2\underset{\displaystyle NH_2}{\underset{|}{C}}HCOOH \text{ (homoserine)}$$

Up to 60% of total Se in marine waters is organically bound (Siu and Berman, 1989). This results in surface depletion of dissolved Se species during the growing season, as paralleled by nutrients such as phosphorus and silica. A large portion of organic Se eventually sinks to deeper water, where it is ultimately carried to surface waters through vertical mixing.

Selenium is effectively scavenged by biotic and abiotic components of bottom sediments. Selenites, in particular, bind selectively to iron and aluminum sesquioxides, forming stable, relatively insoluble complexes. Under acidic and/or reducing conditions, ferric selenite may be reduced to elemental selenium for mobilization to the water column. That process only occurs under anaerobic winter conditions, or in lakes suffering from acute acidification. Alkaline and oxidizing conditions favor the formation of selenates, which are soluble and easily transported. Deverel and Millard (1988), for example, showed that the movement and concentration of selenium in groundwater in the San Joaquin Valley (California) was significantly correlated with the specific conductance of several samples.

Currently, relatively little is known about the environmental chemistry of selenium and its complexes. This primarily reflects its low concentration and concomitant negligible impact in most surface waters. However, increasingly intense irrigation and use of coal as an energy source will inevitably enhance mobilization of selenium in the aquatic environment. Because there are already demonstrated problems with selenium in some areas (such as the San Joaquin Valley), much more needs to be known about the factors affecting the mobilization and fate of selenium under different environmental conditions.

Bioaccumulation

Plants

Selenium is ubiquitous, occurring in the tissues of all aquatic plants and animals. Residues in algae typically range from 0.1 to >1 mg/kg dry weight, yielding concentrations factors (residue in plants/residue in water—CF) of 150 to >1,000 (Table 19.3) (Lemly, 1985). Demon et al. (1988), working with the green alga *Scenedesmus pannonicus* and the fungus *Aureobasidium pullulans*, noted that the rate of uptake of ^{75}Se was independent of pH within the range 5 to 7. The same was also noted for Cd^{2+}, Zn^{2+}, and La^{3+}.

Invertebrates

Concentration factors for invertebrate species are highly variable, reflecting the diversity of species within that group of animals. Most reports place the CF at 300 to 5,000 (Lemly, 1985), so total Se residues in most species are relatively low (Table 19.3). Appreciable uptake occurs through food and water. Schultz et al. (1980), working with the cladoceran *Daphnia pulex*, showed that uptake was greatest in unfed animals and that depuration was practically complete after 96 h.

Relatively little is known about the accumulation of different species of selenium. Pelletier (1986) noted that the mussel *Mytilis edulis* sorbed inorganic Se (Na_2SeO_3) at a relatively slow rate (120 μg Se/g/day) whereas organic Se was not accumulated. When inorganic Hg was added to the culture waters, the rate of accumulation of inorganic Se doubled, and when methyl–Hg was added, the accumulation rate tripled. The presence of methyl Hg resulted in uptake of organic Se at a rate of 150 fg Se/g/day. Other studies have also noted that the uptake of selenium and mercury are closely tied to one another, and that selenium may detoxify mercury and cadmium (Micallef and Tyler, 1987).

Table 19.3. Concentration (mg/kg dry weight) of total selenium in aquatic plants and invertebrates.

Species	Average (range)	Location
Submersed plants, 8 species[1]	NR[a] (ND[b]–2.70)	Saginaw Bay, Lake Michigan
Terrestrial/emergent plants, 15 species[1]	NR (ND–1.6)	Saginaw Bay, Lake Michigan
Submersed plants, 7 species[1]	NR (ND–0.9)	Small lakes, Michigan
Oyster, *Crassostrea virginia*[2]	0.1 (NR)	Lake Pontchartrain, Louisiana
Bivalve mollusc, *Rangia cuneata*[2]	0.3 (NR)	Lake Pontchartrain, Louisiana
Coral, *Pocillopora damicornis*[3]		
tissues	ND (ND)	Intertidal reefs,
skeleton	1.5 (1.4–1.5)	Thailand
Lobster, *Homarus americanus*[4]		
digestive gland	NR (0.3–2.7)[c]	Coastal waters, eastern Canada

[a]Not reported.
[b]Not detected.
[c]Wet weight.
Sources: [1]Estabrook et al. (1985), [2]Byrne and DeLeon (1986), Howard and Brown (1987), [4]Chou and Uthe (1978).

Fish

Total Se in fish muscle tissue generally ranges up to 1.5 mg/kg wet weight in noncontaminated and up to >5 mg/kg in contaminated areas (Table 19.4). Some of the highest levels on record come from the San Joaquin Valley in California, where heavy irrigation and concomitant evaporation has caused an upward movement of selenium-tainted groundwater to the surface. Residues in fish from that area were relatively low during the early 1970s, increasing about twofold between 1973 and 1977 (Saiki and May, 1988). Eventually, die-offs of some species, such as mosquito fish *Gambusia affinis* were observed (Fan et al., 1988). The following health advisories were issued: (1) no consumption of fish where selenium concentrations exceed 5 mg/kg; (2) 4 ounces per 2 weeks where fish contain elevated selenium residues, but not exceeding 5 mg/kg; and no consumption of selenium-tainted fish by women who are pregnant or may soon become pregnant or by children under the age of 15. High residues and die-offs were also noted in several species of aquatic birds (Presser and Ohlendorf, 1987).

Selenium is selectively deposited in the liver of fish. Capelli and Minganti (1987) noted that total Se in the liver of Atlantic bonito *Sarda sarda* from the Gulf of Genoa averaged 5.8 mg/kg wet weight with a range of

Table 19.4. Concentration (mg/kg wet weight) of total Se in the muscle of fish.

Species	Average (range)	Location
Sole, *Solea solea*[1]	0.3 (NR[a])	Coast of Holland
Cod, *Gadus morhua*[1]	0.3 (NR)	Coast of Holland
Herring, *Clupea harengus*[1]	0.3 (NR)	Coast of Holland
Yellowtail kingfish, *Serio lalanland-grandis*[2]	0.3 (0.05–0.7)	Coast of Australia, Sydney
Atlantic bonito, *Sarda sarda*[3]	0.7 (0.2–1.7)	Gulf of Genoa, Italy
Bluegill, *Lepomis macrochirus*[4]	0.5 (0.4–3.2)	San Joaquin Valley, California[b]
Carp, *Cyprinus carpio*[4]	0.7 (0.6–5.5)	San Joaquin Valley, California[b]
Mosquito fish, *Gambusia affinis*[5]	NR (24–98)[c]	San Joaquin Valley, California[b]

[a]Not reported.
[b]Whole body.
[c]Contaminated with selenium from groundwater.
Sources: [1]Hagel (1986), [2]Chvojka (1988), [3]Capelli and Minganti (1987), [4]Saiki and May (1988), [5]Fan et al. (1988).

2.58 to 8.48 mg/kg. The corresponding values for muscle tissue from the same specimens were 0.7 (0.2–1.7) mg/kg.

Both selenate and selenite are concentrated from water by fish. In an experimental study, Bertram and Brooks (1986) showed that residues in fathead minnow *Pimephales promelas* reached an equilibrium within 4 weeks of exposure, yielding a concentration factor of 25. Fish exposed to selenium in the food had not reached an equilibrium after 11 weeks, but equilibrium concentrations were calculated to be only 0.3 of the residue (54 fg Se/g) in food. The accumulation of selenium from both food and water was additive.

Relatively little is known about the uptake of organic Se by fish or about the uptake of inorganic Se under different environmental conditions. Although it now appears that the uptake of mercury by many species is not closely related to that of inorganic Se, there are essentially no data on the effects of mercury and methyl Hg on the uptake of organic Se.

Toxic Effects to Aquatic Organisms

Plants

Inorganic Se is moderately to highly toxic to aquatic plants. Wang (1986) reported that the Effective Concentration $(EC)_{50}$ for duckweed *Lemna minor* was 2.4 mg Se/L whereas the maximum permissible concentration for that same species was 0.24 mg/L. For comparison, the surface water quality standard for the state of Illinois (where the work was done) was 1.0 mg/L. Hutchinson and Stokes (1975) noted that the growth of two green algal species was limited by inorganic Se at a much lower concentration (50 μg Se/L) that found by Wang, while Bennett (1988) noted growth retardation in the green alga *Chlorella pyrenoidsa* at a concentration of only 0.8 μg Se/L. At the opposite end of the spectrum, the EC_{50} of the marine alga *Skeletonema costatum* exposed to selenite acid was 8,200 μg/L.

Invertebrates

Inorganic Se is acutely toxic to invertebrates at the following concentrations:

Cladoceran *Daphnia magna*, 0.4 mg Se/L (water hardness <100 mg/L $CaCO_3$) (LeBlanc, 1984)
Cladoceran *Daphnia magna*, 0.68 mg Se/L (selenite), 0.75 mg Se per liter (selenate) (water hardness, not reported) (Johnston, 1987)
Amphipod *Hyallela azteca*, 0.3 mg Se/L (water hardness 329 mg/L $CaCO_3$) (Halter et al., 1980)
Shrimp *Mysidopsis bahia*, 15.0 mg Se/L (saltwater) (Suter and Rosen, 1988)
Crab *Scylla serrata*, 33.0 mg Se/L (saltwater) (Krishnaja et al. 1987)

Of those species tested, *Daphnia magna* and *Hyallela azteca* were most sensitive and the two marine species were most resistant to inorganic Se. Chronic toxicity has been reported for a number of freshwater and marine species at concentrations below 0.5 mg/L (Mance, 1987; Johnston, 1987).

Fish

Although the LC_{50} of inorganic Se for freshwater fish varies from 0.3 to >5 mg/L, depending on species, life stage, and environmental conditions (Mance, 1987), chronic effects have been noted at much lower concentrations. Hunn et al. (1987) reported that survival of young rainbow trout *Oncorhynchus mykiss* was significantly reduced following exposure to Se^{4+} at 0.04–0.10 mg/L. Those exposures, which lasted for 90 days, also caused a reduction in the length and weight of fry. Similarly, Davies and Goetti (1975) showed that the No-Effect concentration for developing eggs and fry of rainbow trout exposed to inorganic Se was 0.04–0.08 mg/L. Other chronic effects, such as hematological changes, have been noted at concentrations as low as 0.03 mg/L (Hodson et al., 1980).

Relatively little is known about the toxicity of organic Se complexes, and the factors affecting toxicity, to fish (and other aquatic species). In one of the few studies available, Niimi and LaHam (1975) noted that selenomethionine was approximately 10 times more toxic than inorganic Se.

Although that study was restricted to newly hatched zebrafish *Brachydanio rerio*, the same response would likely be observed in many other aquatic species. Because organic Se compounds are present in varying concentrations in surface waters, further studies need to be conducted on the toxicity of such compounds.

Several nations have promulgated surface water guidelines for the protection of aquatic species, as illustrated below.

Canada
  0.001 mg/L (total selenium)
United States
  0.035 mg/L (total selenium as a 24-h average)
  0.260 mg/L (total selenium, maximum).

Lemly (1985) recommended the following water quality objectives: 0.050 mg/L (total selenium, for the protection of salmonids), and 0.010 mg/L (total selenium, for the protection of centrarchids such as sunfish and bass).

Health Effects

Intake

The typical Western diet yields approximately 0.15 mg Se/day in a 70-kg reference man. An additional 0.02–0.03 mg/day comes from water and inhalation for the nonoccupationally exposed population. Once ingested, selenium is efficiently (>75%) absorbed, and generally distributed throughout the body. Greatest residues are found in the liver and kidneys, with lesser amounts in the heart, lung, spleen, and pancreas. Total soft-tissue body burden in the 70-kg reference man is approximately 13 mg. Selenium is eliminated in the urine (0.05 mg/day), feces (0.02 mg/day), and perspiration and other routes (0.08 mg/day).

Selenium is an essential nutrient, the recommended daily intake for adults being 0.05–0.20 mg. A deficiency in selenium typically suppresses the immune system, resulting in the following responses (Kiremidjian-Schumacher and Stotzky, 1987): reduced resistance to microbial and viral infection; reduced neutrophil function; reduced antibody production; proliferation of T and B lymphocytes; cytodestruction by T lymphocytes.

Selenium is also beneficial owing to its antagonistic action toward the toxicity of arsenic, cadmium, mercury, silver, and thallium. Inorganic Se compounds also inhibit tumors, including those of the skin, liver, mammary gland, colon, and lung. In animals, diets containing <0.05 mg Se/kg result in selenium deficiency and a condition known as ''white muscle disease.'' That condition is caused by a deficiency of both vitamin E and selenium.

Acute Toxicity

Industrial exposure to excessive selenium (usually selenious acid, H_2SeO_3) often results in an allergenic and/or irritative response, primarily in the mucous membranes and eyes (Carson et al., 1987). Acute exposure of livestock to selenium results in a condition known as "blind staggers," which is characterized by impairment of vision, abdominal pain, and respiratory failure. Acute exposure of mallard ducks *Anas platyrhynchos* to selenomethionine resulted in embryotoxic and teratogenic effects (Hoffman and Heinz, 1988).

Chronic Toxicity

Chronic effects in humans are rare and have not been reported following the ingestion of treated water. Livestock may develop "alkali disease" following ingestion of excessive amounts of selenium over the long term. That condition is characterized by emaciation, atrophy and decompensation of the heart, liver cirrhosis, renal glomerulonephritis, and anemia (Carson et al., 1987).

Carcinogenicity

The International Agency for Research on Cancer considers that there are insufficient data on animals to allow an evaluation of the carcinogenicity of selenium compounds. In man, there is no clinical or epidemiological evidence to suggest that any selenium compound is carcinogenic.

Drinking Water

Residues

Selenium is often detected in drinking water, but typically in low concentrations (<0.01 mg/L). In a survey of the American Water Works Association (1985) of drinking water systems in 39 states and three territories, there were 88 cases of noncompliance with the Maximum Contaminant Limit of 0.01 mg/L. For comparison, fluoride and nitrate were in noncompliance in 907 and 369 cases, respectively. In those cases of noncompliance, selenium residues averaged 0.026 mg/L with a range of 0.011–0.07 mg/L. The World Health Organization (1984), summarizing a number of studies from West Germany, Australia, Canada, and the United States, also noted that selenium concentrations only rarely exceeded 0.01 mg/L; the highest levels were typically limited to groundwater supplies in seleniferous or heavily irrigated areas.

Consumption Guidelines

Many nations, plus the World Health Organization, use a drinking water guideline of 0.01 mg Se/L. This value was derived from a number of epidemiological and laboratory studies indicating that the adjusted acceptable daily intake of selenium is 0.106 mg/L (US Environmental Protection Agency, 1985). An uncertainty factor of 10 was used in the derivation of the final guideline. In addition, the guideline assumes that the maximum daily intake of selenium from drinking water does not exceed 10% (World Health Organization, 1984).

Treatment

Se^{4+} is effectively removed from drinking water supplies by coagulation with ferric sulfate and by pH adjustment by lime softening. Fixed bed activated alumina adsorption is also effective in the removal of Se^{4+} and Se^{6+} if acidic pH is maintained.

Recommendations

Selenium is a potentially toxic contaminant of surface water and may bioaccumulate to relatively high levels in fish and other aquatic species. Several selenium complexes, with different chemical and toxicologic properties, exist in the environment, and, in several cases, relatively little detailed information is available on the fate of these compounds. Fortunately, episodes of adulteration of surface waters by selenium are relatively rare and limited to seleniferous zones and areas of intensive irrigation. Hence, selenium is not a widespread contaminant of water, but in areas where it abundant (such as the San Joaquin Valley), significant toxicologic impacts may be noted.

As the need for food production increases worldwide, so does the use of irrigation, both in terms of the area serviced and the intensiveness of existing practices. This means that episodes of selenium contamination are likely to increase in frequency, particularly in seleniferous regions of the world. Monitoring of the water in such regions is therefore a priority. There are also several deficiencies in the data base regarding selenium complexes as noted below.

1. Mobilization and fate of selenium complexes under different environmental conditions
2. Uptake of inorganic Se and organic Se by invertebrates and fish under different environmental conditions
3. Effect of methyl Hg on uptake of organic Se
4. Toxicity of organic-Se complexes to fish and other aquatic species

References

American Water Works Association. 1985. An AWWA survey of inorganic contaminants in water supplies. *Journal of the American Water Works Association* 77:67–72.

Bennett, W.N. 1988. Assessment of selenium toxicity in algae using turbidostat culture. *Water Research* 22:939–942.

Bertram, P.E., and A.S. Brooks. 1986. Kinetics of accumulation of selenium from food and water by fathead minnows. *Water Research* 20:877–884.

Byrne, C.J., and I.R. DeLeon. 1986. Trace metal residues in biota and sediments from Lake Pontchartrain, Louisiana. *Bulletin of Environmental Contamination and Toxicology* 37:151–158.

Capelli, R., and V. Minganti. 1987. Total mercury, organic mercury, copper, manganese, selenium, and zinc in *Sarda sarda* from the Gulf of Genoa. *Science of the Total Environment* 63:83–99.

Carson, B.L., H.V. Ellis, and J.L. McCann. 1987. *Toxicology and biological monitoring of metals in humans*. Lewis Publishers, Chelsea, MI. 328 pp.

Chou, C.L., and J.F. Uthe. 1978. *Heavy metal relationships in lobster* (Homarus americanus) *and rock crab* (Cancer irriratus) *digestive glands*. International Council for the Exploration of the Sea, Report E:15/C.M., Copenhagen.

Chvojka, R. 1988. Mercury and selenium in axial white muscle of yellowtail kingfish from Sydney, Australia. *Marine Pollution Bulletin* 19:210–213.

Cooke, T.D., and K.W. Bruland. 1987. Aquatic chemistry of selenium: evidence of biomethylation. *Environmental Science and Technology* 21:1214–1219.

Cutter, G.A. 1989. Freshwater systems. *In: Occurrence and distribution of selenium,* ed. M. Ihnat, 243–262. CRC Press, Boca Raton, FL.

Davies, P.H., and J.P. Goetti. 1975. *Study of the effects of the metallic ions on fish and aquatic organisms*. Water Pollution Studies, Colorado Department of Natural Resources, Wildlife Division, Boulder, CO. pp. 7–29.

Demon, A., M. De Bruin, and H.T. Wolterbeek. 1988. The influence of pH on trace element uptake by an alga *Scenedesmus pannonicus* subsp. *Berlin*) and fungus (*Aureobasidium pullulans*). *Environmental Monitoring and Assessment* 10:165–173.

Deverel, S.J., and S.P. Millard. 1988. Distribution and mobility of selenium and other trace elements in shallow groundwater of the western San Joaquin Valley, California. *Environmental Science and Technology* 22:697–702.

Dutkiewicz, V.A., and L. Husain. 1988. Spatial pattern of non-urban Se concentrations in the northeastern U.S. and its pollution source implications. *Atmospheric Environment* 22:2223–2228.

Estabrook, G.F., D.W. Burk, D.R. Inman, P.B. Kaufman, J.R. Wells, J.D. Jones, and N. Ghosheh. 1985. Comparison of heavy metals in aquatic plants on Charity Island, Saginaw Bay, Lake Huron, U.S.A., with plants along the shoreline of Saginaw Bay. *American Journal of Botany* 72:209–216.

Fan, A.M., S.A. Book, R.R. Neutra, and D.M. Epstein. 1988. Selenium and human health implications in California's San Joaquin Valley. *Journal of Toxicology and Environmental Health* 23:539–559.

Hagel, P. 1986. Monitoring of pollutants in Dutch fishery products. *Environmental Monitoring and Assessment* 7:257–262.

Halter, M.T., W.J. Adams, and H.E. Johnson. 1980. Selenium toxicity to *Daphnia magna, Hyallela azteca*, and the fathead minnow in hard water. *Bulletin of Environmental Contamination and Toxicology* 24:102–107.

Hodson, P.V., D.J. Spry, and B.R. Blunt. 1980. Effects on rainbow trout (*Salmo gairdneri*) of a chronic exposure to waterborne selenium. *Canadian Journal of Fisheries and Aquatic Sciences* 37:233–240.

Hoffman, D.J., and G.H. Heinz. 1988. Embryotoxic and teratogenic effects of selenium in the diet of mallards. *Journal of Toxicology and Environmental Health* 24:477–490.

Howard, L.S., and B.E. Brown. 1987. Metals in *Pocillopora damicornis* exposed to tin smelter effluent. *Marine Pollution Bulletin* 18:451–454.

Hunn, J.B., S.J. Hamilton, and D.R. Buckler. 1987. Toxicity of sodium selenite to rainbow trout fry. *Water Research* 21:233–238.

Hutchinson, T.C., and P.M. Stokes. 1975. *Heavy metal toxicity and algal bioassays*. ASTM STP 573. American Society for the Testing of Materials, Philadelphia, PA. 320 pp.

Johnston, P.A. 1987. Acute toxicity of inorganic selenium to *Daphnia magna* (Straus) and the effect of sub-acute exposure upon growth and reproduction. *Aquatic Toxicology* 10:335–352.

Kiremidjian-Schumacher, L., and G. Stotzky. 1987. Selenium and immune responses. *Environmental Research* 42:277–303.

Krishnaja, A.P., M.S. Rege, and A.G. Joshi. 1987. Toxic effects of certain heavy metals (Hg, Cd, Pb, As and Se) on the intertidal crab *Scylla serrata*. *Marine Environmental Research* 21:109–119.

LeBlanc, G.A. 1984. Interspecies relationships in acute toxicity of chemicals to aquatic organisms. *Environmental Toxicology and Chemistry* 3:47–60.

Lemly, A.D. 1985. Ecological basis for regulating aquatic emissions from the power industry: the case with selenium. *Regulatory Toxicology and Pharmacology* 5:465–486.

Mance, G. 1987. *Pollution threat of heavy metals in aquatic environments*. Elsevier, New York. 372 pp.

Marsalek, J., and H. Schroeter. 1988. Annual loadings of toxic contaminants in urban runoff from the Canadian Great Lakes. *Water Pollution Research Journal of Canada* 23:360–378.

Micallef, S., and P.A. Tyler. 1987. Preliminary observations of the interactions of mercury and selenium in *Mytilus edulis*. *Marine Pollution Bulletin* 18:180–185.

Nevissi, A.E., F.B. Dewalle, J.F.C. Sung, K. Mayer, and R. Dalsey. 1988. Heavy metal variability of different municipal sludges as measured by atomic absorption and inductively coupled plasma emission spectroscopy. *Journal of Environmental Science and Health* 23:823–841.

Niimi, A.J., and Q.N. LaHam. 1975. Relative toxicity of organic and inorganic compounds of selenium to newly hatched zebrafish (*Brachydanio rerio*). *Canadian Journal of Zoology* 54:501–509.

Noetstaller, R. 1988. *Industrial minerals*. World Bank Technical Paper Number 76. World Bank, Washington, DC. 117 pp. Nriagu, J.O. 1989. A global assessment of natural sources of atmospheric trace metals. *Nature* 338:47–49.

Nriagu, J.O., and J.M. Pacyna. 1988. Quantitative assessment of worldwide contamination of air, water and soils by trace metals. *Nature* 333:134–139.

Patterson, J.H., L.S. Dale, and J.F. Chapman. 1987. Trace element partitioning during the retorting of Julia Creek oil shale. *Environmental Science and Technology* 21:490–494.

Pelletier, E. 1986. Modification de la bioaccumulation du selenium chez *Mytilus edulis* en presence du mercure organique et inorganique. *Canadian Journal of Fisheries and Aquatic Sciences* 43:203–210.

Presser, T.S., and H.M. Ohlendorf. 1987. Biogeochemical cycling of selenium in the San Joaquin Valley, California. *Environmental Management* 11:805–821.

Saiki, M.K., and T.W. May. 1988. Trace element residues in bluegills and common carp from the lower San Joaquin River, California, and its tributaries. *Science of the Total Environment* 74:199–217.

Schultz, T.W., S.R. Freeman, and J.N. Dumont. 1980. Uptake, depuration, and distribution of selenium in *Daphnia* and its effects on survival and ultrastructure. *Archives of Environmental Contamination and Toxicology* 9:23–40.

Siu, K.W.M., and S.S. Berman. 1989. The marine environment, *in: Occurrence and distribution of selenium,* ed. M. Ihnat, 263–293, CRC Press, Boca Raton, FL.

Stull, J.K., and R.B. Baird. 1985. Trace metals in marine surface sediments of the Palos Verdes Shelf, 1974–1980. *Journal of the Water Pollution Control Federation* 57:833–840.

Sung, J.F.C., A.E. Nevissi, and F.B. Dewalle. 1986. Concentration and removal efficiency of major and trace elements in municipal wastewater. *Journal of Environmental Science and Health* 21:435–448.

Suter, G.W., and A.E. Rosen. 1988. Comparative toxicology for risk assessment of marine fishes and crustaceans. *Environmental Science and Toxicology* 22:548–556.

Wang, W. 1986. Toxicity tests of aquatic pollutants by using common duckweed. *Environmental Pollution (Series B)* 11:1–14.

World Health Organization. 1984. *Guidelines for drinking-water quality,* Volume 2: *Health criteria and other supporting information.* World Health Organization, Geneva. 335 pp.

20
Silver

Silver is a relatively rare element, with an average concentration in the earth's crust of 0.7 mg/kg. That makes it the 69th most abundant element out of 88 in the crust (Taylor, 1964). Although silver can occur in the elemental form in the environment, several ores also contain silver, including argentite (Ag_2S), cerargyrite ($AgCl$), pyrargyrite ($3Ag_2S \cdot Sb_2S_3$), and proustite ($3AgS \cdot As_2S_3$). Because of its value, silver is recovered to minimize loss in the waste water of most production and user facilities.

Production, Sources, and Residues

Production

World production of silver was 8.6×10^3 metric tons in 1940, decreasing to 7.5×10^3 metric tons in 1960, but rising to 11×10^3 metric tons in 1980 (US Minerals Yearbooks, 1940–1989). Production currently stands at approximately 13×10^3 metric tons/yr. That makes it the 70th most important mineral in terms of worldwide production weight, but in terms of economic value, it currently ranks 10th (Noetstaller, 1988). The world's leading mine producers are Mexico, Peru, the USSR, the USA, and Canada (Table 20.1). The main uses of silver are in photographic materials, jewelry, coinage, dental and medical supplies, some alloys, and electrical and electronic products. The last application reflects the outstanding electrical and thermal conductivities of silver.

Table 20.1. World's leading producers of silver.

Nation	Quantity (metric tons/yr)	Nation	Quantity (metric tons/yr)
Mexico	2159	Peru	1770
USSR	1600	USA	1530
Canada	1327	Australia	1063
Poland	744	Chile	505
Japan	324	South Africa	244

Source: Cordero (1988).

Sources

Worldwide transport of silver in rivers is relatively small, amounting to only 13×10^3 metric tons/yr (Westall and Stumm, 1980). That value is dwarfed by the amounts of iron and aluminum transported by the same rivers (9.9×10^8 and 17×10^8 metric tons, respectively). The amount of silver mobilized by atmospheric rainout is even smaller, averaging approximately 1.0×10^3 metric tons/yr (Westall and Stumm, 1980).

Loadings of silver to surface waters of heavily industrialized regions have gradually increased over the years. Ayres and Rod (1986), for example, estimated that total discharge to the Hudson–Raritan Basin (New York) was only 0.1 metric tons in 1880. That value is an order of magnitude lower than discharge to the basin in 1940 and 40 times less than that of 1980.

The major sources of anthropogenically derived silver in surface waters are mining and milling of base metal ores, mining and milling of precious metals, use and disposal (particularly in photography), and municipal wastes including sewage sludge. The first two sources are gradually decreasing in importance as wastes, both solid and liquid, are scavenged for residual silver. Hoye (1988) noted that the steps to effectively remove silver from low-grade ores include development of engineered heaps of ore and application of a lixiviant, usually an alkaline cyanide solution.

The No. 1 environmental problem associated with these procedures is the potential discharge of cyanide-bearing wastes to surface waters. These wastes included metallocomplexes such as $Fe(CN)_6^{4-}$, $Fe(CN)_6^{3-}$, $Au(CN)_2^-$, $Cu(CN)_2^-$, $Ni(CN)_4^{2-}$, $Ag(CN)_2^-$, $Zn(CN)_4^{2-}$, $Cd(CN)_3^{2-}$, and $Co(CN)_6^{4-}$. Since the majority of these complexes are subject to dissolution under environmentally relevant conditions, migration is a significant source of contamination to surface waters when cyanide destruction procedures are not implemented (Hoye, 1988). Such procedures center around the use of hypochlorite as a cyanicide and thiourea as an alternate lixiviant.

Silver is a significant contaminant of municipal wastes, largely reflecting the poorly controlled discharge of wastes from end-use industries. For example, total discharge from southern California's four largest municipal outfalls was approximately 10 metric tons in 1988, down from 45 metric tons in 1979 (Schafer, 1989). Similarly, Sung et al. (1986), working on 25 sewage treatment plants in the state of Washington, found that silver residues in secondary sludge were relatively high (0.63 mg/L), but averaged 0.004 mg/L in the final discharge (Table 20.2). A related study (Nevissi et al., 1988) showed that the rate of discharge of silver from two sewage treatment plants in the state of Washington ranged from 0.4–0.8 kg sludge/day.

Residues

Silver residues in surface waters are almost always low, generally <0.01 mg/L and often below 0.001 mg/L. Although comparably low concentrations (<0.5 mg/kg dry weight) are found in surface sediments of both coastal and freshwaters, anthropogenic discharges may ultimately produce much higher concentrations. For example, residues in the sediments of San Francisco Bay averaged 14 mg/kg with a range of 1–66 mg/kg (Luoma and Phillips, 1988). This is a reflection of discharges from user industries.

Chemistry

The principal species of silver in surface waters, Ag^+, occurs in a number of forms including $AgCl$, Ag_2S, Ag_2Se, Ag_3AsS_3, and $Na(AgCl_2)$. The last of these is formed in waters with high sodium chloride content. Silver may also be dissolved as double complexes of sulfur, arsenic, antimony, selenium, and tellurium, and form complexes with humic acids. In a study of the Sava River (Yugoslavia), Huljev (1986) showed that the concentra-

Table 20.2. Concentration (mg/L) of total silver in wastes from 25 treatment plants in the state of Washington.

Sample	Average	Range
Raw sewage	0.020	ND[a]–0.270
Primary effluent	0.008	ND–0.019
Secondary effluent	0.004	ND–0.014
Discharge	0.004	ND–0.017
Primary sludge	1.19	0.13–6.12
Secondary sludge	0.63	0.10–1.15

[a]Not detected.
Source: Sung et al. (1986).

tion of silver in humic acids was 1.1 mg/kg. That same investigation showed the presence of silver in four humic acid hydrolystates in the following concentrations (mg/kg): amino acids, 3.4; phenols and phenolic acids, 0.07; carbohydrates, 0.5; polycyclic aromatics, 0.4.

Sorption is the dominant process controlling the fate of silver in most surface waters. Favorable partition coefficients have been noted (in order of descending magnitude) for complex organic compounds associated with sediments, manganese dioxide, $Fe(OH)_3$, montmorillonite clay, illite clay, kaolinite clay, and Fe_2O_3. Lin et al. (1988) noted that silver is more or less immobile in the presence of sulfide complexes, and that strong associations are formed with some sulfur-bearing minerals such as galena. Other studies have shown that sulfur, and concomitant acidic runoff, are associated with the mining of silver, particularly prior to 1950 (Kelly, 1988).

Other processes such as volatilization and photolysis probably have little effect on the fate of silver complexes in surface waters.

At present, only a few detailed studies are available on the environmental chemistry of silver in marine and freshwaters. Some of the outstanding deficiencies include:

Complexation of Ag^+ with ligands, both organic and inorganic, under different environmental conditions
Effect of changes in redox on the sorption/desorption of Ag^+
Partitioning of Ag^+ in different environmental compartments

Although these are major deficiencies, silver is not considered a significant contaminant of most waterways, so such work is not a high priority at this time.

Bioaccumulation

Plants

Silver is often detected in plant and animal tissues, albeit in relatively low concentrations in most surface waters.Ramelow et al. (1987), working with benthic diatoms in the Calcasieu Estuary (Louisiana), reported that total Ag residues ranged from 0.1 to 3.6 mg/kg dry weight. Those values gave sediment concentration factors (residue in diatoms/residue in sediment) of <10, whereas water concentration factors (residue in diatoms/residue in water) were $>10{,}000$. Estabrook et al. (1985) analyzed a large number of plant species collected from the shoreline of Saginaw Bay, Lake Huron. Although most of the plants collected in that study contained residues below the level of detection (0.1 mg/kg), relatively high maximum concentrations were noted as follows: *Cladophora* sp. (a submerged species), 25.7 mg/kg; *Scirpus acutus* (a floating species with emergent leaves), 56.3 mg/kg;*Scirpus validus* (a floating species with emergent leaves), 66.9 mg/kg. Because there was no apparent anthropogenic and widespread source of silver in Saginaw Bay, it must be assumed that the high residues were associated with localized concentrations of silver in the sediments.

Invertebrates

Total Ag does not concentrate through the food chain, so residues are typically low in invertebrates from most surface waters. Byrne and DeLeon (1986) noted that oysters *Crassostrea virginica* from Lake Pontchartrain (Louisiana) carried average burdens of 5.5 mg/kg dry weight compared to sediment residues of 0.004 mg/kg dry weight (concentration factor 1,375). Lake Pontchartrain, an estuary of the Mississippi River, receives an enormous amount of municipal and agricultural runoff and overflows from a population of >1.5 million. The corresponding average residues for the clam *Rangia cuneata* collected from two locations in the lake were 0.4 and 2.4 mg/kg (concentration factors 100 to 600).

Much higher residues were reported for invertebrates from San Francisco Bay, which receives extensive discharges from end-user industries. The following concentrations (mg/kg dry weight) were reported by Luoma and Phillips (1988): mollusc *Ilyanassa obsoleta*, 181 (range 46–320); mollusc *Macoma balthica*, 18 (range 1–67); mollusc *Macoma nasuta*, 3.5 (range 2–5); polychaete *Marphysa sanguinea*, 4 (range 2–6). These values reflect soft-tissue concentrations.

Fish

There do not appear to be any cases of significant adulteration of fish tissues. In laboratory experiments, a condition known as argyria can be induced when fish are exposed to high (but nontoxic) concentrations of silver sulfide. This disorder, characterized by discoloration of skin and eye tissue, has not been reported for fish from natural waters.

Toxic Effects to Aquatic Organisms

Plants and Invertebrates

Silver, particularly the free ion, is highly toxic to most plant and animal species, both marine and freshwater. In his review of several studies, Mance (1987) noted that the LC_{50} of silver nitrate is typically <0.01 mg/L in a number of species, including the crustacean *Daphnia magna* and the ephemeropteran nymph *Ephemerella grandis*. Similarly, Martin and Holdich (1986), working with silver nitrate in soft water (50 mg/L $CaCO_3$), reported that the 96-h LC_{50} in the amphipod *Crangonyx pseudogracilis* was 0.005 mg Ag/L. In another study, Holcombe et al. (1987) exposed five invertebrate species to silver nitrate under standard test conditions and determined the following LC_{50}s (mg/L; average, 95% confidence limits):

Crustacean *Daphnia magna* (48-h LC_{50}): 0.0009 (0.0008–0.0010)
Chironomid larva *Tanytarus dissimilis* (48-h LC_{50}): 0.42 (0.28–0.64)
Crayfish *Orconectes immunis* (96-h LC_{50}): 0.56 (0.45–0.69)
Gastropod *Aplexa hypnorum* (96-h LC_{50}): 0.083 (0.067–0.103)
Leech *Nephelopsis obscura* (96-h LC_{50}): 0.029 (0.020–0.042)

These experiments were conducted at an average water hardness of 45 mg/L as $CaCO_3$ and a temperature of 17° C. The LC_{50}s listed above were lower than those obtained for any of the other 13 chemicals tested, including acrolein and trichlorophenol. In another multi–agent study, Khangarot et al. (1987) reported that toxicity of heavy metals to the cladoceran *Daphnia magna* followed the order Hg > Ag > Cu > Zn > Cd > Co > Cr > Pb > Ni > Sn.

Fish

Silver nitrate is also highly toxic to freshwater fish. The work of Holcombe et al. (1987), referred to above, yielded the following 96-h LC_{50}'s (mg/L; average, 95% confidence limits):

Fathead minnow *Pimephales promelas*: 0.009 (0.008–0.010)
Rainbow trout *Oncorhynchus mykiss*: 0.006 (0.005–0.007)
Bluegill *Lepomis macrochirus*: 0.013 (0.009–0.020)

Similarly, Khangarot and Ray (1988), working on three Asian freshwater species held in moderately hard water (180 mg/L as $CaCO_3$), gave the following 96-h LC_{50}s (average): *Channa punctatus*, 0.018 mg/L; *Puntius sophore*, 0.008 mg/L; *Lebistes reticulatus*, 0.006 mg/L. In another study, the sensitivity of fathead minnow to different silver species was determined under soft water conditions (38 mg/L $CaCO_3$) (LeBlanc et al., 1984):

Silver species	Relative toxicity
Free silver	1
Silver chloride	0.0033
Silver sulfide	0.000067
Silver thiosulfate	0.000057

Silver is generally less toxic to marine species, reflecting the formation of silver chloride. In his review of a number of studies, Mance (1987) noted that the LC_{50} of silver to a number of invertebrate and fish species ranged from 0.1 to 0.3 mg/L—1 or 2 orders of magnitude greater than that reported under freshwater conditions.

Relatively little is known about the chronic effects of different species of silver to plants and animals, both marine and freshwater.

Many nations have promulgated standards/guidelines for the protection of fish and other aquatic life, as exemplified by the following:

Canada
  0.1 μg/L
United States
  1.2 μg/L (water hardness 50 mg/L as $CaCO_3$)
  4.1 μg/L (water hardness 100 mg/L as $CaCO_3$)
  13 μg/L (water hardness 200 mg/L as $CaCO_3$)

Health Effects

Intake

The typical Western diet yields approximately 0.07 mg Ag/day in a 70-kg reference man (Carson et al., 1987). Essentially no silver comes from inhalation in the nonoccupationally exposed population. Once absorbed, silver is inefficiently absorbed, and in most cases, absorption efficiencies of applied dose do not exceed 10%. Daily loss amounts to 0.06 mg in feces, 0.009 mg in urine, 0.0006 mg in hair, and 0.0004 in sweat. There are multiple primary sites of deposition including the liver, skin, adrenal glands, pancreas, muscle, kidney, spleen, and heart.

Acute Toxicity

Suicide provides the No. 1 example of the acute effects of silver intoxication. Large oral doses lead to violent abdominal pain, convulsions, and shock. Patients dying from intravenous doses of silver compounds develop necrosis of the bone marrow, liver, and kidney.

Chronic Toxicity

The primary chronic effect of silver is characterized by a series of symptoms collectively known as argyria. This condition is manifested by a slate-gray pigmentation of the hair, skin, and internal organs. There is a decrease in the activity of glutathione peroxidase and metabolism of cyclic adenosine monophosphate.

Carcinogenicity

Silver compounds are not considered to be teratogenic or mutagenic. There is no evidence of carcinogenicity in any species following the ingestion of silver compounds in water or food.

Drinking Water

Residues

Total Ag residues are typically low and often go undetected in drinking water. In a survey of the American Water Works Association (1985) of 39 states and three territories, silver was in noncompliance with the maximum contaminant limit (0.05 mg/L) in only four cases. For comparison, fluoride and nitrate were in noncompliance in 907 and 369 cases, respectively. In those cases in which silver was in noncompliance, residues ranged from 0.22 to 0.87 mg/L. Much lower levels (maximum 0.08 mg/L) were found in three surveys of American groundwater (US Environmental Protection Agency, 1985). Point-of-use water treatment devices containing silver-impregnated carbon may be the primary source of contamination in some drinking waters (World Health Organization, 1984).

Consumption Guidelines

Many nations currently use a drinking water guideline/standard of 0.05 mg Ag/L to prevent argyria, but the World Health Organization has recommended no guideline for drinking water. The value of 0.05 mg/L was based upon several clinical reports in which humans developed argyria as a result of intravenous and oral exposure to silver (reviewed by US Environmental Protection Agency, 1985). The World Health Organization did not establish a guideline because silver is relatively nontoxic to humans and is normally found in low concentrations in drinking water.

Treatment

Since silver is not a significant contaminant of most drinking waters, treatment process evaluations are limited. It is generally assumed that routine sedimentation will remove any silver sorbed to suspended particles. Coagulation using alum and/or ferric chloride is also effective in silver removal, as is lime softening. Reverse osmosis and ion exchange should also be effective in treatment, particularly for small-scale operations.

Recommendations

Remarkably little is known about the environmental chemistry of silver and its various complexes. Although some general information has been published on sorption, no replicated data are available on the following areas:

Complexation of Ag^+ with ligands, both organic and inorganic, under different environmental conditions

Effect of changes in redox on the sorption/desorption of Ag^+
Partitioning of Ag^+ in different environmental compartments
Effect of acidification on mobilization and speciation

The acute effects of silver nitrate to invertebrates and fish have been adequately studied, but practically nothing is known about the effects of other compounds. It appears that the free silver ion is one of the most toxic agents potentially found in surface waters. The influence of changing environmental conditions on the chronic toxicity of silver has gone almost untouched by aquatic toxicologists.

Fortunately, silver is relatively rare in the earth's crust, so the amount weathered to surface waters is small. In addition, anthropogenic discharges are closely controlled (for economic reasons), and there is no reason to suppose that such discharges will disproportionately expand in future years.

Because residues in surface waters are likely to remain low for the foreseeable future, none of the aforementioned deficiencies can be considered research priorities at this time. There is also no pressing need for routine monitoring of silver in most surface waters. However, determinations should be made of the free ion in areas where silver is released (albeit in small amounts) to the environment.

References

American Water Works Association. 1985. An AWWA survey of inorganic contaminants in water supplies. *Journal of the American Water Works Association* 77:67–72.

Ayres, R.U., and S.R. Rod. 1986. Patterns of pollution in the Hudson-Raritan Basin. *Environment* 28:14–20.

Byrne, C.J., and I.R. DeLeon. 1986. Trace metal residues in biota and sediments from Lake Pontchartrain, Louisiana. *Bulletin of Environmental Contamination and Toxicology* 37:151–158.

Carson, B.L., H.V. Ellis, and J.L. McCann. 1987. *Toxicology and biological monitoring of metals in humans*. Lewis Publishers, Chelsea, MI. 328 pp.

Cordero, R. 1988. *Metal Bulletin's prices and data 1988*. Metal Bulletin Books, Surrey, England, 375 pp.

Estabrook, G.F., D.W. Burk, D.R. Inman, P.B. Kaufman, J.R. Wells, and J.D. Jones, and N. Ghosheh. 1985. Comparison of heavy metals in aquatic plants on Charity Island, Saginaw Bay, Lake Huron, U.S.A., with plants along the shoreline of Saginaw Bay. *American Journal of Botany* 72:209–216.

Holcombe, G.W., G.L. Phipps, A.H. Sulaiman, and A.D. Hoffman. 1987. Simultaneous multiple species testing: acute toxicity of 13 chemicals to 12 diverse freshwater amphibian, fish, and invertebrate families. *Archives of Environmental Contamination and Toxicology* 16:697–710.

Hoye, R. 1988. *Gold/silver heap leaching and management practices that minimize the potential for cyanide releases*. US Environmental Protection Agency, EPA/600/2–88/002, Cincinnati, OH. 103 pp.

Huljev, D.J. 1986. Trace metals in humic acids and their hydrolysis products. *Environmental Research* 39:258–264.

Kelly, M. 1988. *Mining and the freshwater environment*. Elsevier, New York. 231 pp.

Khangarot, B.S., and P.K. Ray. 1988. The acute toxicity of silver to some freshwater fishes. *Acta Hydrochimica Hydrobiologica* 16:541–545.

Khangarot, B.S., P.K. Ray, and H. Chandra. 1987. *Daphnia magna* as a model to assess heavy metal toxicity: comparative assessment with mouse system. *Acta Hydrochimica Hydrobiologica* 15:427–432.

LeBlanc, G.A., J.D. Mastone, A.P. Paradice, B.F. Wilson, H.B. Lockhart, and K.A. Robillard. 1984. The influence of speciation on the toxicity of silver to fathead minnow (*Pimephales promelas*). *Environmental Toxicology and Chemistry* 3:37–46.

Lin, Y., G.W. Bailey, and A.T. Lynch. 1988. *Metal interactions at sulphide mineral surfaces. III. Metal affinities in single and multiple ion adsorption reactions*. US Environmental Protection Agency, EPA/600/D–88/247, Athens, GA. 45 pp.

Luoma, S.N., and D.J.H. Phillips. 1988. Distribution, variability, and impacts of trace elements in San Francisco Bay. *Marine Pollution Bulletin* 19:413–425.

Mance, G. 1987. *Pollution threat of heavy metals in aquatic environments*. Elsevier, London. 372 pp.

Martin, T.R., and D.M. Holdich. 1986. The acute lethality of heavy metals to peracarid crustaceans (with particular reference to fresh-water asellids and gammarids). *Water Research* 20:1137–1147.

Nevissi, A.E., F.B. DeWalle, J.F.C. Sung, K. Mayer, and R. Dalsey. 1988. Heavy metal variability of different municipal sludges as measured by atomic absorption and inductively coupled plasma emission spectroscopy. *Journal of Environmental Science and Health* 23:823–841.

Noetstaller, R. 1988. *Industrial minerals*. World Bank Technical Paper Number 76. World Bank, Washington, DC. 117 pp.

Ramelow, G.J., R.S. Maples, R.L. Thompson, C.S. Lueller, C. Webre, and J.N. Beck. 1987. Periphyton as monitors for heavy metal pollution in the Calcasieu River estuary. *Environmental Pollution* 43:247–261.

Schafer, H. 1989. Improving southern California's coastal waters. *Journal of the Water Pollution Control Federation* 61:1395–1401.

Sung, J.F.C., A.E. Nevissi, and F.B. Dewalle. 1986. Concentration and removal efficiency of major and trace elements in municipal wastewater. *Journal of Environmental Science and Health* 21:435–448.

Taylor, S.R. 1964. Abundance of chemical elements in the continental crust: a new table. *Geochimica Cosmochimia Acta* 28:1273–1285.

Westall, J., and W. Stumm. 1980. The hydrosphere. *In: The natural environment and the biogeochemical cycles,* ed. O. Hutzinger, 17–49. Springer-Verlag, New York.

World Health Organization. 1984. *Guidelines for drinking-water quality,* Volume 2: *Health criteria and other supporting information*. World Health Organization, Geneva, 335 pp.

US Environmental Protection Agency. 1985. National primary drinking water regulations. *Federal Register* 50:46935–47022.

US Minerals Yearbooks. 1940–1989. Bureau of Mines, US Department of the Interior, Washington, DC.

21
Sulfur

Sulfur has enormous environmental significance because it (1) complexes with many toxic agents, organic materials, and hydrogen in surface waters, and (2) is the primary agent of acidification in many lakes and rivers. Sulfur, with a crustal abundance of approximately 260 mg/kg, is one of the four ore-forming elements (chalcogens) in Group VIA of the periodic chart. Valency states of 2− to 6+ occur in the environment, so sulfur and its complexes are stable over a wide range of conditions. More than 2,000 sulfur-bearing ores have been identified, and some of these, such as bravoite [$(Ni,Fe)S_2$], chalcopyrite (Cu_2S), cubanite ($CuFe_2S_3$), greigite (Fe_3S_4), molybdenite (MoS_2) and pyrite (FeS_2), may contain more than 35% by weight of sulfur.

Production, Sources, and Residues

Production and Sources

Natural Sources. Sulfur occurs abundantly in surface waters, a result of natural weathering and emissions from volcanoes, sea salt aerosols, forest fires, and microbial decomposition of organic material. On a worldwide basis, volcanoes emit 2–30 $\times$ 10^6 metric tons/yr to the atmosphere. Emissions consist primarily of SO_2, with lesser amounts of H_2S, SO_3, SO_4^{2-}, elemental S, and OCS.

Sea salt aerosols contribute substantially to the global sulfur cycle. Total emissions come to 44–175 $\times$ 10^6 metric tons S/yr, but only about 10%

of this quantity is deposited on land, the remainder falling back into the ocean (Meszaros, 1982; Brown, 1982). At least two other sulfur species, from non-sea-salt origins, exist over oceans (Savoie and Prospero, 1989): (1) reduced sulfur gases produced by biological activity in the oceans, and (2) sulfur derived from continental anthropogenic sources.

Total annual production of sulfur from microbial decomposition amounts to 100–280 × 10^6 metric tons (Brown, 1982). Approximately 50% of this quantity originates in coastal marine waters, mudflats, swamps, and sediments. The primary species produced by decomposition are dimethyl disulfide [$(CH_3)_2S_2$], hydrogen sulfide (H_2S), carbon disulfide (CS_2), dimethyl sulfide [$(CH_3)_2S$], and methane thiol (CH_3SH).

Anthropogenic Emissions. Anthropogenic emissions to the atmosphere now dominate the sulfur cycle in many parts of the world. In fact, at least 80% of global SO_2 emissions and >45% of total riveborne sulfate sulfur come from man-made sources (Zehnder and Zinder, 1980). Asia is the leading producer of SO_2, followed by Europe and North America (Table 21.1). There is considerable variable in production estimates for the year 2000, simply because no one knows if control measures will be implemented by different countries. In addition, if the public can accept increased use of nuclear energy, sulfur production will be greatly reduced.

Residues

Residues in Precipitation. Conversion of SO_2 to sulfate in the atmosphere results from photooxidation and heterogenous reactions. In polluted atmospheres, the rate of photooxidation is relatively fast, a result of collisions with strong oxidizing radicals, including HO, HO_2, and CH_3O_2. Heterogeneous reactions consist of (1) catalytic oxidation by transition metals in water droplets, (2) surface-catalyzed oxidation of SO_2 subsequent to collision with elemental carbon and other solid particles, and (3) oxidation of SO_2 in the liquid phase by ozone (O_3) and hydrogen peroxide (H_2O_2).

Table 21.1. Global production of sulfur dioxide (× 10^6 metric tons S per year).

Continent	1930	1940	1950	1960	1970	1980	2000
Asia	5	9	12	34	43	57	30–90
Europe	21	25	21	30	30	30	12–30
North America	22	17	25	24	34	29	25
Africa	0.5	0.7	1	2	3	4	6
South America	0.4	0.5	1	2	3	4	6
Oceania	0.4	0.5	1	1	1	2	2
Total	49	53	61	93	114	126	81–159

Sources: Dignon and Hameed (1989), Hordijk (1988), Moller (1984).

SO_4^{2-} levels are greatly elevated in polluted atmospheres in several regions of Europe, North America, and other areas where coal, oil, and other sulfur-bearing fuels are used, as shown in the following examples:

Czechoslovakia, rainfall (1983–84), average ~120 μEq/L (range 100–205 μEq/L) (Moldan and Vesely, 1987)
Southwest China, rainfall (1981–84), average 251 μEq/L (range 112–411 μEq/L) (Zhao et al., 1988)
Michigan, snow melt (1984), average ~40 μEq/L (range 5–120 μEq/L) (Cadle et al., 1987)
Ontario, rain and snowfall (1980–84), average 52 μEq/L (range 30–66 μEq/L) (Chan et al., 1987)
Pacific Northwest (U.S.), snow (1983), average 0.14 mg/L (range 0.04–0.32 mg/L) (Laird et al., 1986)

Deposition of such high levels greatly affects the sulfur cycle in many surface waters, as discussed under Chemistry.

Other Sources Acid mine drainage, a result of the mining and milling of sulfur-bearing ores, has played a dominant role in surface water chemistry in many areas of the world including the Appalachia, Wales, and the Canadian Shield. The following series of reactions results in the production of acidified runoff from poorly regulated mine sites:

$$2FeS_2 + 2H_2O + 7O_2 \rightleftharpoons 2FeSO_4 + 2H_2SO_4 \quad (1)$$

$$4FeSO_4 + 2H_2SO_4 + O_2 \rightleftharpoons 2Fe_2(SO_4)_3 + 2H_2O \quad (2)$$

$$Fe_2(SO_4)_3 + 6H_2O \rightleftharpoons 2Fe(OH)_3 + 3H_2SO_4 \quad (3)$$

$$FeS_2 + 14Fe^{3+} + 8H_2O \rightleftharpoons 15Fe^{2+} + 2SO_4^{2-} + 16H^+ \quad (4)$$

In equation 1, iron disulfide is oxidized to sulfate by oxygen, while, in equation 2 ferrous iron is oxidized to ferric iron. In equation 3, ferric iron is hydrolyzed to ferric hydroxide, and in equation 4, ferric iron is oxidized to pyrite. These reactions have been associated with coal and iron mining, and other base metal mining.

The ferrous iron reacts with atmospheric oxygen and hydrolyzes according to the equation:

$$4Fe^{2+} + O_2 + 10\,H_2O \longrightarrow 4Fe(OH)_3 + 8H^+ \quad (5)$$

Reactions 1 to 5 ultimately lead to an enormous production of acidified water owing to the presence of sulfur in the ore/coal body.

Water. Sulfur concentrations in many surface waters have increased greatly in historic and recent times, a result of SO_2 emissions and acid mine runoff. In a study of 768 lakes in the northeastern United States, Brakke et al. (1988) reported that sulfate averaged 118 μEq/L and that approximately 2% of the lakes contained concentrations > 200 μEq/L.

A related study (Ellers et al., 1988a) of 592 lakes in the upper midwest of the United States found that sulfate averaged 61 μEq/L and that approximately 2% of the lakes contained concentrations > 200 μEq/L.

The final study (Ellers et al., 1988b) of the series on 2,424 lakes in the southeastern United States found that sulfate averaged 63 μEq/L and that in Florida, more than 35% of the lakes contained sulfate at >200 μEq/L.

These relatively high levels reflect atmospheric deposition of sulfate, typically 10–30 kg SO_4/ha/yr in affected areas (Wright, 1988).

One of the difficulties in determining the magnitude of the flux in sulfate residues in surface waters is the lack of historical data. Epstein (1988) estimated the historical sulfate concentrations of natural freshwaters by the following equation:

$$[SO_4^{2-}] = (\text{specific conductance} - 0.102\,[HCO_3^-] - 0.134\,[Cl\text{-}] - 0.129\,[NO_3^-] - 0.292\,[H^+] + k/0.138$$

where specific conductance is expressed in μomhos/cm, all concentrations are in μEq/L, and k is a constant reflecting the concentrations of sodium and potassium found in the water. This equation is based on extensive data from 23 lakes in the northeastern United States.

Acid mine drainage has the potential to contribute an enormous amount of sulfate to freshwaters on both a local and regional basis. Barton (1978), for example, reported that waste waters from three mines contained the following residues (mg/L): coal mine (Pennsylvania), 1,474; zinc mine (Idaho), 63,000; —uranium mine (Ontario), 7,440.

In past years, such high residues would be discharged to surface waters, resulting in a large drop in pH and increase in waterborne sulfate levels. However, some jurisdictions now control these discharges using relatively simple treatment technology, so the concomitant impact on surface waters is reduced.

Chemistry

Sulfur Species

The dominant sulfur species under the pH and Eh conditions commonly encountered in surface waters are SO_4^{2-}, HSO_4^-, sulfides (H_2S, HS^-, and S^{2-}), and elemental sulfur (Figure 21.1). In most natural waters, the actual species is variably regulated by the presence of iron (Figure 21.2) S^{6+} and S^{2-} are the dominant, stable oxidation states but, under reducing conditions, thionates, thiosulfate, polysulfides, and sulfites may be detected. There are several unstable sulfur species which typically play a key role in the biological sulfur cycle (Figure 21.3).

Sulfide oxidation inevitably produces large amounts of H^+, as illustrated by the following equation:

$$HS^- + 1.5O_2 \rightleftharpoons SO_3^{2-} + H^+$$

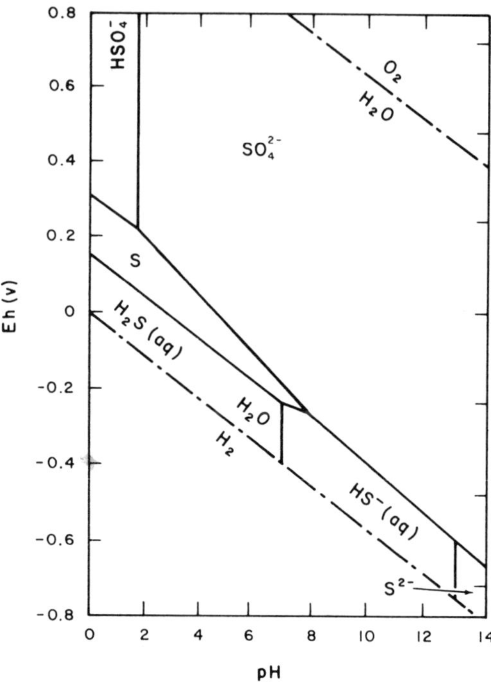

Figure 21.1. Equilibrium distribution of sulfur in water at 25°C and 1 atm of total pressure (Zehnder and Zinder, 1980).

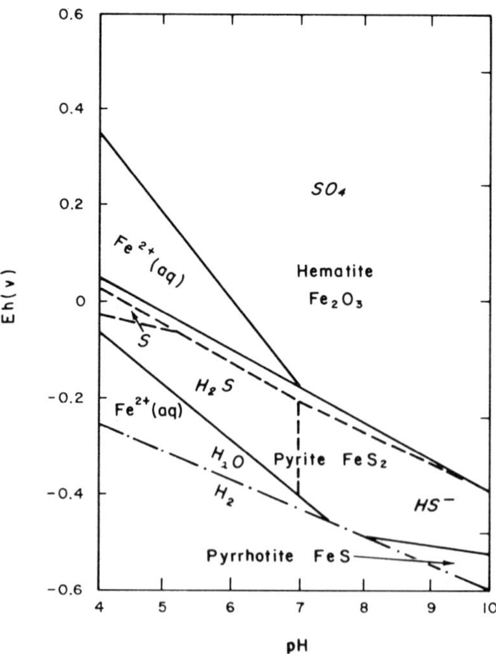

Figure 21.2. Equilibrium distribution of total dissolved sulfur species in the presence of iron (Zehnder and Zinder, 1980).

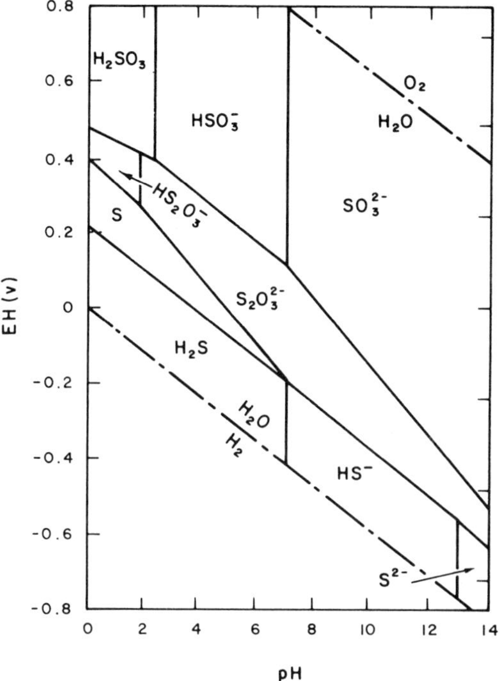

Figure 21.3. Equilibrium distribution of unstable sulfur species in the absence of sulfate (Zehnder and Zinder, 1980).

The rate reaction, which follows first- (or near to first-) order dynamics, increases greatly in the presence of heavy metals such as Ni, Co, Mn, and Cu. Some organic compounds, including phenol and aldehydes, also enhance the rate of oxidation, whereas others inhibit oxidation. Compounds within this latter group include EDTA, NTA, and cyanide.

Atmospherically derived sulfur dioxide is oxidized in water according to the following reactions (Nriagu and Hem, 1978):

$$SO_2\ (aq) + H_2O \rightleftharpoons H_2O{\cdot}SO_2 \tag{6}$$

$$H_2O{\cdot}SO_2 + H_2O \rightleftharpoons H_3O^+ + HSO_3^- \tag{7}$$

$$HSO_3^- + H_2O \rightleftharpoons H_3O^+ + SO_3^{2-} \tag{8}$$

$$2HSO_3^- \rightleftharpoons S_2O_5^{2-} + H_2O \tag{9}$$

The following oxidation of the sulfite is metal catalyzed and proceeds slowly by the radical chain mechanism:

$$SO_3^- + O_2 \rightleftharpoons SO_5^- \tag{10}$$

$$SO_3^{2-} + Me^{n+} \rightleftharpoons SO_3^- + Me^{(n-1)+} \tag{11}$$

$$SO_5^- + SO_3^{2-} \rightleftharpoons SO_4^- + SO_4^{2-} \tag{12}$$

$$SO_4^- + SO_3^{2-} \rightleftharpoons SO_4^{2-} + SO_3^- \quad (13)$$

$$SO_4^- + SO_4^- \rightleftharpoons S_2O_8^{2-} \quad (14)$$

$$SO_3^- + SO_3^- \rightleftharpoons S_2O_6^{2-} \quad (15)$$

Metal-catalyzed oxidation, invoking a metal sulfite complex as the active intermediate, also occurs in most waters.

Effects on Surface Water Chemistry

Sulfur has pronounced effect on the chemistry of surface waters.

1. Increase in the production of H^+, as per the reactions listed preceding
2. Change in Eh through several reactions, including:

$$HS^- \rightleftharpoons S^o + H^+ + 2e$$

3. Decrease in alkalinity through the oxidation of H_2S
4. Mobilization of metals and nutrients from sediments due to changes in pH and Eh. This effect has also been noted for naturally and anthropogenically derived radionuclides (Sheppard and Sheppard, 1988).

All of these effects have been noted in response to atmospheric deposition of sulfur and release of acid mine runoff.

Bioaccumulation

Sulfur is an essential element, representing approximately 0.5% of the dry weight of plants and 1.3% of the dry weight of animals. Some of the more important compounds of biologic origin are listed in Table 21.2. Metallothionein, present in many aquatic species, binds numerous toxic metals in the liver, kidney, and other tissues. Although sulfur compounds are readily transferred through the aquatic food chain, there is no evidence of bioaccumulation.

Table 21.2. Some biological sulfur compounds.

Name	Occurrence	Name	Occurrence
Cysteine	all species	methionine	all species
Thiamine	all species	biotin	all species
Coenzyme A	all species	lipoic acid	all species
Chrondroitin	animals	sulfolipid	photosynthetic bacteria/plants
Metallothionein	animals	ferredoxin	plants/bacteria
Coenzyme M	methane-forming bacteria	phosphatidyl sulfocholine	diatoms

Toxic Effects to Aquatic Organisms

The main problem associated with increased sulfur deposition is acidification of surface waters. Other potential problems, such as the buildup of toxic levels of hydrogen sulfide, pale in magnitude compared to acidification.

An enormous body of information is available on the effects of acidification on plants and animals, so there is no need to review that information here. Virtually everyone agrees that a vast number of lakes and rivers has been adversely affected by acidification. Mitigation, particularly liming, may greatly reduce the toxic effects of acidification (reviewed by Brocksen and Wisniewski, 1988). However, the technique is expensive and can be applied to only a small number of lakes and rivers. The only solution to the toxic effects of acidification is the control of sulfur (and other agents) at the source, as discussed in the Recommendations.

Health Effects

Sulfate

Intake. The typical Western diet yields 450–500 mg sulfate/day in a 70-kg reference man. Although intake from air is insignificant (<0.5 mg/day) in the nonoccupationally exposed population, much larger doses (>500 mg/day) can be ingested through water.

Acute Toxicity. The minimum reported lethal dose of magnesium sulfate in mammals is 200 mg/kg body weight (reviewed by World Health Organization, 1984). There are no reports of mortality in humans following ingestion of high doses of sulfates.

Chronic Toxicity. In adult humans, a sulfate dose of 1–2 g will induce a cathartic effect, whereas magnesium sulfate at concentrations >1,000 mg/L generally act as a purgative agent.

Taste threshold concentrations for several sulfate salts are sodium sulfate 200–500 mg/L; calcium sulfate 250–900 mg/L; magnesium sulfate 400–600 mg/L.

Carcinogenicity. Sulfates are not considered carcinogenic in humans or experimental animals.

Sulfide

Intake. No detailed information is available on the intake of sulfides in the typical Western diet. Apart from their occurrence in water, sulfides are also found in raw and cooked foods, beer, candy, soft drinks, and creams.

Acute Toxicity. Most sulfides are absorbed rapidly from the intestine (World Health Organization, 1984). An oral dose of 10–15 g of sodium sulfide is fatal in adult humans; death from inhalation of hydrogen sulfide occurs at 700–1,000 mg/m^3.

Chronic Toxicity. Hydrogen sulfide inhibits the activity of several enzymes, including succinic dehydrogenase, ATPase, dopa oxidase, carbonic anhydrase, and some iron-containing enzymes. The taste threshold for hydrogen sulfide in water is 0.05–0.10 mg/L. No one is likely to consume a harmful dose of sulfide because of its disagreeable taste.

Carcinogenicity. Sulfides in water are not considered carcinogenic to humans or experimental animals.

Acidic Pollutants

Acidification of surface waters has yet to have a demonstrable effect on human health despite increased mobilization of toxic heavy metals, particularly copper, cadmium, lead, mercury, and aluminum (McDonald, 1985). This is largely a reflection of the effectiveness of traditional water treatment systems in removing metals from source waters. Although acidification also increases the rate of corrosion of pipes, most regulatory agencies now require the installation of acid-resistant materials in the delivery system.

Carcinogenicity

Exposure to acid aerosols may result in increased risk of laryngeal cancer, lung cancer, and alteration of enzyme and mitotic regulation (Soskolne et al., 1989). Until recently, such health impacts have been restricted to occupational exposure, but increased acidification of urban air has increased the risk of chronic respiratory disease in the general population. The extent of acidification has increased to the point where criteria may need to be established for acid aerosols (Lipfert et al., 1989).

Drinking Water

Sulfate

Residues. Sulfate is almost always present in drinking water, often in relatively high concentrations. In a survey of the American Water Works Association (1985) of 39 states and three territories, sulfate was in noncompliance with the Maximum Contaminant Level (250 mg/L) in 1,466 cases. For comparison, iron, fluoride and nitrates were in noncompliance in 4,394, 907, and 369 cases, respectively. Canadian drinking waters contained residues in the <10–1,795 mg/L range (Canadian Water Quality Guidelines, 1987) whereas sulfate in drinking water from 23 European cities averaged 64 mg/L, range 9–125 mg/L (World Health Organization, 1984).

Consumption Guidelines. The World Health Organization recommends a drinking water guideline for sulfate of 400 mg/L. In Canada, the corresponding level is 500 mg/L, and in the United States, it is 250 mg/L. Most other nations maintain a guideline within these limits. The development of the guideline was based on aesthetic factors, particularly bad taste. No health-related problem is likely to develop at concentrations <500 mg/L.

Treatment. Because most sulfate salts are highly soluble, conventional water treatment methods are ineffective in sulfate removal. Hence, ion exchange, reverse osmosis, and electrodialysis must be used, but only on a limited scale (owing to high costs). In areas where sulfate is high, water treated by such methods may be diluted with water from conventional systems, thereby reducing sulfate to the guideline level.

Sulfide

Residues. Sulfides, including hydrogen sulfide, are not normally detected in finished drinking water. This reflects the rapid oxidation of these compounds under aerobic conditions.

Consumption Guidelines. The World Health Organization has proposed a drinking water guideline of "not detectable by consumer," equivalent to approximately 0.05 mg/L (as hydrogen sulfide). This guideline is again based on aesthetic factors and not health criteria.

Treatment. Most sulfides can be easily removed from potable water through aeration, metal precipitation, or chemical oxidation. Chlorine, potassium permanganate, ozone, and ferrate have all been effectively used as chemical oxidants.

Recommendations

Sulfur is enormously important in most, if not all, surface waters because it complexes with many toxic metals and organic materials. This process limits the movement of such agents in the environment, so concomitant toxicologic effects are also minimized. Paradoxically, excess sulfur ultimately leads to acidification of freshwaters, resulting in the mobilization of these same agents. Many thousands of lakes and rivers have been devastated by acidification, and the potential for adulteration of drinking water increases with acidification.

Numerous studies have already been conducted on many facets of acidification. One example is the plethora of information on the effect of low pH on fish and other aquatic species; work in such areas cannot be considered a priority at the present time. On the other hand, the issue of mobilization of heavy metals and other toxic agents in the presence of excess H^+ continues to be important in many areas of the world. Routine monitoring of metals and pH in both surface and drinking waters must be conducted in areas susceptible to acidification.

So much has been written on acid deposition that most scientific issues have been resolved. This does not discount the need for more subtle research to hone our understanding of, for example, source/receptor relationships. Thus, governments can now turn their attention to the implementation of various abatement schemes.

References

American Water Works Association. 1985. An AWWA survey of inorganic contaminants in water supplies. *Journal of the American Water Works Association* 77:67–72.

Barton, P. 1978. The acid mine drainage. *In: Sulfur in the environment. Part II: Ecological impacts,* ed. J.O. Nriagu 313–358. Wiley, New York.

Brakke, D.F., D.H. Landers, and J.M. Ellers. 1988. Chemical and physical characteristics of lakes in the northeastern United States. *Environmental Science and Technology* 22:155–163.

Brocksen, R.W., and J. Wisniewski. 1988. Restoration of aquatic and terrestrial systems. *Water, Air, and Soil Pollution* 41:1–501.

Brown, K.A. 1982. Sulphur in the environment: a review. *Environmental Pollution (Series B)* 3:47–80.

Cadle, S.H., J.M. Dasch, and R.V. Kopple. 1987. Composition of snowmelt and runoff in northern Michigan. *Environmental Science and Technology* 21:295–299.

Canadian Water Quality Guidelines. 1987. Canadian Council of Resource and Environment Ministers. Environment Canada, Ottawa.

Chan, W.H., A.J.S. Tang, D.H.S. Chung, and N.W. Reid. 1987. An analysis of precipitation chemistry measurements in Ontario. *Environmental Science and Technology* 21:1219–1224.

Dignon, J., and S. Hameed. 1989. Global emissions of nitrogen and sulfur oxides from 1860–1980. *Journal of the Air and Waste Management Federation* 39:180–186.

Ellers, J.M., D.F. Brakke, and D.H. Landers. 1988a. Chemical and physical characteristics of lakes in the upper midwest, United States. *Environmental Science and Technology* 22:164–172.

Ellers, J.M., D.H. Landers, and D.F. Brakke. 1988b. Chemical and physical characteristics of lakes in the southeastern United States. *Environmental Science and Technology* 22:172–177.

Epstein, C.B. 1988. A method for the estimation of historical sulfate concentrations in natural freshwaters. *Environmental Science and Technology* 22:146–1463.

Hordijk, L. 1988. A model approach to acid rain. *Environment* 30:16–41.

Laird, L.B., H.E. Taylor, and V.C. Kennedy. 1986. Snow chemistry of the Cascade–Sierra Nevada mountains. *Environmental Science and Technology* 20:275–290.

Lipfert, F.W., S.C. Morris, and R.E. Wyzga. 1989. Acid aerosols: the next criteria air pollutants? *Environmental Science and Technology* 23:1316–1322.

McDonald, M.E. 1985. Acid deposition and drinking water. *Environmental Science and Technology* 19:772–776.

Meszaros, E. 1982. On the atmospheric input of sulphur into the ocean. *Tellus* 34:277–282.

Moldan, B., and M. Vesely. 1987. Chemical composition of atmospheric precipitation in Czechoslovakia, 1976–1984. I. Monthly samples. *Atmospheric Environment* 21:2383–2395.

Moller, D. 1984. Estimation of the global man-made sulphur emission. *Atmospheric Environment* 18:19–27.

Nriagu, J.O., and J.D. Hem. 1978. Chemistry of pollutant sulfur in natural waters. *In: Sulfur in the environment. Part II: Ecological impacts,* eds. J.O. Nriagu and J.D. Hem. 211–270. Wiley, New York.

Savoie, D.L., and J.M. Prospero. 1989. Comparison of oceanic and continental sources of non-sea-salt sulphate over the Pacific Ocean. *Nature* 339:685–687.

Sheppard, S.C., and M.I. Sheppard. 1988. Modeling estimates of the effect of acid rain on background radiation dose. *Environmental Health Perspectives* 78:197–206.

Soskolne, C.L., G. Pagano, M. Cipollaro, J.J. Beaumont, and G.G. Giordano. 1989. Epidemiologic and toxicologic evidence for chronic health effects and the underlying biologic mechanisms involved in sub-lethal exposures to acidic pollutants. *Archives of Environmental Health* 44:180–191.

World Health Organization. 1984. *Guidelines for drinking-water quality,* Volume 2: *Health criteria and other supporting information.* World Health Organization, Geneva. 335 pp.

Wright, R.F. 1988. Acidification of lakes in the eastern United States and southern Norway: a comparison. *Environmental Science and Technology* 22:178–182.

Zehnder, A.J.B., and S.H. Zinder. 1980. The sulfur cycle. In: *The natural environment and the biogeochemical cycles,* ed. O. Hutzinger, 105–145. Springer, New York.

Zhao, D., J. Xiong, Y. Xu, and W.H. Chan. 1988. Acid rain in southwestern China. *Environmental Science and Technology* 22:349–358.

22
Thallium

Thallium, with an average crustal abundance of approximately 0.5 mg/kg, occurs in a only a small number of independent minerals. These include lorandite ($TlAsS_2$), vrbaite ($Hg_3Tl_4As_8Sb_2S_{20}$), hutchinsonite [Pb, Tl$)_2$(Cu, Ag)As_5S_{10}], crookesite [Cu, Tl, Ag$)_2$Se], and avicennite (Tl_2O_3). Thallium is considered a soft acid, so it prefers to complex with sulfur rather than oxides or nitrogen.

Production, Sources, and Residues

Production

World production of thallium is relatively small, standing at less than 0.1 $\times$ 10^3 metric tons/yr. That makes it the 85th most important metal in terms of production weight (Noetstaller, 1988). The major uses of thallium are relatively restricted and include electrical cells and lamps, electronic conductors, catalysts in organic synthesis, specialized optical uses, and formation of alloys with numerous metals.

Sources

Worldwide environmental mobilization of thallium through natural sources comes to 2,400 metric tons/yr (Canadian Water Quality Guidelines, 1987). Most of this total is the result of weathering of thallium-bearing geological materials. Anthropogenic mobilization is slightly smaller, amounting to 2,000 metric tons/yr (Ewers, 1988).

The No. 1 anthropogenic source of thallium is the combustion of fossil fuels. Worldwide emissions from this source amount to 600 metric tons/yr, of which 80 metric tons/yr comes from Europe and 140 metric tons/yr from the United States (Ewers, 1988). Another 600 metric tons/yr enters the environment from the smelting of lead, copper, and zinc. Part of that total goes to the atmosphere, while the remainder is discharged to surface waters or treated as solid waste. Iron and steel production is another significant source, coming to >200 metric tons/yr.

There are a number of other relatively small, anthropogenic sources of thallium. The most important of these is the roasting of pyrite during cement production. Thallium residues of up to 50,000 mg/kg have been found in the fly dust of cement plants (Ewers, 1988).

Residues

Thallium is usually detected in surface waters, albeit in relatively low concentrations. A survey of lakes and rivers in western Canada yielded residues in the 5–100 μg/L range (Canadian Water Quality Guidelines, 1987). Concentrations of up to 14 μg/L were reported for marine and freshwaters by Ewers (1988).

Chemistry

Thallium exists in three oxidation states in the environment: Tl^{o}, Tl^{+}, and Tl^{3+}. The thallous ion (Tl^{+}), the principal form of thallium under the pH and Eh conditions commonly encountered in surface waters, readily combines with sulfur and, to a lesser degree, halogens and oxygen. Although Tl^{3+} is less stable than Tl^{+}, it does form organometallic compounds. Tl^{o} is found principally in reducing environments.

Although relatively little is known about the environmental fate of thallium compounds, it appears that Tl^{+} readily sorbs to clay and hydrous metal oxides. Under reducing conditions, insoluble Tl_2S is formed in the sediments, whereas the soluble chloride, carbonate, and hydroxy salts of Tl^{+} are formed in aerobic waters. Tl^{+} and Tl^{3+} readily complex with humic acids under basic conditions. Other fate processes such as photolysis and volatilization appear to have little or no effect on thallium compounds.

Bioaccumulation

Almost nothing is known about the accumulation of total Tl through the aquatic food chain. It is assumed, however, that sorption of low-molecular-weight, soluble compounds is rapid, particularly at the lower trophic levels. Zitko et al. (1975) showed that the concentration factor (tissue level/water level) for the clam *Mya arenia* was approximately 18, compared to 27–1430 for Atlantic salmon *Salmo salar*. Exposure of bluegill

Lepomis macrochirus to 0.08 mg Tl/L as Tl_2SO_4 for 28 days produced a concentration factor of 34 (Barrows et al., 1980); the resulting residues had a half-life of >4 days in muscle tissue. Bowhead whales *Balaena mysticetus*, collected from Alaskan waters, contained total Tl residues of <0.01 mg/kg wet weight in all samples (Byrne et al., 1985).

There does not appear to be any replicated information available on uptake of thallium under different water quality conditions, such as pH and hardness. Similarly, the importance of complexation with organic and inorganic ligands on uptake is not known. Essentially nothing is known about the importance of the age and size of test species on thallium accumulation.

Toxic Effects to Aquatic Organisms

Plants

Tl^+ is generally more toxic than Tl^{3+}. Puddu et al. (1988), for example, noted that the EC (effective concentration)$_{50}$ of the marine alga *Dunaliella tertiolecta* exposed to Tl^+ was 0.4×10^{-6} *M* compared to 0.9×10^{-6} *M* for Tl^{3+}. The corresponding concentrations for another alga, *Phaeodactylum tricornutum*, were 0.7×10^{-6} and 1.1×10^{-6} *M*, respectively.

Invertebrates and Fish

In general, thallium compounds are only moderately toxic to most aquatic species. The LC_{50} for the cladoceran *Daphnia magna*, a common test species, typically ranges from 0.9 to 2.5 mg Tl/L. Much higher LC_{50}s (120–130 mg/L) have been found in bluegill *Lepomis macrochirus*, whereas the corresponding value for sheepshead minnow *Cyprinodon variegatus* is >20 mg/L (Suter and Rosen, 1988). That last species can tolerate a Maximum Acceptable Concentration of 6.0 mg/L following long-term exposure to thallium (Suter and Rosen, 1988). The lowest reported concentration of thallium inducing adverse effects in fish (fathead minnow *Pimephales promelas*) is 0.04 mg/L (US Environmental Protection Agency, 1980).

Essentially nothing is known about the effects of low levels of thallium compounds on blood parameters, enzyme activity, reproductive potential, or other indices of chronic intoxication.

Health Effects

Intake

Although thallium intake/loss data are scarce, it is assumed that the typical Western diet yields approximately 0.002 mg Tl/day in a 70-kg reference man (Carson et al., 1987). Inhalation by the nonoccupationally exposed population is essentially nil. Losses occur principally through the feces (0.001 mg/day) and urine (0.0005 mg/day).

Acute Toxicity

Murder and suicide at doses >14 mg Tl/kg provide the primary examples of acute intoxication by thallium compounds. Following ingestion, thallium is initially concentrated in the kidneys, particularly the renal medulla (De Groot and Heijst, 1988). Highest residues then appear in the liver and blood, and, in the final stages of poisoning, thallium is detected in essentially all tissues. The usual cause of death is respiratory depression, followed by pneumonia and respiratory paralysis.

Chronic Toxicity

The only examples of chronic intoxication in humans come from occupationally exposed workers. In those cases, a common symptom is alopecia areata, characterized by hair loss and complete epilation. Polyneuritis and particularly retrobulbar neuritis have been reported in some cases of occupational exposure.

Carcinogenicity

There is inadequate information available to assess the potential carcinogenicity of thallium compounds to humans. Certainly, no cancers in humans have been linked to the ingestion of thallium-tainted waters.

Drinking Water

Residues and Consumption Guidelines

Many nations, plus the World Health Organization, have yet to establish a guideline for total Tl in drinking water. This reflects the low or nondetectable concentrations of thallium in drinking water and the relatively low toxicity of most thallium compounds to humans.

The Acceptable Daily Intake for thallium is approximately 0.037 mg in a 70-kg reference man (Ewers, 1988). Assuming that drinking water accounts for approximately two thirds of the daily dose of thallium, a potential drinking water criterion is 0.013 mg Tl/L (US Environmental Protection Agency, 1980). This value also assumes that the daily consumption of drinking water is 2 L.

Treatment

Because thallium does not pose a threat to most drinking water, relatively little is known about advanced treatment/removal methods. It is, however, widely assumed that application of alum, or a related agent, plus conventional filtration will remove particulate-bound thallium.

Recommendations

There are many deficiencies in our knowledge of thallium, particularly its chemistry, environmental fate, and toxicity. In fact, the data base for thallium is among the weakest of any environmentally relevant inorganic agent.

The primary deficiencies include:

1. Effect of pH, hardness, organic ligands, inorganic ligands on uptake by plant, invertebrate, and fish species
2. Effect of age, size, and other biological conditions on uptake by aquatic plants and animals
3. Sublethal effects of Tl^{+} and Tl^{3+} to fish and other aquatic species
4. Establishment of a drinking water guideline

These points should be considered a priority, simply because replicated data on the environmental fate and toxicity of thallium and its compounds are not generally available.

References

Barrows, M.E., S.R. Petrocelli, K.J. Macek, and J.J. Carroll. 1980. Bioaccumulation and elimination of selected water pollutants by bluegill sunfish (*Lepomis macrochirus*). *In: Exposure and hazard assessment of toxic chemicals,* ed. R. Haque, 379–392. Ann Arbor Science, Ann Arbor, MI.

Byrne, C., R. Balasubramanian, E.B. Overton, and T.F. Albert. 1985. Concentrations of trace metals in the bowhead whale. *Marine Pollution Bulletin* 16:497–498.

Canadian Water Quality Guidelines. 1987. Canadian Council of Resource and Environment Ministers. Environment Canada, Ottawa.

Carson, B.L., H.V. Ellis, and J.L. McCann. 1987. *Toxicology and biological monitoring of metals in humans*. Lewis Publishers, Chelsea, MI. 328 pp.

De Groot, G., and A.N.P. Van Heijst. 1988. Toxicokinetic aspects of thallium poisoning. Methods of treatment by toxin elimination. *Science of the Total Environment* 71:411–418.

Ewers, U. 1988. Environmental exposure to thallium. *Science of the Total Environment* 71:285–292.

Noetstaller, R. 1988. *Industrial minerals*. World Bank Technical Paper Number 76. World Bank, Washington, DC. 117 pp.

Puddu, A., M. Pettine, T. La Noce, R. Pagnotta, and F. Bacciu. 1988. Factors affecting thallium and chromium toxicity to marine algae. *Science of the Total Environment* 71:572.

Suter, G.W., and A.E. Rosen. 1988. Comparative toxicology for risk assessment of marine fishes and crustaceans. *Environmental Science and Technology* 22:548–556.

US Environmental Protection Agency. 1980. *Ambient water quality criteria for thallium*. US Environmental Protection Agency, EPA 440/5–80–074, Washington, DC.

Zitko, V., W.V. Carson, and W.G. Carson. 1975. Thallium: occurrence in the environment and toxicity to fish. *Bulletin of Environmental Contamination and Toxicology* 13:23–30.

23
Tin

The environmental significance of tin has risen enormously in recent years owing to the increased use of organic antifouling agents in marine and, to a lesser degree, freshwaters. Numerous environmental impacts, including the elimination or contamination of fish and invertebrates, have been observed in and around marinas and estuaries. Tin occurs in the earth's crust at an average concentration of approximately 2 mg/kg, and is concentrated in the following commercially important minerals: cassiterite (SnO_2), stannite (Cu_2FeSnS_4), and teallite ($PbZnSnS_2$). Tin is a borderline metal, meaning that it exhibits equal preference for complexation with sulfur and oxygen/nitrogen.

Production, Sources, and Residues

Production

World production of tin was 179 $\times$ 10^3 metric tons in 1930, increasing to only 183 $\times$ 10^3 metric tons in 1960 and 251 $\times$ 10^3 metric tons in 1980 (US Minerals Yearbooks, 1930–1989). Production in recent years has remained relatively steady, and has averaged 200–230 $\times$ 10^3 metric tons/yr (Cordero, 1988). The world's leading producers of refined tin are Malaysia, Mexico, Brazil, the United Kingdom, and Indonesia; the leading consumers are the United States, Japan, the Soviet Union, West Germany, and China (Table 23.1).

Tin's physicochemical properties make it useful in the prevention of

Table 23.1. World's leading producers and consumers of refined tin.

Nation	Production ($\times$ 10^3 metric tons/yr)	Nation	Consumption ($\times$ 10^3 metric tons/yr)
Malaysia	46	USA	52
Mexico	46	Japan	32
Brazil	25	USSR	30
UK	24	FRG	17
Indonesia	21	China	14
USSR	19	UK	9
China	17	France	7
Spain	15	Italy	5
Bolivia	14	The Netherlands	5

Source: Cordero (1988).

corrosion and other chemical reactions with various metals.Tin plating is widely used as a protective coating in food packaging, foil, and wires. Metallic tin also goes into the formation of alloys such as solder, bronze, and brass. Organotin compounds are widely used as antimicrobials, dyes, pigments, and ceramics.

Sources

The primary anthropogenic sources of inorganic tin to surface waters are: base metal mining and smelting; municipal wastewater and sewage sludge; combustion of fossil fuels, particularly oil and, to a lesser degree, coal; steel manufacture; and atmospheric deposition, particularly near municipal incinerators. The No. 1 source of organotins is the dissolution of tributyl tin, and related compounds, used as antifouling agents.

Approximately 93% of atmospheric tin in the northern hemisphere originates from anthropogenic sources while the corresponding value for the southern hemisphere is 70% (Byrd and Andreae, 1986). The most important anthropogenic source is the combustion of oil, followed by combustion of coal and emissions from base metal mines and smelters (Table 23.2). Most of the tin in rainfall is in the particulate rather than dissolved form. Total residues in rainfall typically range from 0.1 to 10 ng/L (Byrd and Andreae, 1986).

Residues

Until recently, the concentration of inorganic Sn in surface water was not routinely measured, so the data base on residues is relatively limited. Uncontaminated freshwaters generally carry residues in the 0.001–0.005 mg/L range, but higher concentrations (up to 0.05 mg/L) have been found

Table 23.2. Worldwide anthropogenic emissions of tin to the atmosphere.

Source	Emission ($\times 10^3$ metric tons/yr)
Oil combustion	
industry and domestic	0.29–3.58
electric utilities	0.35–2.32
Coal combustion	
industry and domestic	0.10–0.99
electric utilities	0.16–0.76
Pyrometallurgical metal production	
copper/nickel production	0.43–1.70
Refuse incineration	
municipal	0.14–1.40
sewage sludge	0.02–0.06
Total	1.49–10.81

Source: Nriagu and Pacyna (1988).

in some lakes (Maguire et al., 1982). Much lower levels (0.01 μg/L) are found in seawater (Forstner and Wittmann, 1979).

Tributyltin is generally detected in surface waters receiving wastes or runoff from municipalities, sewage treatment plant, and marinas. Residues are typically low, ranging from 1 to 70 ng/L (Muller et al., 1989; Goldberg, 1986), but may exceed 500 ng/L around marinas (where antifouling paint is used). In extreme cases in England and Denmark, concentrations of up to 3,000 ng/L have been detected (Muller et al., 1989).

Most of the environmental concern about tin in sediments centers around anthropogenic inputs of tributyltin. Residues of that compound may exceed 120 μg/kg dry weight in heavily contaminated areas (Table 23.3). Related compounds such as dibutyltin, butyltin, and methylated tins typically occur in much lower concentrations, generally <0.1 μg/kg dry weight (Table 23.3).

Chemistry

Although the prevalent oxidation state in aquatic environments is Sn^{4+}, tin may also occur as Sn^{4-}, Sn^0, and Sn^{2+}. The Sn^{2+}, which can be oxidized to Sn^{4+} in surface waters, occurs principally in anaerobic sediments and waters with a low redox potential. The principal forms of Sn^{4+} in surface waters include $Sn(OH)_4$, $SnO(OH)_2$, and $SnO(OH)_3^-$.

All forms of tin are readily sorbed to suspended solids, clay, and (probably) sulfides and hydroxides. Accordingly, tin residues are generally

Table 23.3. Concentration (μg/kg dry weight) of organotins in sediments.

Location	Species	Concentration average (range)
Sarah Creek, Chesapeake Bay (USA)[1]	Tributyltin	62 (23–120)
Boston Harbor (USA)[2]	Monobutyltin	0.017 (ND[a]–0.126)
	Dibutyltin	0.040 (ND–0.316)
	Tributyltin	0.119 (ND–0.518)
Boston Harbor (USA)[3]	Monobutyltin	0.014 (ND–0.028)
	Dibutyltin	0.053 (0.036–0.070)
	Tributyltin	0.093 (0.084–0.104)
Great Bay Estuary, New Hampshire (USA)[4]	Methyltin	0.014 (ND–0.080)
	Butyltin	0.019 (ND–0.049)
Coastal waters, Alexandria (USA)[5]	Methylbutyltin	0.175 (0.0–1.200)
	Dimethyltin	0.035 (0.0–0.135)
	Trimethyltin	0.007 (0.0–0.080)
	Butyltin	0.089 (0.0–0.450)
	Dibutyltin	0.080 (0.0–0.425)
	Tributyltin	0.230 (0.030–1.375)
Harbors, New Brunswick (Canada)[6]	Methyltin	9.0 (2.5–17.2)[b]
	Dimethyltin	ND
	Trimethyltin	ND
Vancouver Harbor (Canada)[6]	Methyltin	0.015 (0.01–0.02)[b]
	Dimethyltin	<0.01
	Trimethyltin	0.025 (0.02–0.03)[b]

[a]Not detectable.
[b]mg/kg wet weight.
Sources: [1]Unger et al. (1988), [2]Makkar et al. (1989), [3]Cooney et al. (1988), [4]Randall et al. (1986), [5]Dahab (1988), [6]Maguire et al. (1986).

greatest in the sediments and, to a lesser degree, the sediment surface microlayer. Concentration factors (microlayer concentration/surface water concentration) may exceed 6,000, but are typically <100 (Cleary and Stebbing, 1987).

Organotins

Butyltins and inorganic Sn can be methylated in the environment to form methylbutyltins and methyltins, respectively. There are a number of possible agents and routes of methylation, including (Cooney, 1988; Craig and Rapsomanilis, 1985) methylcobalamin (methyl-B_{12}), CH_3I, $(CH_3)_3S^+I^-$, and $(CH_3)_3N^+CH_2COO^-$.

Although the rate of methylation is slow in surface waters (Maguire et al., 1986), the resulting compounds accumulate in sediments and are biologically active. The presence of fulvic acids in the sediments limits the rate of methylation of Sn^{2+} (Ring and Weber, 1988). The influence of other environmental factors such as pH, salinity, and temperature on methylation has not been adequately documented.

Tributyltin undergoes relatively rapid biodegradation ($T_{1/2}$ = approximately 20 days) in the water column (Seligman et al., 1988; Clark et al., 1988). The process is mediated by fungi, bacteria, and algae, yielding the primary breakdown product dibutyltin and, to a lesser degree, butyltin (Cooney, 1988; Seligman et al., 1986). The $T_{1/2}$ of tributyltin in sediments is longer (16–23 weeks) than in water, reflecting the toxic effects of tin and other heavy metals that are concentrated in the sediments. The ultimate product of biodegradation in both water and sediments is inorganic Sn. Tributyltin undergoes slow ($T_{1/2}$ = 89 days) sunlight photolytic degradation, at least partially by stepwise debutylation to inorganic Sn (Maguire et al., 1983).

Bioaccumulation

Plants

Under most circumstances, uptake of inorganic Sn by aquatic plants is initially rapid, but then slows to eventually reach a plateau. The rate of uptake is dose-dependent and is retarded by the presence of ligands, both organic and inorganic. Wong et al. (1984), working with the green alga *Ankistrodesmus falcatus*, achieved a concentration factor of 91,000 when the alga was exposed for 80 min to 46 μg Sn/L of medium. Most (85%) of the tin was stored in polysaccharides, with smaller amounts in protein (18.8%), lipid (0.1%), and low-molecular-weight metabolites (0.1%).

Uptake of tributyltin by *Ankistrodesmus falcatus* is also rapid, producing a concentration factor of approximately 30,000 (Maguire et al., 1984). In Maguire's study, approximately 50% of the original burden of tributyltin was converted, after 4 weeks, to dibutyltin. Francois et al. (1989), working with marine eelgrass *Zostera marina*, also noted that successive debutylation was the primary means of degradation. The process followed the model:

$$Bu_3Sn^+ \xrightarrow{k_1} Bu_2Sn^{2+} \xrightarrow{k_2} BuSn^{3+} \xrightarrow{k_3} Sn^{4+}$$

The rate constants were first-order and calculated as follows:

$$d[Bu_3Sn^+]/dt = k_1[Bu_3Sn^+] \quad (1)$$

$$d[Bu_2Sn^{2+}]/\mathrm{dt} = k_1[Bu_3Sn^+] - k_2[Bu_2Sn^{2+}] \quad (2)$$

$$d[BuSn^{3+}]/dt = k_2[Bu_2Sn^{2+}] - k_3[BuSn^{3+}3] \quad (3)$$

Debutylation in eelgrass was then modeled.

$$
\begin{array}{ccccccc}
Bu_3Sn^{+}\ (eel) & \xrightarrow{k_1} & Bu_2Sn^{2+}\ (eel) & \xrightarrow{k_2} & BuSn^{3+}\ (eel) & \longrightarrow & Sn^{4+} \\
\uparrow k_{ads} & & & & \downarrow & & \\
Bu_3Sn^{+}\ (aq) & & & & BuSn^{3+}\ (aq) & &
\end{array}
$$

where "eel" refers to association with eelgrass, k_{ads} is the rate of uptake of tributyltin by the plant, and aq is aqueous concentration.

Invertebrates

Tributyltin and dibutyltin are readily accumulated by many invertebrate species. In some cases, residues of well over 1,000 μg/kg have been recorded (Table 23.4), yielding concentration factors (tissue level/water

Table 23.4. Concentration (μg/kg dry weight) of organic tin in invertebrate species.

Species	Compound[a]	Average (range)	Location
Mussel, *Mytilus edulis*[1]	TBT	425 (100–1,540)	estuaries, USA
	DBT	239 (<5–870)	estuaries, USA
	BT	221 (<5–1,240)	estuaries, USA
Oyster, *Crassostrea virginica*[1]	TBT	329 (<5–790)	estuaries, USA
	DBT	99 (<5–270)	estuaries, USA
	BT	58 (<5–120)	estuaries, USA
Mussel, *Mytilus edulis*[2]	TBT	310 (190–470)[b]	San Diego Bay, USA
	TBT	200 (<10–360)[b]	Monterey Bay, USA
	TBT	75 (<10–120)[b]	Puget Sound, USA
	TBT	73 (<10–220)[b]	Puget Sound marina, USA
	TBT	65 (<10–170)[b]	
	TBT	67 (<10–1,080)[b]	Auke Bay, USA
			Kodiak Harbor, USA
Polychaete, *Nereis diversicolor*[3]	TBT	NR[c] (30–1,590)	Poole Harbor, UK
	DBT	NR (150–5,650)	Poole Harbor, UK
Gastropod, *Littorina littorea*[3]	TBT	NR (100–1,120)	Poole Harbor, UK
	DBT	NR (230–3,220)	Poole Harbor, UK
Bivalve, *Mya arenaria*[3]	TBT	NR (5,550–11,450)	Poole Harbor, UK
	DBT	NR (1,780–9,950)	Poole Harbor, UK

[a]TBT, tributyltin; DBT, dibutyltin; BT, butyltin.
[b]Wet weight.
[c]Not reported.
Sources: [1]Wade et al. (1988), [2]Short and Sharp (1989), [3]Langston et al. (1987).

level) of 3,000–133,000 (Langston et al., 1987). Such high levels are usually limited to sites of anthropogenic contamination, whereas invertebrates collected from noncontaminated areas contain residues that are typically below the analytical detection limit (Bailey and Davies, 1988a). Uptake of tributyltin is strongly concentration-dependent, and a steady state in residues is generally achieved within 2 weeks (Laughlin and French, 1988). Tissue concentrations in molluscs are often greatest in the gills, followed by those in the viscera, adductor muscle, and mantle.

Most regulatory agencies routinely monitor the concentrations of tributyltin, dibutyltin, and butyltin in molluscs and other invertebrate species. Monitoring is usually limited to coastal and estuarine waters that receive input of tributyltin and other organic Sn compounds. In some cases, organic Sn residues can be so high as to threaten important shellfish fisheries. Tanabe (1988) reported that, in Japan, the tentative acceptable level of tributyltin in seafood is 890 μg/kg, a concentration that was exceeded in a number of samples.

Fish

Tributyltin residues are often high in fish collected from harbors and other coastal waters. Tanabe (1988), for example, noted that sea bass (species not identified) contained levels of 1.1–1.3 mg/kg wet weight in muscle tissue. Comparably high levels (0.5–1.5 mg/kg) were found in the muscle tissue of Atlantic salmon *Salmo salar* kept in nets treated with tributyltin antifoulants (Davies and McKie, 1987) whereas chinook salmon *Oncorhynchus tshawytscha* held under the same conditions contained residues of 0.28–0.90 mg/kg (Short and Thrower, 1986).

Sorption of tributyltin follows an exponential path, increasing enormously during the first hours of exposure, followed by a plateau in the rate of sorption. Martin et al. (1989), working with rainbow trout *Oncorhynchus mykiss*, noted that tributyltin partitioned favorably to peritoneal fat (mean concentration 9.2 mg/kg), the kidney/liver/gallbladder (mean concentration 3.1–3.7 mg/kg), and all other tissues (mean concentration 0.5–1.5 mg/kg). Tsuda et al. (1988), working with carp *Cyprinus carpio*, noted that the concentration of butyltin compounds decreased in the order:

tributyltin: kidney > gallbladder > liver > muscle
dibutyltin: liver > gallbladder > kidney > muscle
butyltin: liver ⩾ gallbladder > kidney > muscle

Sorption of phenyltins also follows an exponential path, increasing rapidly during the first 24 h of exposure, followed by a plateau in the rate of sorption. Tsuda et al. (1987) reported that concentrations of triphenyltin in the muscle of carp *Cyprinus carpio* ranged up to 1.6 mg/kg wet weight, whereas the primary metabolites (diphenyltin, phenyltin) were found at <0.1 mg/kg. The same study showed that residues of the metabolites were much greater in the liver and kidney, both ranging up to 1 mg/kg.

Toxic Effects to Aquatic Organisms

Plants

Most species of aquatic plants are highly sensitive to tributyltin and related compounds. Thain (1983), for example, reported that the LC_{100} (lethal concentration) of tributyltin oxide to the diatom *Skeletonema costatum* was 18 μg/L whereas Walsh et al. (1985) found a 48 h LC_{50} of 14 μg/L when the same species was exposed to tributyltin. A much lower 48-h LC_{50} (5 μg/L) was noted by Beaumont and Newman (1986) when *Skeletonema costatum*, *Dunaliella tertiolecta*, and *Pavlova lutheri* were exposed to tributyltin oxide. The 72-h EC_{50} (Effective Concentration) of tributyltin oxide to *Skeletonema costatum* is approximately 0.3 μg/L (Walsh et al., 1985).

Such low concentrations (all derived from laboratory experiments) suggest that the productivity of some algal species may be significantly affected by elevated levels of organotins in harbors, marinas, and other areas where antifoulants are used. However, some algal species are probably less sensitive to organotins than those described preceding. In fact, Goldberg (1986) recorded the presence of algae in two California marinas where tributyltin residues ranged from 0.50 to 0.57 μg/L. No invertebrate species were found in the waters of those two marinas.

Invertebrates

Tributyltin is highly toxic to many marine and estuarine invertebrates, with LC_{50}s typically ranging from <0.5 to 5 μg/L (Hall et al., 1988a). These values make tributyltin among the most toxic of all organotins, and in fact, some compounds are more than 1,000 times less toxic to certain invertebrates than tributyltin (Table 23.5).

Tributyltin generally acts as an anionophore of inorganic ions in several invertebrate species, resulting in the collapse of the $Cl^-:OH^-$ gradient across mitochondrial membranes.Inhibition of ion translocation, which then influences osmotic pressure and hemolymph concentrations, is another potential toxic mechanism following exposure to organotins. A few investigators (e.g., Bokman and Laughlin, 1989) have demonstrated that tributyltin is not always a strong anionophore in some common species such as the oyster *Crassostrea virginica*. Hence, the interaction between salinity and organotin toxicity needs to be explored more fully.

Table 23.5. Relative acute toxicity (expressed as LC_{50}) of organotins to zoeal mud crabs (*Rhithropanopeus harrisii*).[a]

Compound	Relative toxicity
Bis(tributyltin) oxide	1
Bis(triisobutyltin) oxide	0.84
Tricyclohexyltin bromide	0.36
Triphenyltin hydroxide	1.8
Bis(triisopropyltin) oxide	3.4
Bis(tripropyltin) oxide	3.5
Dicyclohexyltin dichloride	6.7
Triethyltin hydroxide	7.1
Trimethyltin hydroxide	10.2
Diphenyltin dichloride	46.7
Dibutyltin dichloride	51.1
Dipropyltin dichloride	254.3
Diethyltin dichloride	265.2
Dibenzyltin dichloride	488.5
Dimethyltin dichloride	1677.7

[a]Increasing values indicate decreasing toxicity.
Source: Laughlin (1987).

Numerous chronic effects have been reported for marine invertebrates exposed to organotins, as follows:

1. Decrease in the growth of juvenile *Mytilus edulis* exposed to tributyltin oxide at 0.4 μg TBTO/L (Stromgren and Bongard, 1987).
2. Reduction in oxygen consumption, feeding rate, and growth of the spat of Pacific oyster *Crassostrea gigas* exposed to tributyltin oxide at 0.02–0.05 μg TBTO/L (Lawler and Aldrich, 1987).
3. Increase in frequency of shell deformities of Sydney rock oysters *Saccostrea commercialis* exposed to tributyltin at <0.03–0.30 μg/L (Ellis, 1988).
4. Increased frequency of imposex of the common dogwhelk *Nucella lapillus* exposed to tributyltin in coastal waters of the United Kingdom (Bailey and Davies, 1988b). Imposex is the superimposition of male characters, notably a penis and a vas deferens, on the female.

Fish

Tributyltin is also highly toxic to many fish. Martin et al. (1989), for example, reported that the 96-h LC_{50} for lake trout *Salvelinus namaycush* was 5.2 μg/L whereas the corresponding values for rainbow trout *Oncorhynchus mykiss* and silversides *Menidia beryllina* were only 1.4 and 3 μg/L, respectively (Hall, 1988). Higher LC_{50}s (23–26 μg/L) were found for larval sheepshead minnow *Cyprinodon variegatus* and larval-subadult mummichog *Fundulus heteroclitus* (Hall, 1988). When larval fathead minnow *Pimephales promelas* were exposed to triphenyltin hydroxide, the 96-h LC_{50} was 7 μg/L (Jarvinen et al., 1988).

Chronic effects are typically reported following exposure to extremely low concentrations of organotins. Wester and Canton (1987), for example, reported that the No-Observed-Effect Concentration of tributyltin oxide in the guppy *Poecilia reticulata* was only 0.01 μg/L. That concentration was based on absence of thymus atrophy, liver vacuolation, and hyperplasia of hemopoietic tissue. In another study, Jarvinen et al. (1988) noted that the growth of larval fathead minnow was retarded at 0.2 μg/L, following exposure to triphenyltin hydroxide; the same study also recorded reduced survival of larvae at 2 μg/L.

Relatively little is known about the effect of salinity, inorganic ligands, and other water quality parameters on the toxicity of organotin compounds to fish.

Health Effects

Intake

The typical Western diet yields approximately 4 mg S/day in a 70-kg reference man (reviewed by Carson et al., 1987). Inhalation is an insignificant route in the nonoccupationally exposed population. Once ingested, inorganic Sn is poorly absorbed, ranging from 0.6–7.7% (Schafer and Femert, 1984), whereas organotins are absorbed at a rate of approximately 10%. The primary sites of accumulation of inorganic Sn are the kidney, liver, and bones; for organotins, the main sites are the liver, kidney, and brain. Total Sn in blood serum of nonoccupationally exposed people averages 30–103 mg/kg, ranging up to 350 mg/kg (Schafer and Femert, 1984).

Daily losses from a 70-kg reference man are feces 3.5 mg, sweat 0.5 mg, urine 0.02 mg. Inorganic Sn leaves the body in feces and urine, whereas organotin is primarily excreted in the feces and bile.

Acute Toxicity

Acute exposure to inorganic Sn generally comes from contaminated food. The most frequently reported symptoms are nausea, abdominal cramps, vomiting, headache, diarrhea, and fever. Inorganic Sn appears to interfere with protein digestion, and the uptake and metabolism of metals such as copper, iron, and zinc (Schafer and Femert, 1984). Such effects are seen only after exposure to high doses (150–1400 mg Sn/kg).

Chronic Toxicity

Neurotoxicity is the salient effect of exposure to toxic levels of organotins. Triethyltin and trimethyltin are particularly neurotoxic, showing a different series of effects, typically when dose exceeds 1 mg/kg body weight (Table 23.6).

Tributyltin is thought to act as an agent in cellular depletion of ATP and delipidation of anionic phospholipids (Gray et al., 1987a); it is also possible that tributylstannylperoxy radicals are formed, resulting in lipid peroxidation as illustrated below:

Initiation:

$$(C_4H_9)_3\ Sn^+ + O_2^{\dot{-}} \longrightarrow (C_4H_9)_3\ SnOO^{\cdot} \qquad (4)$$

Propagation:

$$(C_4H_9)_3SnOO^{\cdot} + R \diagup\!\!=\!\!\vee\!\!=\!\!\diagdown R' \longrightarrow (C_4H_9)_3SnOOH + R\text{—}C\text{—}C\text{—}\dot{C}\text{—}C\text{—}C\text{—}R' \qquad (5)$$

$$R\text{—}C\text{—}C\text{—}\dot{C}\text{—}C\text{—}C\text{—}R' + O_2 \longrightarrow R\text{—}C(\text{—}O\text{—}\dot{O})\text{—}C{=}C\text{—}C{=}C\text{—}R' \qquad (6)$$

$$R\text{—}C(\text{—}O\text{—}\dot{O})\text{—}C{=}C\text{—}C{=}C\text{—}R' + R + R \diagup\!\!=\!\!\vee\!\!=\!\!\diagdown R' \longrightarrow R\text{—}C(\text{—}O\text{—}O\text{—}H)\text{—}C{=}C\text{—}C{=}C\text{—}R' + R\text{—}C\text{—}C\text{—}\dot{C}\text{—}C\text{—}C\text{—}R' \qquad (7).$$

Organotins also induce in vitro hemolysis in human erythrocytes in the following order of potency (Gray et al. 1987b): tributyltin > tripropyltin > tetrabutyltin > triphenyltin chloride > triethyltin bromide > dibutyltin chloride > stannous chloride > trimethyltin chloride = butyl chloride dihydroxide.

Table 23.6. Neurologic effects of triethyltin and trimethyltin.

Triethyltin
- A distinct myelinopathy
- A generalized edema of the nervous system
- A disturbance of cerebral energy metabolism
- A general depression of behavioral processes with concomitant sensory impairments

Trimethyltin
- Degeneration of neurons in the hippocampus, the limbic forebrain, the brainstem, and primary sensory structures
- Change in the regional concentration of dopamine and aminobutyric acid
- Alterations in visual, auditory, and somatosensory processing
- Hyperactivity and cognitive impairment

Sources: Walsh and DeHaven (1988), Cockerill et al. (1987), Messing et al. (1988), O'Callaghan and Miller (1987), Snoeij et al. (1987).

Carcinogenicity

Tin is not considered carcinogenic to humans or experimental animals.

Drinking Water

Residues

Total Sn is rarely detected in finished drinking water, largely reflecting its scarcity in feed water. Although organotins may leach into drinking water from new polyvinyl chloride pipe, the rate of extraction reaches a steady state within 12 h (Wu et al., 1989). There are almost no data available on the relative proportions of inorganic versus organic compounds in drinking water samples.

Consumption Guidelines

Most nations, plus the World Health Organization, have yet to establish a drinking water guideline for tin. This reflects the low toxicity of most tin compounds to human consumers. The National Sanitation Foundation (USA) has developed a guideline of 50 μg/L for tin (Wu et al., 1989).

Because organotin usage continues to rise in many industrial and municipal sectors, residues in drinking water will also likely increase in future years. This reflects the contamination of source water by antifoulants and dissolution of distribution system materials. Hence, a multinational drinking water guideline should be developed in future years.

Treatment

A primary means of controlling organotins in drinking water is to limit the usage of tin-containing distribution lines. Otherwise, conventional coagulation and filtration are effective in removing particulate-bound tin.

Recommendations

The environmental significance of tin has risen enormously in recent years, a reflection of the expanded use of organotins in and around surface waters. Although this increase in usage has not had major, direct effect on human consumers of water, there have been many impacts on aquatic plants, invertebrates, and fish. The majority of these effects have occurred in and around marinas, estuaries, and other coastal waters. Some nations have now moved to limit, in a small way, the use of organotins in such areas.

Despite the increasing importance of tin compounds, relatively little basic research has been conducted on their environmental fate and toxicity. Several of the most important deficiencies are listed below.

1. Partitioning of tin compounds in different components of sediments.
2. Effect of temperature, pH, salinity, and other environmental factors on the rate of methylation of inorganic Sn and butyltins.
3. Factors influencing the toxicity of organotins to algae, invertebrates, and fish, particularly the effects of changing salinity.
4. Development of a drinking water guideline for organotins.

References

Bailey, S.K., and I.M. Davies. 1988a. Tributyltin contamination around an oil terminal in Sullom Voe (Shetland). *Environmental Pollution* 55:161–172.

Bailey, S.K., and I.M. Davies. 1988b. Tributyltin contamination in the Firth of Forth (1975–87). *Science of the Total Environment* 76:185–192.

Beaumont, A.R., and P.B. Newman. 1986. Low levels of tributyl tin reduce growth of marine micro-algae. *Marine Pollution Bulletin* 17:457–461.

Bokman, E., and R.B. Laughlin. 1989. A study of steady state and kinetic regulation of chloride ion and osmotic pressure in hemolymph of oysters, *Crassostrea virginica*, exposed to tri-n-butyltin. *Archives of Environmental Contamination and Toxicology* 18:832–838.

Byrd, J.T., and M.O. Andreae. 1986. Concentrations and fluxes of of tin in aerosols and rain. *Atmospheric Environment* 20:931–939.

Carson, B.L., H.V. Ellis, and J.L. McCann. 1987. *Toxicology and biological monitoring of metals in humans*. Lewis Publishers, Chelsea, MI. 328 pp.

Clark, E.A., R.M. Sterritt, and J.N. Lester. 1988. The fate of tributyltin in the aquatic environment. *Environmental Science and Technology* 22:600–604.

Cleary, J.J., and A.R.D. Stebbing. 1987. Organotin in the surface microlayer and subsurface waters of southwest England. *Marine Pollution Bulletin* 18:238–246.

Cockerill, D., L.W. Chang, A. Hough, and F. Bivins. 1987. Effects of trimethyltin on the mouse hippocampus and adrenal cortex. *Journal of Toxicology and Environmental Health* 22:149–161.

Cooney, J.J. 1988. Interactions between microorganisms and tin compounds. *In: The biological alkylation of heavy elements,* eds. P.J. Craig, and F. Glockling, 92–104. Royal Society of Chemistry, London.

Cooney, J.J., A.T. Kronick, G.J. Olson, W.R. Blair, and F.E. Brinckman. 1988. A modified method for quantifying methyl and butyltins in estuarine sediments. *Chemosphere* 17:1795–1802.

Cordero, R. 1988. *Metal Bulletin's prices and data 1988.* Metal Bulletin Books, Surrey, England, 375 pp.

Craig, P.J., and S. Rapsomanilis. 1985. Methylation of tin and lead in the environment: oxidative methyl transfer as a model for environmental reactions. *Environmental Science and Technology* 19:726–730.

Dahab, O.A. 1988. Speciation of tin compounds in sediments of the Alexandria coastal belt. *Water, Air, and Soil* 40:433–441.

Davies, I.M., and J.C. McKie. 1987. Accumulation of total tin and tributyltin in muscle tissue of farmed Atlantic salmon. *Marine Pollution Bulletin* 18:405–407.

Ellis, D. 1988. First Australian TBT ban. *Marine Pollution Bulletin* 19:652.

Forstner, U., and G.T.W. Wittmann. 1979. *Metal pollution in the aquatic environment.* Springer, New York. 486 pp.

Francois, R., F.T. Short, and J.H. Weber. 1989. Accumulation and persistence of tributyltin in eelgrass (*Zostera marina* L.) tissue. *Environmental Science and Technology* 23:191–196.

Goldberg, E.D. 1986. TBT, an environmental dilemma. *Environment* 28:17–44.

Gray, B.H., M. Porvaznik, C. Flemming, and L.H. Lee. 1987a. Tri-n–butyltin: a membrane toxicant. *Toxicology* 47:35–54.

Gray, B.H., M. Porvaznik, C. Flemming, and L.H. Lee. 1987b. Organotin-induced hemolysis, shape transformation and intramembranous aggregates in human erythrocytes. *Publications in Cell Biology and Pathology* 3:23–38.

Hall, L.W. 1988. Tributyltin environmental studies in Chesapeake Bay. *Marine Pollution Bulletin* 19:431–438.

Jarvinen, A.W., D.K. Tanner, E.R. Kline, and M.L. Knuth. 1988. Acute and chronic toxicity of triphenyltin hydroxide to fathead minnows (*Pimephales promelas*) following brief or continuous exposure. *Environmental Pollution* 52:289–301.

Langston, W.J., G.R. Burt, and Z. Mingjiang. 1987. Tin and organotin in water, sediments, and benthic organisms of Poole Harbour. *Marine Pollution Bulletin* 18:634–639.

Laughlin, R.B. 1987. Quantitative structure-activity studies of di–and triorganotin compounds. *In: QSAR in environmental toxicology,* ed., K.L.E. Kaiser, 189–206. Reidel Publishing Co., Dordrecht, Holland.

Laughlin, R.B., and W. French. 1988. Concentration dependence of bis(tributyl) tin oxide accumulation in the mussel, *Mytilus edulis*. *Environmental Toxicology and Chemistry* 7:1021–1026.

Lawler, I.F., and J.C. Aldrich. 1987. Sublethal effects of bis(tri-n-butyltin)oxide on *Crassostrea gigas* spat. *Marine Pollution Bulletin* 18:274–278.

Maguire, R.J., Y.K. Chau, G.A. Bengert, E.J. Hale, P.T.S. Wong, and O. Kramer. 1982. Occurrence of organic compounds in Ontario lakes and rivers. *Environmental Science and Technology* 16:698–702.

Maguire, R.J., J.H. Carey, and E.J. Hale. 1983. Degradation of the tri-n-butylin tin species in water. *Journal of Agricultural and Food Chemistry* 31:1060–1065.

Maguire, R.J., P.T.S. Wong, and J.S. Rhamey. 1984. Accumulation and metabolism of tri-n-butyltin cation by a green alga, *Ankistrodesmus falcatus*. *Canadian Journal of Fisheries and Aquatic Sciences* 41:537–540.

Maguire, R.J., R.J. Tkacz, Y.K. Chau, G.A. Bengert, and P.T.S. Wong. 1986. Occurrence of organotin compounds in water and sediment in Canada. *Chemosphere* 15:253–274.

Makkar, N.S., A.T. Kronock, and J.J. Cooney. 1989. Butyltins in sediments from Boston Harbor, USA. *Chemosphere* 23:2043–2050.

Martin, R.C., D.G. Dixon, R.J. Maguire, P.V. Hodson, and R. J. Tkacz. 1989. Acute toxicity, uptake, depuration and tissue distribution of tri-n-butyltin in rainbow trout, *Salmo gairdneri*. *Aquatic Toxicology* 15:37–52.

Messing, R.B., G. Bollweg, Q. Chen, and S.B. Sparber. 1988. Dose-specific effects of trimethyltin poisoning on learning and hippocampal corticosterone binding. *Neurotoxicology* 9:491–502.

Muller, M.D., L. Renberg, and G. Rippen. 1989. Tributyltin in the environment—sources, fate and determination. An assessment of present status and research needs. *Chemosphere* 18:2015–2042.

Nriagu, J.O., and J.M. Pacyna. 1988. Quantitative assessment of worldwide contamination of air, water and soils by trace metals. *Nature* 333:134–139.

O'Callaghan, J.P., and D.B. Miller. 1987. Acute exposure of the neonatal rat to triethyltin results in persistent changes in neurotypic and gliotypic proteins. *Journal of Pharmacology and Experimental Therapeutics* 244:368–378.

Randall, L., J.S. Han, and J.H. Weber. 1986. Determination of inorganic tin, methyltin, and butyltin compounds in sediments. *Environmental Technology Letters* 7:571–576.

Ring, R.M., and J.H. Weber. 1988. Methylation of tin (II) by methyl iodide under simulated estuarine conditions in the absence and presence of fulvic acid. *Science of the Total Environment* 68:225–239.

Schafer, S.G., and U. Femert. 1984. Tin—a toxic heavy metal? A review of the literature. *Regulatory Toxicology and Pharmacology* 4:57–69.

Seligman, P.F., A.O. Valkirs, and R.F. Lee. 1986. Degradation of tributyltin in San Diego Bay, California, waters. *Environmental Science and Technology* 20:1229–1235.

Seligman, P.F., A.O. Valkirs, P.M., and R.F. Lee. 1988. Evidence for rapid degradation of tributyltin in a marina. *Marine Pollution Bulletin* 19:531–534.

Short, J.W., and J.L. Sharp. 1989. Tributyltin in bay mussels (*Mytilus edulis*) of the Pacific coast of the United States. *Environmental Science and Technology* 23:740–743.

Short, J.W., and F.P. Thrower. 1986. Accumulation of butyltins in muscle tissue of chinook salmon reared in sea pens treated with tri-n-butyltin. *Marine Pollution Bulletin* 17:542–545.

Snoeij, N.J., A.H. Penninks, and W. Seinen. 1987. Biological activity of organotin compounds—an overview. *Environmental Research* 44:335–353.

Stromgren, T., and T. Bongard. 1987. The effect of tributyltin oxide on growth of *Mytilus edulis. Marine Pollution Bulletin* 18:30–31.

Tanabe, S. 1988. TBT contamination in Japan. *Marine Pollution Bulletin* 19:500.

Thain, J.E. 1983. *The acute toxicity of bis(tributyltin) oxide to the adults and larvae of some marine organisms.* ICES Paper CM1983/E:13. International Council for the Exploration of the Sea, Copenhagen.

Tsuda, T., H. Nakanishi, S. Aoki, and J. Takebayashi. 1987. Bioconcentration and metabolism of phenyltin chlorides in carp. *Water Research* 21:949–953.

Tsuda, T., H. Nakanishi, S. Aoki, and J. Takebayashi. 1988. Bioconcentration and metabolism of butyltin compounds in carp. *Water Research* 22:647–651.

Unger, M.A., W.G. MacIntyre, and R.J. Huggett. 1988. Sorption behavior of tributyltin on estuarine and freshwater sediments. *Environmental Toxicology and Chemistry* 7:907–915.

US Minerals Yearbooks. 1930–1989. Bureau of Mines, US Department of the Interior, Washington, DC.

Wade, T.L., B. Garcia-Romero, and J.M. Brooks. 1988. Tributyltin contamination in bivalves from United States coastal estuaries. *Environmental Science and Technology* 22:1488–1493.

Walsh, G.E., L.L. McLaughlan, E.M. Lores, M.K. Louie, and C.H. Deans. 1985. Effects of organotins on growth and survival of two marine diatoms *Skeletonema costatum* and *Thalassiosira pseudonana. Chemosphere* 14:383–392.

Walsh, T.J., and D.L. DeHaven. 1988. Neurotoxicity of the alkyltins. *In: Metal neurotoxicity.* eds., S.C. Bondy and K.N. Prasad, 87–107. CRC Press, Boca Raton, FL. 87–107.

Wester, P.W., and J.H. Canton. 1987. Histopathological study of *Poecilia reticulata* (guppy) after long-term exposure to bis(tri-n-butyltin)oxide (TBTO) and di-*n*-butyltindichloride (DBTC). *Aquatic Toxicology* 10:143–165.

Wong, P.T.S., R.J. Maguire, Y.K. Chau, and O. Kramer. 1984. Uptake and accumulation of inorganic tin by a freshwater alga, *Ankistrodesmus falcatus. Canadian Journal of Fisheries and Aquatic Sciences* 41:1570–1574.

Wu, W., R.S. Roberts, Y.-C. Chung, W.R. Ernst, and S.C. Havlicek. 1989. The extraction of organotin compounds from polyvinyl chloride pipe. *Archives of Environmental Contamination and Toxicology* 18:839–843.

24
Vanadium

Vanadium is a ubiquitous element in the earth's crust, ranking 22nd in abundance with a mean concentration 150 mg/kg. Although igneous rock, carbonatite complexes, titaniferous magnetite complexes, and deposits of iron, uranium, chromium, and manganese all contain vanadium, high-content ores include vanadinite, descloizite, partonite, roscoelite, and carnotite. Elevated concentrations can also be found in many coals and other fossil fuels. Vanadium is a hard acid, so it prefers to complex with oxides and nitrogen.

Production, Sources, and Residues

Production

World production of vanadium was 3.1×10^3 metric tons in 1940, increasing to 6.4×10^3 metric tons in 1960 and 35×10^3 metric tons/yr in 1980 (US Minerals Yearbooks, 1940–1989). Production in recent years has continued to remain relatively low, generally $<40 \times 10^3$ metric tons/yr. Vanadium ranks 64th in terms of world production (by weight) of minerals and ores, but is 42nd in terms of value of output (Noetstaller, 1988). The primary uses of vanadium and its compounds are (1) metallurgy, particularly steel for use in combustion, tools, dies, and springs; (2) catalysts in the production of polymeric plastics, and sulfuric and nitric acids; and (3) dyes, inks, and paints.

Sources

The total amount of vanadium discharged from worldwide anthropogenic sources to freshwaters amounts to 2.1–21 × 10^3 metric tons/yr (Table 24.1). The No. 1 source is atmospheric fallout, principally due to the combustion of fossil fuels and, to a lesser degree, windblown dusts. Other significant sources include the dumping of sewage sludge and the discharge of domestic wastewater. Vanadium is also solublized from stabilized oil ash waste in seawater (Breslin and Duedall, 1988).

The total amount of vanadium emitted to the atmosphere from worldwide anthropogenic sources ranges up to 142 × 10^3 metric tons/yr (average 86 × 10^3 metric tons/yr), compared to a maximum of 54 × 10^3 metric tons/yr (average 28 × 10^3 metric tons/yr) for natural sources (Nriagu, 1989). Hence, the ratio of natural emissions to total emissions is approximately 0.25. The primary natural sources of vanadium to the atmosphere are windblown soil particles, volcanoes, and seasalt spray; the major anthropogenic source is the combustion of oil and coal (Table 24.2).

Residues

Water. Total V residues in freshwaters generally range from <0.5 μg/L to 50 μg/L. In one study from Sri Lanka, Dissanayake et al. (1987) reported that concentrations in an urban canal averaged 18 μg/L with a range of 2–45 μg/L. The canal received waste water from both industrial and municipal sources. In a Canada-wide survey of freshwaters, total V residues generally ranged from nondetectable (<0.5 μg/L) to 19 μg/L (Na-

Table 24.1. Worldwide anthropogenic input of vanadium to freshwaters.

Source	Input (thousand metric tons/yr)
Atmospheric fallout	1.4–9.1
Dumping of sewage sludge	0.7–4.3
Domestic wastewater	
central	0–2.7
noncentral	0–1.8
Smelting and refining	
nonferrous metals	0–1.2
Coal-burning power plants	0–0.6
Manufacturing processes	
metals	0–0.75
chemicals	0–0.35

Source: Nriagu and Pacyna (1988).

Table 24.2. Worldwide anthropogenic and natural emissions of vanadium to the atmosphere.

Source	Input (thousand metric tons/yr)
Anthropogenic sources	
Oil combustion	
industrial and domestic	21.5–71.6
electrical utilities	7.0–52.2
Coal combustion	
industrial and domestic	1.0–9.9
electrical utilities	0.3–4.7
Incineration	
sewage sludge	0.3–2.0
Steel and iron production	0.1–1.4
Natural Sources	
Windborne soil particles	1.2–30
Volcanoes	0.2–11
Sea-salt spray	0.1–7.2
Forest fires	0.02–3.6
Continental biogenic particles	0.04–1.8

Sources: Nriagu (1989), Nriagu and Pacyna (1988).

quadat, 1985). However, a maximum value of 900 μg/L was found in a Pacific Coast river (no apparent anthropogenic source), whereas extremely low levels (<0.5 μg/L) were recorded from Lake Ontario.

Sediments. Total V in surficial sediments typically ranges from 20 to 150 mg/kg dry weight. Mudroch and Duncan (1986), working on sediment from the Niagara River on the Canada-USA border, reported relatively high concentrations (101–139 mg/kg) in the fine sediment fraction (Table 24.3). Although residues ranged from 31 to 88 mg/kg in the 31–53 μm fractions, much higher concentrations were found in the 54–150 μm size fraction (Table 24.3), presumably reflecting the presence of a vanadium-containing mineral in the sediments.

Table 24.3. Concentration (mg/kg dry weight) of vanadium in different sediment size fractions from the Niagara River.

Size fraction (μm)	Concentration average (range)	Size fraction (μm)	Concentration average (range)
54–150	158 (36–734)	40–53	48 (31–88)
27–39	52 (37–86)	19–26	62 (47–86)
13–18	71 (55–88)	<13	123 (101–139)

Sources: Mudroch and Duncan (1986).

Araujo et al. (1988) reported that total V in the clay/silt fraction of sediments from the North Sea and Scheldt Estuary ranged from 91 to 106 mg/kg, compared to 48–49 mg/kg for the bulk sediments. Slightly lower residues, 29–48 mg/kg, were found for bulk sediments from the Arabian Gulf near Kuwait (Samhan et al., 1987) whereas vanadium from the highly contaminated Ganges Estuary (India) ranged from 42 to 180 mg/kg (Subramanian et al., 1988).

Chemistry

Vanadium has a number of possible oxidation states: V^{o}, V^{+}, V^{2+}, V^{3+}, V^{4+} and V^{5+}. The pentavalent form is the most soluble, and is the primary agent of transport in surface waters. Vanadium can be oxidized to the pentavalent form from the tetravalent form, so becoming soluble in water.

Vanadium often appears in association with humic acids. The vanadyl ion forms stable complexes with these substances, a reflection of its hard acid classification. Mangrich and Vugman (1988) suggested that the vanadyl ion-binding site in a humic acid from a Brazilian estuary was of the salicylate type. Under this configuration, vanadium was bound to oxygen atoms only. The susceptibility of vanadium–humic acid complexes to photolysis and microbial degradation is poorly understood at this time.

Bioaccumulation

Plants

Total V is only infrequently determined in aquatic plants, so the corresponding data base is relatively small. Soderlund et al. (1988) reported that residues in *Fucus vesiculosus* collected from the east coast of Sweden ranged from 0.10 to 2.15 mg/kg dry weight with an average of 0.45 mg/kg. There was no significant difference in residues for plants collected at different depths. The same study showed that an aquatic moss *Fontinalis* sp. carried burdens of 3.7–5.7 mg/kg dry weight (average 4.6 mg/kg). Lee (1983), in a review of a number of studies, noted that some marine algae, particularly *Thalassia testudinum*, *Pontedaria cordata*, and *Phyllospadix iwatensis*, can accumulate vanadium.

Invertebrates

Total V in invertebrate tissues typically ranges from 0.1 to 5 mg/kg dry weight, but higher levels, ranging up to 9.5 mg/kg, have been found under some conditions (Table 24.4). Fowler (1986) noted that the vanadium content of microplankton (principally copepods, larval crustaceans, chaetognaths, phytoplankton, and detritus) from the Mediterranean Sea averaged 1.45 mg/kg dry weight. This was the highest concentration in the

Table 24.4. Concentration (mg/kg dry weight) of total V in the soft tissues of invertebrates.

Species	Average (range)	Location
Coral, *Pocillopora damicornis*[1]	5.2 (2.6–9.5)	Coastal waters, Thailand
Coral, *Fungia fungites*[2]	2.7 (1.3–7.2)	Coastal waters, Thailand
Pearl oyster, *Pinctada radiata*[3]	0.28 (0.0–1.20)	Arabian Gulf, Saudi Arabia
Asiatic clam, *Corbicula fluminea*[4]	1.1 (<0.1–5.0)	Shatt al-Arab River, Iraq
Mollusc, *Arctica islandica*[5]	0.94 (0.02–4.0)	Georges Bank, USA

Sources: [1]Howard and Brown (1987), [2]Howard and Brown (1986), [3]Sadig and Alam (1989), [4]Abaychi and Mustafa (1988), [5]Phillips et al. (1987).

pelagic invertebrate food chain. Residues at the other levels of the food chain were:

Euphasiid *Meganyctiphanes norvegica*	0.23 mg/kg
Amphipod *Phronima sedentaria*	0.45 mg/kg
Shrimp *Pasiphaea sivado*	0.07 mg/kg

Essentially nothing is known about the factors influencing the uptake of vanadium by invertebrate species.

Fish

Total V is only infrequently determined in marine and freshwater fish. The limited data base available to date indicates that (1) residues are generally low, and (2) there is no evidence of bioaccumulation. In the pelagic food chain study of the Mediterranean Sea referred to in the previous section (Fowler, 1986), the fish *Myctophum glaciale* contained an average vanadium burden of only 0.08 mg/kg dry weight. This was equivalent to approximately 0.02 mg/kg wet weight.

There do not appear to be any cases in which the consumption of fish has been limited by high vanadium concentrations.

Toxic Effects to Aquatic Organisms

Plants and Invertebrates

Remarkably little is known about the toxic effects of vanadium to aquatic plants and invertebrates. In one study, Miramand and Unsal (1978) reported that the LC_{50} for a number of algal and invertebrate species ranged from 10 to 65 mg/L. The species, which were exposed to $NaVO_3$ in seawater, included the annelid *Nereis diversicolor*, the mollusc *Mytilus edulis*, and the crustacean *Carcinus maenas*. Because vanadium appears to be relatively nontoxic to plants and invertebrates, there has been no detailed assessment of the factors influencing toxicity.

Fish

Vanadium is relatively nontoxic to fish species, both marine and freshwater. LC_{50}s typically range from 1 to >30 mg/L, depending on species, life stage, and water chemistry (Taylor et al., 1985; Mance, 1987). Stendahl and Sprague (1982) reported that the 7-day LC_{50} of rainbow trout *Oncorhynchus mykiss* increased from 2.4–3.0 mg/L in soft water (30 mg/L as $CaCO_3$) to 2.9–5.6 mg/L in hard water (360 mg/L as $CaCO_3$). The same study also showed that the LC_{50} was lowest at pH 5.5 and greatest at pH 7.8.

Because of the low toxicity of vanadium to fish, there are few guidelines aimed at protecting the quality of surface waters. The European Community does, however, maintain an objective of 1.0 mg/L for both marine and freshwater species.

Health Effects

Intake

The typical Western diet yields approximately 2.0 mg V/day in a 70-kg reference man. An additional 0.2 μg V/day comes from inhalation in the nonoccupationally exposed population. Once absorbed, vanadium is concentrated in the lungs of humans, with lesser amounts going to the small and large intestines, omentum, and skin. Snyder et al. (1975) estimated that the soft-tissue body burden is <18 mg. Vanadium is eliminated almost exclusively in the feces.

Acute Toxicity

Most cases of acute vanadium poisoning come from occupational exposure through the inhalation route. The primary symptoms are upper and lower respiratory tract irritation, chronic coughing, and chronic bronchitis. The conjunctiva may be irritated and eczema may develop. Vanadium salts exhibit low oral toxicity; for example, the LD_{50} of V_2O_3 in the mouse is 130 mg V/kg.

Chronic Toxicity

The primary examples of chronic exposure come from the inhalation of industrial-derived emissions. Symptoms include chronic respiratory disorders, irritation of the skin and eyes, and discoloration of the tongue and oral mucosa (National Institute for Occupational Safety and Health, 1977). Such effects have not been recorded following the consumption of tainted drinking water.

Carcinogenicity

Although the number of studies is relatively small, there does not appear to be any sustainable evidence that vanadium salts are carcinogenic in animals or man.

Drinking Water

Vanadium can be detected at concentrations of 4–222 μg/L in many finished drinking waters (reviewed by Carson et al., 1987). However, because of the low oral toxicity of vanadium salts, most nations (plus the World Health Organization) have not developed any guideline for the protection of drinking water. Advanced treatment technology for vanadium removal has not been developed, again because of the low toxicity of vanadium salts.

Recommendations

Vanadium does not constitute a threat to surface waters except in highly localized situations. There is no evidence of concentration of vanadium through the food chain, and all compounds appear to be relatively nontoxic to aquatic species. Although vanadium is found in drinking waters, most regulatory agencies do not recommend a consumption guideline, reflecting the low oral toxicity of vanadium salts.

There are several outstanding areas of potential research on vanadium. These include subjects in environmental fate, partitioning into different biotic/abiotic compartments, and chronic toxicity to aquatic species. However, since vanadium does not hold much environmental significance at the present time, such topics cannot be considered a priority.

References

Abaychi, J.K., and Y.Z. Mustafa. 1988. The Asiatic clam, *Corbicula fluminea*: an indicator of trace metal pollution in the Shatt al-Arab River, Iraq. *Environmental Pollution* 54:109–122.

Araujo, M.F.D., P.C. Bernard, and R.E. Van Grieken. 1988. Heavy metal contamination in sediments from the Belgian coast and Scheldt Estuary. *Marine Pollution Bulletin* 19:269–273.

Breslin, V.T., and I.W. Duedall. 1988. Vanadium release from stabilized oil ash waste in seawater. *Environmental Science and Technology* 22:1166–1170.

Carson, B.L., H.V. Ellis, and J.L. McCann. 1987. *Toxicology and biological monitoring of metals in humans*. Lewis Publishers, Chelsea, MI. 328 pp.

Dissanayake, C.B., J.M. Niwas, and S.V.R. Weerasooriya. 1987. Heavy metal pollution of the mid-canal of Kandy: an environmental case study from Sri Lanka. *Environmental Research* 42:24–35.

Fowler, S.W. 1986. Trace metal monitoring of pelagic organisms from the open Mediterranean Sea. *Environmental Monitoring and Assessment* 7:59–78.

Howard, L.S., and B.E. Brown. 1986. Metals in tissues and skeleton of *Fungia fungites* from Phuket, Thailand. *Marine Pollution Bulletin* 17:569–570.

Howard, L.S., and B.E. Brown. 1987. Metals in *Pocillopora damicornis* exposed to tin smelter effluent. *Marine Pollution Bulletin* 18:451–454.

Lee, K. 1983. Vanadium in the aquatic ecosystem. *In: Aquatic toxicology,* ed. J.O. Nriagu, 155–187. Wiley, New York.

Mance, G. 1987. *Pollution threat of heavy metals in aquatic environments*. Elsevier, London. 372 pp.

Mangrich, A.S., and N.V. Vugman. 1988. Bonding parameters of vanadyl ion in humic acid from the Jucu River estuarine region, Brazil. *Science of the Total Environment* 75:235–241.

Miramand, P., and M. Unsal. 1978. Toxicité aigue du vanadium vis à vis de quelques espêces benthiques et phytoplanctoniques marines. *Chemosphere* 10:827–832.

Mudroch, A., and G.A. Duncan. 1986. Distribution of metals in different size fractions of sediment from the Niagara River. *Journal of Great Lakes Research* 12:117–126.

Naquadat. 1985. *National water quality data bank*. Environment Canada, Ottawa.

National Institute for Occupational Safety and Health. 1977. *Criteria for a recommended standard: occupational exposure to vanadium*. National Institute for Occupational Safety and Health, Pub. No. 77–222, NIOSH, Washington, DC.

Noetstaller, R. 1988. *Industrial minerals,* World Bank Technical Paper Number 76. World Bank, Washington, D.C.

Nriagu, J.O. 1989. A global assessment of natural sources of atmospheric trace metals. *Nature* 338:47–49.

Nriagu, J.O., and J.M. Pacyna. 1988. Quantitative assessment of worldwide contamination of air, water and soils by trace metals. *Nature* 333:134–139.

Phillips, C.R., J.R. Payne, J.L. Lambach, G.H. Farmer, and R.R. Sims. 1987. Georges Bank monitoring program: hydrocarbons in bottom sediments and hydrocarbons and trace metals in tissues. *Marine Environmental Research* 22:33–74.

Sadig, M., and I. Alam. 1989. Metal concentrations in Pearl oyster, *Pinctada radiata*, collected from Saudi Arabian coast of the Arabian Gulf. *Bulletin of Environmental Contamination and Toxicology* 42:111–118.

Samhan, O., M. Zarba, and V. Anderlini. 1987. Multivariate geochemical investigation of trace metal pollution in Kuwait marine sediments. *Marine Environmental Research* 21:31–48.

Snyder, W.S., M.J. Cook, E.S. Nasset, L.R. Karhausen, G.P. Howells, and I.H. Tipton. 1975. *Report of the task group on reference man,* ICRP Publication 23. International Commission on Radiological Protection, New York.

Soderlund, S., A. Forsberg, and M. Pedersen. 1988. Concentrations of cadmium and other metals in *Fucus vesiculosus* L. and *Fontinalis dalecarlica* Br. Eur. from the northern Baltic Sea and the southern Bothnian Sea. *Environmental Pollution* 51:197–212.

Stendahl, D.H., and J.B. Sprague. 1982. Effects of water hardness and pH on vanadium lethality to rainbow trout. *Water Research* 16:1479–1488.

Subramanian, V., P.K. Jha, and R. Van Grieken. 1988. Heavy metals in the Ganges Estuary. *Marine Pollution Bulletin* 19:290–293.

Taylor, D., B.G. Maddock, and G. Mance. 1985. The acute toxicity of nine "grey list" metals (arsenic, boron, chromium, copper, lead, nickel, tin, vanadium and zinc) to two marine fish species: dab (*Limanda limanda*) and grey mullet (*Chelon labrosus*). *Aquatic Toxicology* 7:135–144.

US Minerals Yearbooks. 1940–1989. Bureau of Mines, US Department of the Interior, Washington, DC.

25
Zinc

Zinc occurs in the earth's crust at an average concentration of 70 mg/kg, making it the 24th most abundant element. The principal ores are sulfides, such as sphalerite, wurtzite (cubic and hexagonal ZnS, respectively), carbonate (known as smithsonite or calamine, $ZnCO_3$), and silicate (willemite, Zn_2SiO_4). Although an enormous amount of zinc is mobilized each year from natural and anthropogenic sources, the number of environmental problems caused by zinc has declined during the latter half of this century, reflecting the implementation of sound waste control practices.

Production, Sources, and Residues

Production

World production of zinc was 1,394 × 10^3 metric tons in 1930, increasing to 3,286 × 10^3 metric tons in 1960 and 5,229 × 10^3 metric tons in 1980 (US Minerals Yearbooks, 1930–1989). Production in recent years has continued to increase and is now near 7,000 × 10^3 metric tons/yr. The major producers are Canada, the USSR, Australia, Peru, and China; the leading consumers are the USA, Japan, the USSR, the FRG and China (Table 25.1).

The major uses of zinc are in coatings to protect iron and steel (35% of global consumption), in alloys for die casting (25%), and in brass (20%).

Table 25.1. World's leading producers and consumers of zinc.

Producing nation	Quantity (10^3 metric tons/yr)	Consuming nation	Quantity (10^3 metric tons/yr)
Canada	1,294	USA	999
USSR	810	USSR	990
Australia	662	Japan	753
Peru	598	FRG	434
China	396	China	360
Mexico	285	France	261
USA	226	Italy	232
Spain	223	UK	182
Japan	222	Belgium	172
Sweden	214	Poland	157

Source: Cordero (1988).

Sources

The total amount of zinc discharged to freshwaters from anthropogenic sources comes to 77–373 × 10^3 metric tons/yr (Nriagu and Pacyna, 1988). There are several major sources including the discharge of domestic wastewater, coal-burning power plants, manufacturing processes involving metals, and atmospheric fallout (Table 25.2). Approximately 34% of all emissions of zinc to the atmosphere come from natural sources, the remainder originating from metal production, burning of coal and oil, and

Table 25.2. Worldwide anthropogenic input of zinc to freshwaters.

Source	Input (thousand metric tons/yr)
Manufacturing processes	
metals	25–138
chemicals	0.2–5
pulp and paper	0.1–1.5
Domestic wastewater	
central	9–45
noncentral	6–36
Smelting and refining	
iron and steel	6–24
nonferrous metals	2–20
Atmospheric fallout	21–58
Coal-burning power plants	6–30
Base metal mining and dressing	<0.1–6

Source: Nriagu and Pacyna (1988).

Table 25.3. Annual input of zinc to The Netherlands' part of the North Sea in 1980 (actual) and 1990 (projected).

Source	Input 1980 (metric tons/yr)	Input 1990 (metric tons/yr)
Total input	10,000	6,200–7,300
Atmospheric deposition	2,300	2,300
Rivers	5,900	3,200–4,300
Coastal discharges	180	39
Dredging sludges	1,600	620
Industrial wastes	40	0
Incineration at sea	4	0
Offshore mining	3.5	3.6

Source: Beukema et al. (1986).

fertilizer and cement production (Nriagu, 1989, Nriagu and Pacyna, 1988).

Total zinc discharges to coastal and estuarine waters have decreased in many areas in recent years. In 1972, for example, discharges from the four largest southern California municipal outfalls amounted to approximately 1,200 metric tons, whereas in 1980 and 1988, the corresponding rates were 700 and 200 metric tons, respectively (Schafer, 1989). Similarly, discharge to The Netherlands' part of the North Sea was 10,000 metric tons in 1980, but is projected to decline to 6,200–7,300 metric tons for the year 1990 (Table 25.3). The major sources for both years are atmospheric deposition and input from rivers.

Residues

Water. The concentration of extractable zinc in freshwaters usually ranges from <0.001 to approximately 0.05 mg/L. Urban et al. (1987), working on 24 bogs in northeastern North America, reported residues of 0.007 to 0.022 mg/L. These levels were only 52–60% of those found in precipitation (the major source of zinc in the bogs), implying that sorption by peat controlled residues in water. Similarly, Sridhar (1986) reported that total zinc in the influent of Awba Lake (Nigeria) ranged from 2.4 to 6.4 μg/L, but declined to only 0.4–5.1 μg/L after passing through a growth of the water lettuce *Pistia stratiotes*. That gave a reduction of 20–83%. In contrast, an elevation in zinc levels has been observed in a number of lakes receiving acidic deposition (White and Driscoll, 1987).

Sediments. Sediments are a primary sink for zinc. Residues in excess of 1,000 mg/kg dry weight have been found in the vicinity of mines and smelters. Relatively high concentrations can also be found near major municipalities and coal-burning power stations (Table 25.4). Poulton (1987), for example, reported that a core sample from Hamilton Harbor

Table 25.4. Concentration (mg/kg dry weight) of total zinc in sediments of depositional basins of the Great Lakes.

Lake	Surface	Background
Ontario	87–3,507	83–163
Erie	18–536	8–128
Huron	8–233	60–88
Michigan	40–350	40–50
Superior	143–195	53–137

Source: Mudroch et al. (1988).

(a major steel-producing region in Canada) contained 5,700 mg Zn/kg. By comparison, sediments from uncontaminated waters typically contain residues of 5 to 50 mg/kg.

Chemistry

Zinc is classified as a borderline metal, meaning that it forms bonds with oxygen as well as nitrogen and sulfur donor atoms. Under aerobic conditions, Zn^{2+} is the predominant species at acidic pH, but is replaced by $Zn(OH)_2$ at pH 8–11, and $Zn(OH)_3^-/Zn(OH)_4^{2-}$ at pH >11 (Vymazal, 1985). Anaerobic conditions lead to the formation of ZnS regardless of pH within the range 1–14. Samanidou and Fytianos (1987), working on two rivers in northern Greece, noted that zinc partitioned primarily into the sulfide and Fe–Mn hydrous oxide fractions of the sediments (Table 25.5).

Zinc binds readily with many organic ligands, particularly in the presence of nitrogen or sulfur donor atoms. Stability constants (K_o) for zinc/humic acid complexes are variable, typically ranging from 2.3 to 5.9. This means that the fate of zinc will vary among waterways, depending on the

Table 25.5. Partitioning of zinc in the sediments of two rivers in Greece.

	Distribution (%)	
Fraction	Axios River	Aliakmon River
Cation-exchangeable	0.8	1.8
Carbonates	8.4	3.8
Fe–Mn hydrous oxides	28.4	13.4
Organic sulfides	32.0	52.1
Residual	30.4	28.9

Source: Samanidou and Fytianos (1987).

type of humic material present in the system. Zinc also shows variable behavior in binding to suspended particulates, depending on pH and Eh conditions, and the input of anthropogenically derived zinc.

Bioaccumulation

Plants

Zinc is an essential component of the enzyme carbonic anhydrase, which catalyzes the dehydration of carbonic acid and participates in the flux of carbon dioxide within the cell. Hence, zinc is found in all green plants, and is essential for their growth.

Total Zn residues in aquatic plants collected from polluted surface waters generally range from 100 to 500 mg/kg dry weight, but considerably higher levels have been found in the vicinity of base metal mines. Mason and Macdonald (1988), for example, reported maximum concentrations of up to 2,810 mg/kg in the aquatic moss *Fontinalis squamosa*. Similarly, collections of *Fucus vesiculosus* along the east coast of Sweden yielded a maximum residue of 877 mg/kg (Soderlund et al., 1988). The elevated levels in those collections were due to the discharge of municipal and industrial wastes into coastal waters. In nonpolluted areas total Zn in plants typically ranges from 5 to 35 mg/kg.

A comprehensive review of factors influencing the uptake of zinc by aquatic plants was prepared by Vymazal (1986).

Invertebrates

Although total Zn in the soft tissues of invertebrates generally ranges from 50 to 500 mg/kg dry weight (Table 25.6), residues of over 3,000 mg/kg may be found in exceptional cases. Maximum concentrations in molluscs are usually associated with the digestive gland, kidney, gill, and gonad, rather than the foot or muscle tissue. This differential deposition is likely due to the presence of metal-binding proteins (metallothionein) in the various tissues and the concomitant effect on biological half-life.

Rate of uptake is linear or exponential in many species. Under natural conditions, the rate may be greatest in the winter, spring, or summer, or show no consistent seasonal trend. The presence of ligands, either organic or inorganic, significantly reduces uptake in most species studied to date.

Fish

Total Zn in the muscle tissue of fish is generally much lower than residues in plants or invertebrates. Most reports place total Zn at <50 mg/kg wet

Table 25.6. Concentration (mg/kg dry weight) of total Zn in the soft tissues of invertebrates.

Species	Average (range)	Location
Euphausiids, *Euphausia* sp.[1]	92 (58–140)	Mediterranean Sea
	104 (NR[a])	NE Atlantic
	69 (53–83)	NE Pacific
	100 (59–195)	NW Pacific
Oysters, *Crassostrea angulata*[2]	466 (296–651)	Huelva Estuary, Spain
Gastropod, *Lymnaea stagnalis*[3]	75 (62–118)	Lake Balaton, Hungary
Mollusc, *Elliptio complanata*[4]	259 (132–535)	Lakes, northern Quebec

[a]Not reported.
Sources: [1]Fowler (1986), [2]Lopez-Artiguez et al. (1989), [3]V.-Balogh et al. (1988), [4]Tessier et al. (1984).

weight, with a range of <1–100 mg/kg. Much higher levels (100–300 mg/kg) can be found in the liver and kidney of many fish species. This reflects the presence of metallothionein in such tissues (Overnell et al., 1987).

With the exception of nonfeeding periods, fish generally sorb the majority of zinc from food rather than water. The rate of uptake is typically exponential, prior to the attainment of a steady state in residues. The presence of either organic or inorganic ligands in solution may significantly reduce the rate of uptake in most (if not all) species.

Toxic Effects to Aquatic Organisms

Plants

Toxicity of zinc to aquatic plants is highly variable, with EC_{50}s (Effective Concentration) ranging from <0.01 to >100 mg/L. This extreme variability is due to (1) the effect of different physicochemical conditions on uptake, and (2) the ability of many species to adapt to high zinc levels. As early as 1980, Allen et al. noted that algal (*Microcystis aeruginosa*) growth decreased by more than an order of magnitude as the quantity of chelator in solution declined; in fact toxicity was not related to total Zn but to predicted free metal concentration.

Other investigations (Starodub et al., 1987) showed that extracellular ligands produced by *Scenedesmus quadricauda* bound zinc, thereby reducing its toxicity. The same study also showed that the toxicity of a combination of Zn/Cu/Pb was significantly greater at pH 4.5 than at pH 8.5 or pH 6.5.

The toxic effects of zinc are, at least in part, related to interference with phosphorus metabolism (Kuwabara, 1985). Apparently, as the concentration of cellular polyphosphate declines with increasing culture age, a portion of the previously bound intracellular zinc is released into the cell (Bates et al., 1985). When the quantity of intracellular zinc exceeds a critical threshold, phosphorus metabolism is disrupted, interfering with cell division and decreasing cell yield.

Invertebrates

Acute toxicity of zinc to invertebrates is relatively low, with 48- to 96-h LC_{50}s generally ranging from 0.5 to 5 mg/L (but periodically >100 mg/L; Table 25.7). Chronic effects, such as reduced production in multiple generation studies, may be induced at 10% of the 48-h LC_{50} (Verriopoulos and Hardouvelis, 1988).

Mortality of marine invertebrates following exposure to zinc is likely related to a progressive decrease in the ability to osmoregulate. McLusky and Hagerman (1987), for example, showed that changes in salinity above or below the isosmotic point led to reduced median survival time when the mysid *Praunus flexuosus* was exposed to zinc sulfate. Mortality also decreased as temperature increased from 5 to 15° C. De March (1988), working with the amphipod *Gammarus lacustris*, noted that K or Mg, in combination with Zn, had an additive effect on toxicity, whereas combinations Cu/Zn and Cd/Zn had more than additive effects. Other studies have shown that calcium ions offer better protection from intoxication than magnesium ions. In addition, increasing water hardness decreases toxicity to most invertebrate species.

Fish

Zn^{2+} is moderately toxic to most species of fish, both marine and freshwater. Although the 48- to 96-h LC_{50} for Zn^{2+} usually falls within the range 0.5–5 mg/L, physicochemical and biological factors may extend the limits to 0.1 to >100 mg/L.

Table 25.7. Acute toxicity (LC_{50}) of Zn^{2+} to two crustacean species.

48-h LC_{50} (mg/L)		96-h LC_{50} (mg/L)	
Average	95% Confidence limits	Average	95% Confidence limits
Asellus aquaticus			
52	32–85	18	12–25
Crangonyx pseudogracilis			
121	95–171	20	15–25

Source: Martin and Holdich (1986).

Numerous effects have been noted in fish following long-term exposure to zinc. These effects include (1) reduced mineral uptake, (2) reduced skeletal calcium deposition, (3) teratogenic effects on embryo-larval stages, and (4) reduced growth. The minimum concentration causing such effects is approximately 0.07 mg/L in soft water.

Bradley and Sprague (1985), working with rainbow trout *Oncorhynchus mykiss*, reported that changes in pH from 5.5 to 7.0 increased zinc toxicity by factors of 2 to 5, depending on total hardness. A decrease in hardness from 386 to 31 mg $CaCO_3$/L increased zinc toxicity by more than an order of magnitude at both pH levels. Anadu et al. (1989) noted that preexposure of the same species to zinc increased tolerance to higher levels of zinc by 0.6–5 times. This is likely due to increased concentrations of metallothionein (Hobson and Birge, 1989), which partitions zinc away from critical enzymes paths. Complexation with humic acids (and other ligands/chelators) reduces acute toxicity in most species studied to date (Hutchinson and Sprague, 1987).

Several nations have promulgated guidelines for the protection of fish and other aquatic life:

Canada
- 0.03 mg/L

United States
- 0.065 mg/L (water hardness 50 mg/L as $CaCO_3$, 1-h average)
- 0.120 mg/L (water hardness 100 mg/L as $CaCO_3$, 1-h average)
- 0.210 mg/L (water hardness 200 mg/L as $CaCO_3$, 1-h average)
- 0.059 mg/L (water hardness 50 mg/L as $CaCO_3$, 4-day average)
- 0.110 mg/L (water hardness 100 mg/L as $CaCO_3$, 4-day average)
- 0.190 mg/L (water hardness 200 mg/L as $CaCO_3$, 4-day average)
- 0.095 mg/L (salt water, 1-h average)
- 0.086 mg/L (salt water, 4-day average)

European Community
- 0.01 mg/L (protection of salmonids, water hardness 0–50 mg/L as $CaCO_3$)
- 0.075 mg/L (protection of salmonids, water hardness 100–250 mg/L as $CaCO_3$)
- 0.125 mg/L (protection of salmonids, water hardness >250 mg/L as $CaCO_3$)
- 0.04 mg/L (protection of marine fish and shellfish)

Health Effects

Intake

The typical Western diet yields 10–15 mg Zn/day in a 70-kg reference man. Inhalation is an insignificant source in the nonoccupationally exposed population. Sorption of zinc from food is relatively low, amounting to 10–40% (reviewed by Carson et al., 1987). The total body burden is 1.4–3 g in a 70-kg reference man; maximum residues occur in the muscle, liver, and kidney.

Zinc is essential in (1) several enzymes and enzyme functions; (2) DNA, RNA, and protein synthesis; (3) carbohydrate metabolism; and (4) cell division and growth. The minimum amount of zinc required per day ranges from 1 to 5.5 mg, depending on age and pregnancy/lactation in females.

Acute Toxicity

Ingestion of >2 g Zn produces toxic symptoms (fever, diarrhea, vomiting, and other gastrointestinal tract irritation) in humans. The most common episodes of poisoning come from the ingestion of acidic beverages made in galvanized containers. Levels of 675–2,300 mg/L in drinking water are emetic.

Chronic Toxicity

Medical patients given high doses of zinc (>15 mg/day) generally show no adverse reaction, with the possible exception of reduced copper absorption. Occupationally exposed workers may show gastrointestinal disturbance and clinically latent liver dysfunction.

The taste threshold for zinc in drinking water is approximately 15 mg/L.

Carcinogenicity

Zinc and its compounds are not considered carcinogenic in humans or experimental animals.

Drinking Water

Residues

Zinc is routinely detected in drinking water, reflecting its presence in raw water and dissolution of solder and other material in the distribution system. In a survey of the American Water Works Association (1985) of drinking water in 39 states and three territories, there were 10 cases of noncompliance with the Maximum Contaminant Limit of 5 mg/L. For comparison, fluoride and nitrates were in noncompliance in 907 and 369 episodes, respectively. Total Zn in municipal water from the city of Aligarh (India) ranged from 0.002 to 0.27 mg/L (Ajmal and Uddin, 1986a); the corresponding concentrations for drinking water from Rio de Janeiro (Brazil) was 0.007–0.32 mg/L (Azcue *et al.*, 1988).

Ajmal and Uddin (1986b) noted that the concentration of zinc in standing water from the Aligarh Muslim University (India) was appreciably higher than that of running water: 0.029–0.36 mg/L versus 0.010–0.18 mg/L. This was likely due to the dissolution of zinc from distribution line materials. Similarly, Schock and Neff (1988) showed that the amount of zinc in the first draw of water samples decreased appreciably after 60 days (Figure 25.1). The same trend was also apparent for second- and third-draw samples.

Consumption Guidelines

Many nations, plus the World Health Organization, use a drinking water guideline of 5.0 mg/L. This is based on aesthetic rather than toxicologic factors. Zinc at 5.0 mg/L should impart no offensive taste or appearance to drinking water.

Treatment

Zinc removal is generally ancillary or incidental to other treatment objectives. Based on waste water studies, lime softening at pH 9.5–10 is effec-

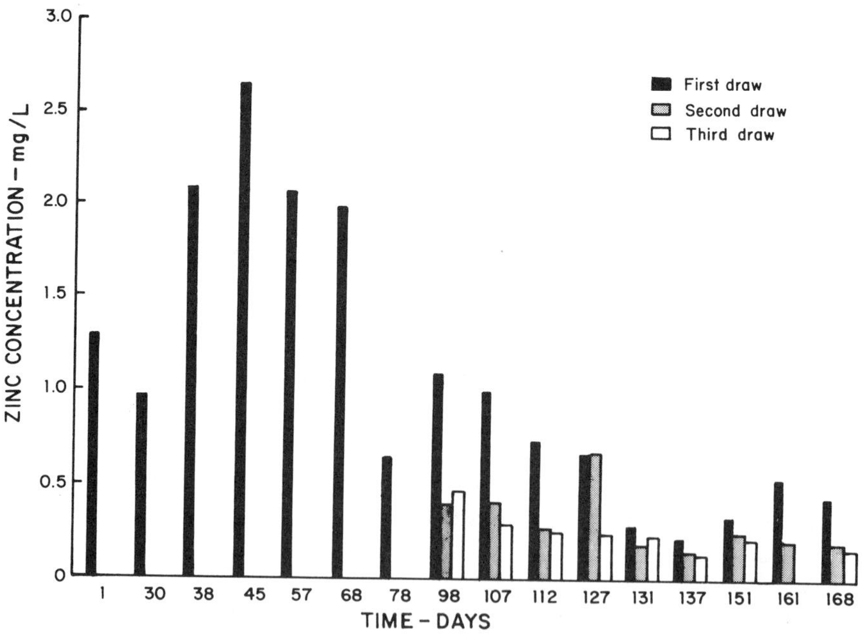

Figure 25.1. Concentration of zinc in successive unfiltered standing water samples from test valves located in experimental pipe loops (Schock and Neff, 1988). Reprinted from *Journal American Water Works Association*, Vol. 80, No. 11 (November 1988), by permission. Copyright © 1988, American Water Works Association.

tive in zinc removal, whereas alum coagulation at pH 6.5–7 provides <30% removal. Reduction in the use of zinc-containing distribution system materials can also control residues in the final product.

Recommendations

At one time, zinc posed a significant threat to the users of water. There are many early accounts of extensive mortality and/or contamination of fish and other aquatic species from the discharge of zinc-containing wastewater. These problems were exacerbated at the same time by the discharge of other heavy metals and acidic water. In some cases, drinking water supplies were tainted.

Today, zinc is relatively innocuous in the aquatic environment. Discharges are closely controlled, so the number of reports of adulteration of aquatic resources is small. An enormous amount of research has also been conducted on the environmental chemistry, accumulation, and toxicity of zinc. In fact, there do not appear to be any major data gaps on the environmental significance of zinc. This makes any research on zinc a low priority.

Many nations have now established surface water quality objectives for the protection of aquatic resources. Although there is some variation in the numerical value placed on these objectives, all appear to be adequate for the protection of fish and other aquatic species. Obviously, routine monitoring of waste water and surface water is important to ensure compliance with these objectives.

Zinc also poses no significant toxicologic threat to drinking waters. The drinking water guideline is based on aesthetic factors, particularly taste and appearance. Although it is important to monitor zinc residues in drinking waters, there is no need to establish a toxicologically based guideline.

References

Ajmal, M., and R. Uddin. 1986a. Studies of heavy metals in the ground waters of the City of Aligarh U.P. (India). *Environmental Monitoring and Assessment* 6:181–194.

Ajmal, M., and R. Uddin. 1986b. Quality of drinking water in the Aligarh Muslim University campus, Aligarh, U.P. (India) with respect to heavy metals. *Environmental Monitoring and Assessment* 6:195–205.

Allen, H.E., R.H. Hall, and T.D. Brisbin. 1980. Metal speciation. Effects on toxicity. *Environmental Science and Technology* 14:441–443.

American Water Works Association (1985). An AWWA survey of inorganic contaminants in water supplies. *Journal of the American Water Works Association* 77:67–72.

Anadu, D.I., G.A. Chapman, L.R. Curtis, and R.A. Tubb. 1989. Effect of zinc exposure on subsequent acute tolerance to heavy metals in rainbow trout. *Bulletin of Environmental Contamination and Toxicology* 43:329–336.

Azcue, J.M.P., W.C. Pfeiffer, M. Fiszman, and O. Malm. 1988. Heavy metal removal by different water treatment plants, in Rio de Janeiro State, Brazil. *Environmental Technology Letters* 9:429–436.

Bates, S.S., A. Tessier, P.G.C. Campbell, and M. Letourneau. 1985. Zinc-phosphorus interactions and variation in zinc accumulation during growth of *Chlamydomonas variabilis* (Chlorophyceae) in batch culture. *Canadian Journal of Fisheries and Aquatic Sciences* 42:86–94.

Beukema, A.A., G.P. Hekstra, and C. Venna. 1986. The Netherlands' environmental policy for the North Sea and Wadden Sea. *Environmental Monitoring and Assessment* 7:117–155.

Bradley, R.W., and J.B. Sprague. 1985. The influence of pH, water hardness, and alkalinity on the acute lethality of zinc to rainbow trout (*Salmo gairdneri*). *Canadian Journal of Fisheries and Aquatic Sciences* 42:731–736.

Carson, B.L., H.V. Ellis, and J.L. McCann. 1987. *Toxicology and biological monitoring of metals in humans*. Lewis Publishing, Chelsea, MI. 328 pp.

Cordero, R. 1988. *Metal Bulletin's prices and data 1988*. Metal Bulletin Books, Surrey, England. 375 pp.

De March, B.G.E. 1988. Acute toxicity of binary mixtures of five cations (Cu^{2+}, $Cd^{2+}2$, Zn^{2+}, Mg^{2+}, and K^{+}) to the freshwater amphipod *Gammarus lacustris* (Sars): alternative descriptive models. *Canadian Journal of Fisheries and Aquatic Sciences* 45:625–633.

Fowler, S.W. 1986. Trace metal monitoring of pelagic organisms from the open Mediterranean Sea. *Environmental Monitoring and Assessment* 7:59–78.

Hobson, J.F., and W.J. Birge. 1989. Acclimation-induced changes in toxicity and induction of metallothionein-like proteins in the fathead minnow following sublethal exposure to zinc. *Environmental Toxicology and Chemistry* 8:157–169.

Hutchinson, N.J., and J.B. Sprague. 1987. Reduced lethality of Al, Zn and Cu mixtures to American flagfish by complexation with humic substances in acidified soft waters. *Environmental Toxicology and Chemistry* 6:755–765.

Kuwabara, J.S. 1985. Phosphorus-zinc interactive effects on growth by *Selenastrum capricornutum* (Chlorophyta). *Environmental Science and Technology* 19:417–421.

Lopez-Artiguez, M., M.L. Soria, and M. Repetto. 1989. Heavy metals in bivalve molluscs in the Huelva Estuary. *Bulletin of Environmental Contamination and Toxicology* 42:634–642.

Martin, T.R., and D.M. Holdich. 1986. The acute lethal toxicity of heavy metals to peracarid crustaceans (with particular reference to fresh-water asellids and gammarids). *Water Research* 20:1137–1147.

Mason, C.F., and S.M. Macdonald. 1988. Metal contamination in mosses and otter distribution in a rural Welsh river receiving mine drainage. *Chemosphere* 17:1159–1166.

McLusky, D.S., and L. Hagerman. 1987. The toxicity of chromium, nickel and zinc: effects of salinity and temperature, and the osmoregulatory consequences in the mysid *Praunus flexuosus*. *Aquatic Toxicology* 10:225–238.

Mudroch, A., L. Sarazin, and T. Lomas. 1988. Summary of surface and background concentrations of selected elements in the Great Lakes sediments. *Journal of Great Lakes Research* 14:241–251.

Nriagu, J.O. 1989. A global assessment of natural sources of atmospheric trace metals. *Nature* 338:47–49.

Nriagu, J.O., and J.M. Pacyna. 1988. Quantitative assessment of worldwide contamination of air, water and soils by trace metals. *Nature* 333:134–139.

Overnell, J., R. McIntosh, and T.C. Fletcher. 1987. The levels of liver metallothionein and zinc in plaice, *Pleuronectes platessa* L., during the breeding season, and the effect of oestradiol injection. *Journal of Fish Biology* 30:539–546.

Poulton, D.J. 1987. Trace contaminant status of Hamilton Harbour. *Journal of Great Lakes Research* 13:193–201.

Samanidou, V., and K. Fytianos. 1987. Partitioning of heavy metals into selective chemical fractions from rivers in northern Greece. *Science of the Total Environment* 67:279–285.

Schafer, H. 1989. Improving southern California's coastal waters. *Journal of the Water Pollution Control Federation* 61:1395–1401.

Schock, M.R., and C.H. Neff. 1988. Trace metal contamination from brass fittings. *Journal of the American Water Works Association* 80:47–56.

Soderlund, S., A. Forsberg, and M. Pedersen. 1988. Concentrations of cadmium and other metals in *Fucus vesiculosus* L. and *Fontinalis dalecarlica* Br. Eur. from the northern Baltic Sea and the southern Bothnian Sea. *Environmental Pollution* 51:197–212.

Sridhar, M.K.C. 1986. Trace element composition of *Pistia stratiotes* L. in a polluted lake in Nigeria. *Hydrobiologia* 131:273–276.

Starodub, M.E., P.T.S. Wong, C.I. Mayfield, and Y.K. Chau. 1987. Influence of complexation and pH on individual and combined heavy metal toxicity to a freshwater green alga. *Canadian Journal of Fisheries and Aquatic Sciences* 44:1173–1180.

Tessier, A., P.G.C. Campbell, J.C. Auclair, and M. Bisson. 1984. Relationships between the partitioning of trace metals in sediments and their accumulation in the tissues of the freshwater mollusc *Elliptio complanata* in a mining area. *Canadian Journal of Fisheries and Aquatic Sciences* 41:1463–1472.

Urban, N.R., S.J. Eisenreich, and E. Gorham. 1987. Aluminum, iron, zinc, and lead in bog waters of northeastern North America. *Canadian Journal of Fisheries and Aquatic Sciences* 44:1165–1172.

US Minerals Yearbooks. 1930–1989. Bureau of mines, US Department of the Interior, Washington, DC.

V.-Balogh, K., D.S. Fernandez, and J. Salanki. 1988. Heavy metal concentrations of *Lymnaea stagnalis* L. in the environs of Lake Balaton (Hungary). *Water Research* 22:1205–1210.

Verriopoulos, G., and D. Hardouvelis. 1988. Effects of sublethal concentration of zinc on survival and fertility in four successive generations of *Tisbe*. *Marine Pollution Bulletin* 19:162–166.

Vymazal, J. 1985. Occurrence and chemistry of zinc in freshwaters—its toxicity and bioaccumulation with respect to algae: A review. Part 1: Occurrence and chemistry of zinc in freshwaters. *Acta Hydrochimica Hydrobiologica* 13:627–654.

Vymazal, J. 1986. Occurrence and chemistry of zinc in freshwaters—its toxicity and bioaccumulation with respect to algae: A review. Part 2: Toxicity and bioaccumulation with respect to algae. *Acta Hydrochimica Hydrobiologica* 14:83–102.

White, J.R., and C.T. Driscoll. 1987. Zinc cycling in an acidic Adirondack lake. *Environmental Science and Technology* 21:211–216.

26
Summary and Conclusions

There is still an enormous amount of work to be done on the environmental fate and toxicity of inorganic contaminants of surface water. Some of the agents evaluated in this book have waned in importance in recent years, whereas others have waxed, reflecting either increased use or mobilization of those agents. In some cases, new toxicological or chemical evidence has come to light, necessitating a reevaluation of the use of the agent.

The following chart summarizes important research and monitoring priorities for those agents described in this book. The monitoring priorities, denoted by an asterisk (*), are supplemental to multielemental, water quality surveys.

Aluminum

National/international standardization of protocols aimed at studying the Al species complex

The formation and/or resolubilization or inorganic and organic complexes under acidic conditions in surface water

The resolubilization of alum sludge and other aluminum-bearing wastes in relation to changes in redox potential and pH of bottom sediments

The toxicity of Al species to standard test organisms such as rainbow trout and *Daphnia*

Chemical and basic scientific evaluation of the role of Al species in the genesis of Alzheimer's and related diseases

Routine monitoring of aluminum in drinking water in relation to alum use*

Arsenic

Physicochemical studies on the mobilization of arsenic into surface and groundwater
Speciation of arsenic in algae, invertebrates, and fish, particularly those used for human consumption
Epidemiological studies on the occurrence of skin cancer in populations exposed to arsenic in water
Role of arsenic's carcinogenic potential in setting drinking water guidelines

Asbestos

Mobilization of heavy metals in surface waters when ambient asbestos residues are high
Asbestos residues in clams and other filter feeders that may be used for human food
Chronic toxicity of asbestos at environmentally relevant levels to freshwater and marine organisms
Sensitive epidemiological surveys relating exposure to neoplasia in the stomach, pancreas, esophagus, and small intestine
Establishment of a widely accepted guideline or maximum recommended level of asbestos in water

Barium

Environmental fate of barium in coastal and freshwaters
Chronic effects of barium exposure to marine and freshwater species, both plant and animal
Revision of drinking water guideline

Beryllium

Atmospheric deposition of beryllium near coal-fired power plants and concomitant effects on surface water loading
Sorption/desorption from sediments and suspended solids under different environmental conditions
Chronic toxic effects to several aquatic species
Guidelines for the protection of freshwater and marine life

Boron

Concentration of boron in a range of aquatic species, both plant and animal

Cadmium

Nonpoint sources of contamination, particularly atmospheric deposition
Toxic effects to estuarine/coastal species under variable salinity regimes

Chromium

Environmental fate of organic chromium complexes
Effects of organic ligands on the toxicity of chromium compounds to marine and freshwater plant and animal species.
Chronic effects of Cr^{3+} and Cr^{6+} to marine and freshwater invertebrate species
Carcinogenicity of Cr^{3+} and Cr^{6+} to fish
Carcinogenicity of water-soluble Cr^{6+} compounds to mammals
Cr^{3+} and Cr^{6+} in drinking water*
Bivalve molluscs as monitors of chromium in surface waters*

Cobalt

Environmental fate and chronic toxicity to multiple species (secondary priority)

Copper

No priorities at this time

Cyanides

Implementation of best available technology in the control of wastewater discharges
Environmental fate and speciation of cyanides in surface waters
Presence of metallic cyanide complexes in surface waters receiving such discharges
Presence of metallic cyanide complexes in fish tissues and other aquatic food products from surface waters receiving such complexes
Chronic toxicity of cyanides to a range of invertebrate species

Iron

Environmental fate of sequestered and remobilized metals in the iron redox cycle
Sorption of toxic metals by iron hydroxides/oxides, and resulting impact on biological uptake
Iron/toxic metal interactions and concomitant effect on chronic toxicity

Lead

Accumulation of lead in deep open sediments
Environmental methylation and demethylation of lead
Metabolism of organic lead compounds by plant and animal species
Chronic toxicity of inorganic and organic lead compounds to marine fish
Carcinogenicity of lead salts via oral administration
Reevaluation of the current drinking water guideline of 0.05 mg/L used in many nations

Manganese

Environmental fate of sequestered and remobilized metals in the manganese redox cycle
Oxidation of metals on the surface of MnO_2
Protective mechanism of Mn^{2+} against toxic metals, using a number of plant and animal species (both marine and freshwater)
Seasonal flux in manganese and concomitant change in the toxicity of manganese-bound metals

Mercury

No research priorities at this time
Methyl mercury in surface waters*
Methyl mercury and total mercury in sediments*
Methyl mercury in fish tissues and other food products*

Nickel

No priorities at this time

Nitrogen

No research priorities at this time (excluding acid-forming atmospheric compounds)
Formation of ammonia-N mixing zones in surface waters*
Nitrate and nitrite levels in drinking waters*

Selenium

Mobilization and fate of selenium complexes under different environmental conditions
Uptake of inorganic and organic selenium by invertebrates and fish under different environmental conditions
Effect of methyl mercury on uptake of organic selenium
Toxicity of organic selenium complexes to fish and other aquatic species

Silver

No priorities at this time

Sulfur

No research priorities at this time (excludes acid-forming compounds)
Heavy metal and pH cycles in acid-stressed freshwaters*

Thallium

Effect of pH, hardness, organic ligands, inorganic ligands on uptake by plant, invertebrate, and fish species

Tin

Partitioning of tin compounds in different components of the sediments
Effect of temperature, pH, salinity, and other environmental factors on the rate of methylation of inorganic tin and butyltins
Factors influencing the toxicity of organotins to algae, invertebrates, and fish, particularly the effects of changing salinity
Development of a drinking water guideline for organotins

Vanadium

No priorities at this time

Zinc

No priorities at this time

Supplementary to these research and monitoring recommendations, I find it necessary to comment on the primary source of the majority of the inorganic contaminants of surface water. That source is the emissions and solid/liquid waste produced from the burning of coal. Popular mythology currently holds that coal is an excellent source of energy, superior to nuclear power on the basis of reliability, safety, and economics. Certainly, events around Chernobyl and Three Mile Island support and even exacerbate this state of mind. Current technology has, however, reduced the risk of a serious radioactive leak to $<10^{-8}$, while reducing the discharge of inorganic contaminants to essentially nil.

If coal (and other fossil fuels) continues to be the primary worldwide energy source for the 1990s and beyond, the mobilization of inorganic contaminants will wax with unforeseen environmental and health effects. Global warming will be further enhanced, as will the production of acid rain. Reduction in the proportionate use of coal can only benefit humankind.

Index